BIOLOGY OF SEX

Biology of Sex

ALEX MILLS

UNIVERSITY OF TORONTO PRESS
Toronto Buffalo London

Library and Archives Canada Cataloguing in Publication

Mills, Alexander Matthew, 1960–, author
Biology of sex / by Alex Mills.

Includes bibliographical references and index.
Issued in print and electronic formats.
ISBN 978-1-4875-9338-4 (hardcover).—ISBN 978-1-4875-9337-7 (softcover).—
ISBN 978-1-4875-9339-1 (EPUB).—ISBN 978-1-4875-9340-7 (PDF).

1. Sex (Biology)—Textbooks. 2. Human reproduction—Textbooks.
3. Sex—Textbooks. 4. Evolution (Biology)—Textbooks. 5. Textbooks.
I. Title.

QP251.M54 2018 612.6 C2017-906263-8
C2017-906264-6

We welcome comments and suggestions regarding any aspect of our publications—please feel free to contact us at news@utphighereducation.com or visit our Internet site at utorontopress.com.

North America
5201 Dufferin Street
North York, Ontario, Canada, M3H 5T8
2250 Military Road
Tonawanda, New York, USA, 14150
ORDERS PHONE: 1–800–565–9522
ORDERS FAX: 1–800–221–9985
ORDERS E-MAIL: utpbooks@utpress.utoronto.ca

UK, Ireland, and continental Europe
NBN International
Estover Road, Plymouth, PL6 7PY, UK
ORDERS PHONE: 44 (0) 1752 202301
ORDERS FAX: 44 (0) 1752 202333
ORDERS E-MAIL: enquiries@nbninternational.com

The University of Toronto Press acknowledges the financial support for its publishing activities of the Government of Canada through the Canada Book Fund.

Canadä

Printed in Canada

Contents

List of Figures xii
List of Boxes xvii
Preface xviii

1 INTRODUCTION 1
Key themes 1
1.1 The wider context: sex, gender, sexuality, and sexual orientation 4
1.1.1 Assigned sex and gender dysphoria 7
1.1.2 Sex as a biological construct and gender as a social construct 8
1.2 A caveat about biological determinism and ideology 8
1.3 The comparative approach 13
1.4 Biology as a part of natural science 14
1.5 Theories and the methodology of science 15
Summary 19
Further reading 24

2 SEX AND REPRODUCTION 25
Key themes 25
2.1 Does sexual motivation reveal its "purpose"? 27
2.1.1 Sex has diverse purposes through co-option 29
2.1.2 Homosexuality could be one manifestation of co-option 31
2.1.3 Sexual features can also be co-opted for non-social purposes 32
2.2 Reproduction: sexual, asexual, and non-sexual 33
2.2.1 Not all organisms reproduce sexually 33
2.2.2 Non-sexual reproductions of human cells is through mitosis and cytokinesis 37
2.2.3 Asexual reproduction in complex organisms 38
2.2.4 Switching between parthenogenesis and sexual reproduction 41

2.3 **Sexual reproduction fundamentals** **43**
2.3.1 Female and male: Eggs and sperm 43
2.3.2 Primary versus secondary sex characteristics 45
2.4 **From fertilization to the production of offspring** **50**
2.4.1 External versus internal fertilization 51
2.4.2 Other variations in patterns of sexual reproduction 56
Summary **57**
Further reading **58**

3 SEX AND INHERITANCE 59
Key themes **59**
3.1 **How an Augustinian friar discovered the rules of sexual inheritance** **62**
3.1.1 Mendel's peas were a "friendly" study system 63
3.2 **Developing Mendel's Law of Segregation** **65**
3.2.1 The second-generation offspring showed a surprising pattern of inheritance 65
3.2.2 The meaning of segregation in the making of gametes 66
3.3 **Mendel's Law of Independent Assortment** **70**
3.4 **Updating Mendelian genetics** **72**
3.4.1 Using modern terminology 73
3.5 **Applying Mendelian genetics to humans** **75**
3.5.1 Mendelian inheritance of blood groups in humans 75
3.6 **Sex, DNA, and chromosomes** **80**
3.6.1 Chromosomes are the hereditary material 80
3.6.2 The human karyotype 82
3.6.3 DNA is the genetic code 84
3.6.4 How DNA encodes information 86
3.7 **What happens to chromosomes in the making of gametes?** **87**
3.7.1 What is crossing-over? 92
3.7.2 The two divisions of meiosis 95
3.7.3 Sex and chromosomes that don't crossover 96
3.7.4 How meiosis relates to Mendel's laws 97
Summary **98**
Further reading **99**

4 SEX AND EVOLUTION 101
Key themes **101**
4.1 **A short history of evolutionary thought** **102**

4.1.1 Evolutionary theory is a natural result of the Enlightenment 104
4.1.2 Exploration, fossils, and a very old Earth also led to evolutionary theory 105
4.1.3 Jean-Baptiste Lamarck proposed a mechanism for evolution in 1809 106
4.1.4 Charles Darwin and the voyage of the *Beagle* 107
4.2 The principles of natural selection 113
4.2.1 The argument for natural selection 113
4.2.2 The concepts of fitness and adaptation 115
4.3 DNA as an evolutionary legacy 116
4.3.1 Mutations, and how alleles differ from one another 117
4.3.2 Chromosomal mutations can also contribute to evolution 118
4.3.3 Mutation rates in sperm and eggs 120
4.4 Thinking of breeding groups as gene pools 121
4.4.1 Human breeding groups can be thought of as gene pools 121
4.4.2 The ABO blood group gene pool 122
4.5 The evolutionary costs of sex 125
4.5.1 The numerical argument that sex is costly 125
4.5.2 Sexual reproduction also involves search costs 127
4.5.3 Sexual reproduction can entail health, injury, and mortality costs 127
4.6 The evolutionary benefits of sex 129
4.6.1 Advantages associated with a diverse gene pool 130
4.6.2 Inbreeding reveals the value of gene mixing through sex 131
4.6.3 The Red Queen hypothesis and Muller's ratchet 133
4.7 The role of sex in the creation of species 134
4.7.1 Two different patterns of species evolution 134
4.7.2 How do gene pools become split at the start of speciation? 137
4.7.3 Pre-zygotic and post-zygotic reproductive barriers 139
Summary 141
Further reading 142

5 SEXUAL SELECTION 143
Key themes 143
5.1 Sexual selection is a sub-category of natural selection 145
5.1.1 Comparing examples of natural selection and sexual selection 146
5.1.2 Re-formulating the natural selection argument for sexual selection 148
5.1.3 Manifestations of sexual selection 151

5.2 Why are females usually the "limiting sex"? 152
5.2.1 Females usually invest more in reproduction 152
5.2.2 Does inter-sexual selection produce payoffs? 155
5.2.3 Does intra-sexual selection produce payoffs? 159
5.3 How mate choice based on ornaments increases female fitness 162
5.3.1 Genetic benefits in the runaway hypothesis: "Sexy sons" 162
5.3.2 Genetic benefits in the good genes hypothesis: Better survival 164
5.3.3 Benefits in the good resources hypothesis: Honest promises 166
5.4 Sex role reversal 166
5.4.1 Sex role reversal: Adjusting investments made by males and females 168
5.5 Not all intra-sexual competition involves fighting 170
5.5.1 Scrambles are a form of indirect competition 171
5.5.2 Endurance and subterfuge are also forms of indirect competition 172
5.5.3 Sperm competition 173
Summary 175
Further reading 176

6 MATING SYSTEMS 177
Key themes 177
6.1 There are five major types of mating systems 179
6.2 Monogamy 181
6.2.1 Distinguishing among types of monogamy 181
6.2.2 Hypotheses for monogamy 184
6.3 Polygyny is the most common form of polygamy 188
6.3.1 Two models for polygyny 188
6.3.2 Human polygyny 191
6.4 The benefits for females of mating with multiple males 193
6.4.1 Non-monogamous females hedge their bets against infertility 195
6.4.2 Non-monogamous females reap genetic benefits 196
6.4.3 Non-monogamous females can reap direct benefits 201
6.5 Polyandry 204
6.5.1 Polyandrous mating systems are much less common than polygynous ones 204
6.6 Polygynandry 207
6.6.1 Reproductive skew 207
6.6.2 Cooperative breeding 209

6.7 **Promiscuity** 210
6.7.1 Factors associated with promiscuity 211
6.7.2 Scramble competitions are usually promiscuous mating systems 211
6.7.3 "Lek polygyny" is a promiscuous system resembling hook-up culture 212
Summary 214
Further reading 215

7 SEXUAL CONFLICT 217
Key themes 217
7.1 **Strategies for sexual success** 219
7.1.1 A comment on terminology 220
7.1.2 Sexual strategies exist in the context of sexual conflict 221
7.2 **Realms of sexual conflict** 221
7.2.1 Sexual conflict before mating 223
7.2.2 Sexual conflict during mating 229
7.2.3 Sexual conflict after mating 234
7.2.4 Sexual conflict during parenting 237
7.3 **Mating strategies can change with circumstances** 239
7.3.1 Making the best of things 239
7.3.2 Life history theory argues for plastic mating strategies 244
7.3.3 Sex allocation 246
7.4 **Same-sex parenting** 248
Summary 252
Further reading 253

8 SEX DETERMINATION AND DIFFERENTIATION 255
Key themes 255
8.1 **Are male and female bodies the only two options in sex determination?** 258
8.2 **The familiar method of sex determination relies on X- and Y-chromosomes** 261
8.2.1 Using insects to discover the role of sex chromosomes 261
8.2.2 The X- and Y-chromosomes in humans 265
8.2.3 Little genetic differences between men and women, but big phenotypic differences 265
8.2.4 The *SRY* gene and transcription 267
8.3 **Not all genetic sex determination relies on the XY system** 269
8.3.1 Genetic sex determination without Y-chromosomes 270
8.3.2 Are females ever the heterogametic sex? 270

8.3.3 Chromosomal sex determination where hermaphroditism is a phenotype 272
8.3.4 Genetic sex determination in honeybees 274
8.4 Environmental sex determination 276
8.4.1 Temperature is a non-social sex-determination system 277
8.4.2 Environmental sex determination can override genetic sex determination 281
8.4.3 Can mothers use temperature to select their offspring's sex? 281
8.4.4 Social systems of environmental sex determination 282
8.4.5 Anomalous sex determination caused by a member of a different species 284
8.4.6 Anomalous sex determination by ecotoxins 286
Summary 287
Further reading 287

9 HUMAN SEXUAL ANATOMY AND REGULATION 289
Key themes 289
9.1 Human sexual differentiation and function is highly dependent on hormones 290
9.1.1 The endocrine system regulates hormones in the bloodstream 292
9.1.2 Major hormones that regulate sexual development, function, and behavior 294
9.2 Sex differentiation is part of development 297
9.2.1 Human sexual differentiation before birth 298
9.3 The male reproductive system 299
9.3.1 Further male differentiation during the prenatal period 299
9.3.2 Male sexual development at puberty 300
9.3.3 Sexual anatomy of the human adult male 301
9.4 The female reproductive system 306
9.4.1 Further female differentiation during the prenatal period 306
9.4.2 Female sexual differentiation at puberty 306
9.4.3 Sexual anatomy of the human adult female 307
9.5 Anomalous sexual phenotypes in humans 309
9.5.1 Consequences of anomalies in the sex chromosomes 310
9.5.2 Intersex conditions that result from dysfunctional alleles 312
9.5.3 When girls become men at puberty 313
9.6 Sex linkage: Why some genetic disorders occur mostly in males 314
9.7 Cancers of sexually differentiated organs and tissues 315
9.7.1 Cancers associated with male organs 318
9.7.2 Cancers associated with female organs 319
Summary 321
Further reading 322

10 HUMAN FERTILITY AND BIRTH 323
Key themes **323**
10.1 Key differences between making sperm and eggs **324**
10.1.1 Spermatogenesis occurs from puberty to old age 325
10.1.2 Oogenesis occurs from the fetal stage to menopause 327
10.2 The menstrual cycle **328**
10.2.1 Most mammals have an estrous cycle instead 330
10.2.2 Is women's fertile period concealed? 331
10.3 Sexual arousal and response in men and women **333**
10.4 Fertilization and the making of a zygote **336**
10.4.1 Infertility 338
10.4.2 Contraception 340
10.5 Pregnancy **343**
10.5.1 The placenta 344
10.5.2 Labor and delivery 345
10.6 Breast-feeding **349**
10.7 Menopause **350**
10.8 Sexually transmitted infections **352**
10.8.1 A diversity of organisms cause STIs 354
10.8.2 Major bacterial STIs 354
10.8.3 Major STIs caused by viruses 355
10.8.4 Non-humans suffer sexually transmitted infections too 358
Summary **360**
Further reading **361**

Glossary **363**
Index **395**

Figures

1.1 Sexual development and regulation is diverse 3
1.2 Culture is a powerful modifier of biological sex differences 6
1.3 Most behaviors reflect both biological and environmental influences 10
1.4 Descriptions and prescriptions of behavior should be distinguished 12
1.5 Making comparisons among species enriches understanding 14
1.6 One classification system is used for all organisms 17
1.7 Line graphs show changes in a dataset connected by a line 20
1.8 Column graphs display values for different categories 21
1.9 Pie charts are used to indicate numerical proportions 22
1.10 Scatterplots are like line graphs but have key differences 22
1.11 More than one set of data can be shown in a single graph 23
2.1 Families tend to become smaller as societies modernize 27
2.2 Exaptation occurs when an older feature is co-opted for a new function 29
2.3 Bacterial conjugation involves a gene transfer but is not sex 34
2.4 Conjugation in the *Paramecium* is much more like sex 35
2.5 Mitotic cell division results in the production of two identical cells 37
2.6 Asexual reproduction is procreation other than by the fusion of gametes 40
2.7 Cyclic parthenogenesis alternates between sexual and asexual reproduction 42
2.8 Sperm are the smaller motile gametes and eggs are the larger immotile ones 44
2.9 Sexual dimorphism is a pervasive pattern in sexually reproducing species 48
2.10 The flower is the sex organ of most plants 53
2.11 Fertilization can be internal or external 54
3.1 The blending inheritance hypothesis was rejected because genetic diversity persists 61
3.2 Mendel manipulated pollination in his plots of experimental peas 64
3.3 Mendel tracked plant attributes over multiple generations of breeding 66
3.4a Punnett squares are used to analyze offspring possibilities for given parents 68
3.4b Mendel's first experiment crossed white-flowered and purple-flowered peas 69
3.4c Mendel also experimented with reciprocal crosses 69
3.4d Mendel's experiments did not stop after just one generation 69

3.5	This is a Punnett square for a dihybrid cross	71
3.6	There are four blood group phenotypes	77
3.7	There are two Punnett square possibilities for a type AB father and a type B mother	79
3.8	Eleanor Carothers was a pioneering chromosome researcher	81
3.9	Chromosomes are most visible when dividing	82
3.10	The human karyotype contains 23 pairs of chromosomes	83
3.11	Transcription and translation form the central dogma of molecular biology	85
3.12	Coded triplets in messenger RNA dictate particular amino acids in proteins	87
3.13	Based on Y-chromosome analysis, Thomas Jefferson probably had two families	89
3.14	There are three alternative sexual life cycles in nature	91
3.15	Crossing-over in gametogenesis creates unique sperm and eggs	93
3.16	Meiotic cell division is the process that creates haploid gametes	94
4.1	*On the Origin of Species* was published in 1859	103
4.2	Progressive changes in Europe led to the discovery of evolution	105
4.3	Charles Darwin traveled the world for five years in the 1830s	108
4.4	Prehistoric *Archaeopteryx* had feathers coupled with reptilian traits	111
4.5	Biogeography patterns reveal regions where different groups have evolved	112
4.6	Fitness is a measure of reproductive success	116
4.7	Small DNA sequence differences can have far-reaching consequences	119
4.8	Punnett squares can be modified for populations	124
4.9	Asexual reproduction is numerically more efficient than sexual reproduction	126
4.10	The pursuit of mating opportunities can be life-shortening	128
4.11	Numerous royal families once suffered from inbreeding depression	132
4.12	Anagenesis and cladogenesis are contrasting patterns of creating new species	135
4.13	The biological species concept relies on a breeding test	136
4.14	Allopatric speciation occurs when an original population is split in two	137
4.15	Hybrids are often sterile due to meiosis challenges	140
5.1	Elaborate traits like the peacock tail challenged early evolutionists	145
5.2	Sexual cannibalism still rewards the male victim	150
5.3	Armaments and fighting behaviors are usually male attributes	151
5.4	Male-female differences in reproductive investment have many consequences	154
5.5	Elongating a male widowbird's tail improves his mating success	157
5.6	Female pronghorn prefer a male that protects them from harassment	159
5.7	Human females use odor to assess attractiveness in a mate	161
5.8	Sexually selected traits can signal genetic quality	165
5.9	Shifting the burdens of reproductive investment can reverse sex roles	167
5.10	Sex role reversal is dependent upon reproductive investment levels	169

5.11	Competition for mates in many species is a scramble	171
5.12	Sperm move more efficiently by aggregating	173
5.13	Some attributes of the human mind are likely sexually selected	174
6.1	Mating systems reflect the structure of sexual relationships	179
6.2	There are five major mating system types	180
6.3	A minority of birds are genetically monogamous	183
6.4	DNA analysis can detect cases of extra-pair mating	185
6.5	Two parents are often better than one	186
6.6	Males will often guard females they have mated with	187
6.7	Some forms of polygyny rely on defending females	189
6.8	Polygyny can be a better choice for females	190
6.9	Human polygyny is widespread but mostly reserved for powerful males	191
6.10	Social status can be a component of extra-pair mating	194
6.11	Numerous models explain female non-monogamy	197
6.12	Female fitness commonly improves when mating is non-monogamous	199
6.13	Nuptial gifts are used in mating transactions	202
6.14	Bateman gradients reveal fitness consequences of having additional mates	205
6.15	Polygynandrous groups may still have a primary breeding pair	209
6.16	In lekking species, females choose males based on status and performance	213
7.1	Sexual coercion is one manifestation of sexual conflict	219
7.2	Extra-pair offspring are rare among humans	225
7.3	When harems are taken over, infanticide by males often follows	226
7.4	Male mating intensity can be sub-optimal for females	228
7.5	Hermaphroditic flatworms compete for the male role	230
7.6	Hermaphroditic snails use love darts to manipulate partners	231
7.7	Longer copulation increases sperm transfer in some invertebrates	233
7.8	Reproductive anatomy sometimes reveals a history of sexual conflict	235
7.9	The male guppy's clawed penis manipulates unreceptive females	237
7.10	Conflict can arise when current and future breeding investment is divided	238
7.11	Brood parasites offload parenting to unsuspecting foster parents	241
7.12	Male mating strategies can be condition-dependent	242
7.13	Among social mammals, subordinate individuals increase mating opportunities by prospecting	244
7.14	Sex allocation is adaptive when having offspring of a particular sex is advantageous	247
7.15	Sex allocation can reflect the body condition of the father	248
7.16	Same-sex parenting is known from a variety of bird species	251
8.1	Henry VIII was determined to have sons	257

8.2	Modern prenatal medical techniques can determine fetal sex	258
8.3	Simultaneous hermaphrodites have functional testes and ovaries	260
8.4	Some species have more than one type of each sex	263
8.5	The XY sex-determination system is found in diverse animals	264
8.6	The X-chromosome has far more genes than the Y-chromosome	266
8.7	Testis-determining factor (TDF) is a decisive gene product	268
8.8	Two key sex-determination systems are named after research insects	271
8.9	Some species have no females	273
8.10	Sex determination in social insects is known as haplodiploidy	275
8.11	Environmental temperature determines sex in many developing reptiles	277
8.12	Reptiles exhibit three patterns of temperature-dependent sex determination	278
8.13	Female spoon worms ensure nearby larvae develop into males	280
8.14	Chromosomal and temperature-dependent sex-determination systems can clash	281
8.15	Sequential hermaphroditism is influenced by social circumstances	283
8.16	Parasitic castration occurs when a host's reproductive system is taken over	285
9.1	Injections of androgen steroids can have serious health consequences	291
9.2	Two primary human steroids are estradiol and testosterone	293
9.3	Hormone regulation relies on feedback systems	296
9.4	Sexual differentiation in humans begins in the second month	299
9.5	The male reproductive system	302
9.6	Intromittent organs are highly diverse in nature	305
9.7	The female reproductive system	308
9.8	Non-disjunction of sex chromosomes leads to anomalies	311
9.9	Children who appear to be girls occasionally develop into men at puberty	314
9.10	Queen Victoria had a hemophilia allele on one of her X-chromosomes	317
9.11	Cancers of sexually differentiated organs vary in incidence and mortality	318
10.1	Fertility has been a perpetual theme in art since prehistoric times	325
10.2	Male and female meiosis differ in several significant ways	326
10.3	A number of features vary with the menstrual cycle	329
10.4	Many female mammals are only briefly receptive once per cycle	331
10.5	The sexual response cycle in humans has four stages	336
10.6	Some infertile couples benefit from in vitro fertilization (IVF)	339
10.7	A baby can have two mothers and one father	341
10.8	Surgery is used to sterilize adults who opt for permanent contraception	343
10.9	The placenta is unique to placental mammals	345
10.10	Thalidomide was a morning-sickness drug that caused developmental abnormalities	346
10.11	Cephalic presentation is the desirable orientation of the baby at term	348

10.12 The human breast is designed for milk production and delivery 349
10.13 Antibiotic resistance is a serious public health problem 353
10.14 HIV incidence varies with sex, ethnicity, and risky behaviors 357
10.15 Wildlife populations also suffer sexually transmitted infections 359

Boxes

1.1	Who's who in the multicellular world?	16
1.2	Interpreting scientific data using graphs	20
2.1	Do single-celled organisms reproduce using sex?	34
2.2	Sexual reproduction in plants	52
3.1	We use Punnett squares to understand patterns of inheritance	68
3.2	Who's the father?	78
3.3	Now who's the father?	88
3.4	There are three life cycle variants of meiosis	90
4.1	Categories of evidence for evolution	110
4.2	Sex and the definition of "species"	136
5.1	Suicide sex: Big cost for big payoff	150
5.2	Choosiness in humans: Females can smell genetically suitable partners	160
5.3	The human mind as a sexually selected trait	174
6.1	Testing for extra-pair paternity in birds	184
6.2	The polygyny threshold model	190
7.1	Do human females show characteristics of a dual mating strategy?	224
7.2	Sexual conflict in hermaphrodites	230
7.3	Off-loading parenting is one way to reduce inter-sexual conflict	240
8.1	Can there be multiple types of one sex?	262
8.2	One lick seals his fate	280
9.1	Intromittent organs are highly variable in nature	303
9.2	Queen Victoria's X-chromosome influenced European history	316
10.1	Two moms and a dad	341
10.2	Teratogens and the placental barrier	346
10.3	Gonorrhea evolution: Genomic responses to antibiotic use	352
10.4	HIV and AIDS	356

Preface

There are many realms of knowledge and ways of knowing, but there is no minimizing the fact that in modern life, the scientific realm is one of utmost importance.

For students to graduate with a breadth of knowledge, many colleges and universities embrace the tradition of general education, which simply means that a component of a student's degree includes enrichment in areas outside the major. At my own university, this is an entrenched practice, and for non-science students it includes a suite of fascinating science courses to choose from created specifically for this demographic. These courses usually have two over-arching purposes. One is to expose students to the scientific enterprise by teaching major methodologies, principles, and concepts. The other is to impart interesting and relevant science-generated knowledge along the way. Given the mandate of universities and colleges to educate students in a manner that prepares them for life in a modern civilization, it's well accepted that graduates will fare better having achieved some real scientific literacy.

In this tradition, many colleges and universities offer a non-majors course in the biology of sex. Such a course is eminently capable of satisfying the requirements of general-education science. For one thing, although sex can be meaningfully considered outside the scientific realm, ignoring the science of sex must impoverish the end result if understanding is to be achieved. For another, sex is also a particularly apt subject for general education because it is not only inherently captivating but it is almost certain to be directly connected to students' lives, because most people have identities that include their sexuality, and most are sexually active at least at some times in their lives. It's plain that sex infuses much of life, from the marketplace to the art gallery and countless points in between.

Yet, although many post-secondary institutions offer a non-majors course in the biology of sex, until now there has been no textbook that has tackled the subject for this group of courses. There are more specialized texts that deal with, for example, human reproductive biology. But for an expansive, non-specialist analysis of the subject in textbook form, this is the first of its kind. Our goal was to provide a resource crafted for the non-majors demographic—students who may not have any senior high school math or science but who nonetheless would benefit from the opportunity to take a science course that delivers interesting and relevant material.

To support the manner in which most versions of biology of sex courses are taught, we've taken the comparative approach. This means that although there is human material throughout, including two final chapters that are almost exclusively on the biology of human sex, there is also a great deal of material about other organisms. This is very purposeful, even if the ultimate goal is to better understand human sex and sexuality. A broad survey of sex across the natural world greatly enriches understanding about nature and about sex, and it assists in underscoring which principles approach the status of being universal and which, in contrast, are more idiosyncratic in their variation from organism to organism.

In preparing for this project, we knew from our prior discussions with instructors that the order in which different themes are taught varies from professor to professor. In recognizing the central position of evolutionary thinking in understanding sex, most argue that evolution should come first. We have almost succeeded in doing this, but there are two interceding chapters that follow the introductory chapter. The first (Chapter 2) is an acknowledgment and summary of a foundational concept—the obvious connection between sex and reproduction. The second is a bit more contentious. However, there is strong research that demonstrates that students succeed much better at understanding evolution if they first understand the basics of genetics. We apply that thinking here, covering the genetics of sex in Chapter 3 before we move on in Chapter 4 to evolution and sex.

Sex, including its biological underpinnings and its almost infinite and surprising permutations, is indeed a rich subject. Its study can be a vehicle for teaching scientific principles and for generating excitement about the power of science to illuminate. The sky is the limit, and we hope that instructors who adopt this book will find that this text helps them in that endeavor.

Alex Mills

1 Introduction

KEY THEMES

Most complex organisms use sex for reproduction, in which the genetic material of two individuals is mixed. Males and females differ fundamentally in their sex organs and in whether they produce sperm or eggs. Human attitudes to sex are shaped to a large degree by cultural forces. In complex organisms like humans, sex is not merely for procreation, even in the biological sense. We understand sex better by studying many species and their diverse sex-related behaviors.

Clownfish are embedded in popular culture by the very engaging and successful Pixar movie *Finding Nemo*. The movie opens with two clownfish, Marlin and Coral, and their brood of soon-to-hatch eggs. They are a sort of traditional, nuclear family. As they survey their brood, Marlin turns to Coral for confirmation that he has fulfilled his role, asking "Did your man deliver?" Marlin suggests naming half of them Coral Jr. and the other half Marlin Jr., but concedes one will be called Nemo in deference to Coral's wishes. When the barracuda strikes and Marlin has failed to save Coral and all their offspring but Nemo, he despairs, but sets out to protect his son from all risks.

As a film that seeks to entertain with a captivating tale rather than to document the biology of the ocellaris clownfish (*Amphiprion ocellaris*, Figure 1.1), we ought not to fault Pixar's mistakes. But it certainly reinforces human stereotypes, not just of sex roles but of the biology of sex too. When a human child is born, sex is almost always immediately announced, based on external male or female sexual apparatus, and parents usually have alternative boy and girl names ready. (Increasingly, the baby's sex is known and announced before birth.) Yet, when clownfish hatch, they are neither male nor female; they have internal immature reproductive organs known as ovotestes that are essentially male but that have egg-producing potential.

Not only that, the clownfish "family" is far from the human stereotype. The largest and dominant individual is the breeding female. The second largest, who is socially subordinate to the female, is the breeding male. There are several other immature, non-reproductive adults who must wait for the opportunity to mature. How does this happen? One of the breeders must first die. If it is the female, the breeding male responds by experiencing changes in the hormone concentrations in his body. This results in his rather rapid conversion into a female, and he then takes her place at the top of the pile. One of the immature individuals matures equally quickly into a breeding male. So, in clownfish, an individual's reproductive success is measured by how far up the social ladder it can go, with the pinnacle being the reproductive female, and an individual's sex is clearly impermanent. Imagine this: the father of one clownfish could be that clownfish's half-sibling's mother! Nor is Marlin's continual fussing over Nemo consistent with what is known

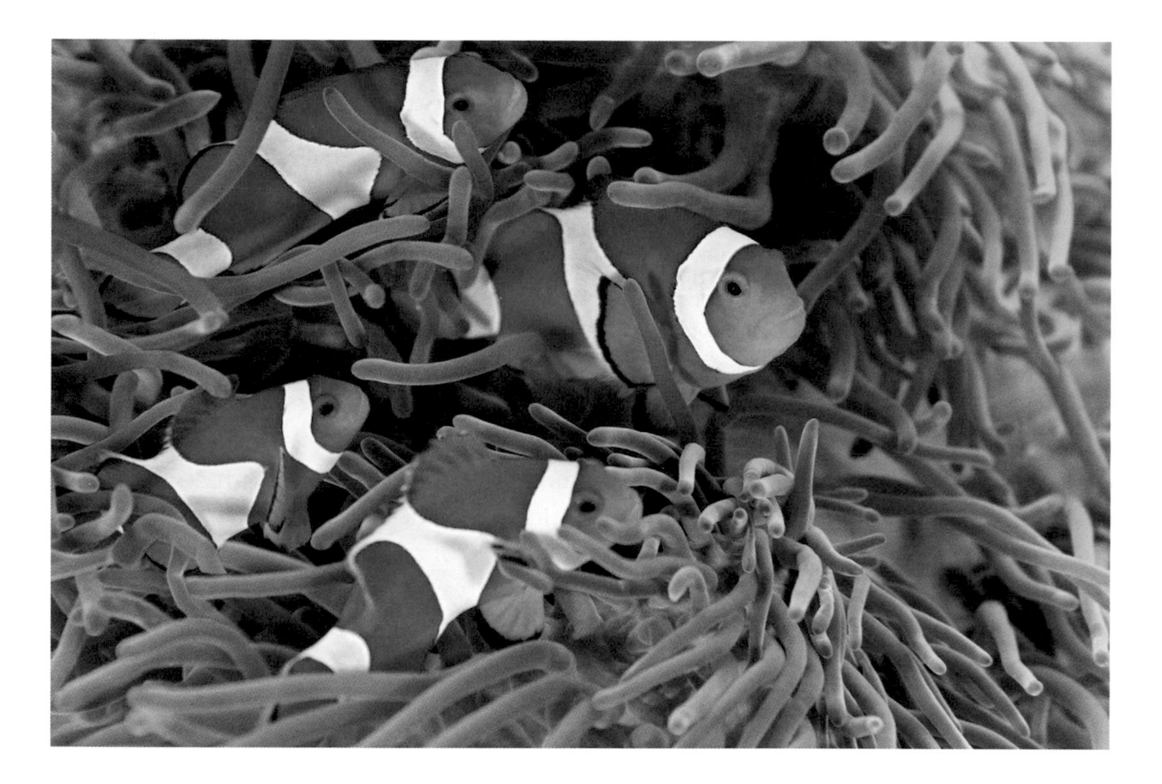

FIGURE I.1 SEXUAL DEVELOPMENT AND REGULATION IS DIVERSE
Clownfish live in groups where the breeding female is dominant, where hormones are used to control members of the group, and where sex change is a common phenomenon, driven by social circumstances.
iStock.com / crisod

about clownfish biology. In the real world, once the eggs hatch, the young fish do indeed float away as Nemo did, and parenting is over until the next set of eggs is laid.

This brief examination of clownfish reproduction introduces some of the most interesting themes in the biology of sex: male–female physical differences, sex determination, sex roles, patterns of dominance, mating systems, and differences in investing in offspring. People are naturally interested in these things. This is partly because for most people sex is a major part of experience and has significant emotional and physical rewards, including the production of children. But our interest is even more pervasive because sex influences such diverse aspects of our lives—relationships with others, health, identity, parenting, ubiquitous expressions in art and other cultural motifs, and, perhaps regrettably, commerce.

Most people, however, know rather little about the subject from a biological perspective. The "facts of life" are learned in Western cultures quite young, and people's pursuit of their own experiences soon follows in most cases, although some communities exert strong barriers. But how it all works and why it works in the way it does is a black box to most people. By this we mean that we essentially know its inputs and outputs but not the processes that go on in between. This is true of the information content (the genetics), the origin and maintenance of some very surprising patterns (evolution, including sexual selection), relationship patterns (mating systems), the sometimes disappointing discord associated with the subject (sexual conflict), and the mechanics and control of how it works in the body (anatomy, physiology, and sex differentiation). In a nutshell, these black box subjects are the themes of this book.

1.1 THE WIDER CONTEXT: SEX, GENDER, SEXUALITY, AND SEXUAL ORIENTATION

There is a lot about sex that is primarily social. Although biologists justifiably contend that many of the social aspects of sex, including gender identity, sexuality, and sexual orientation have biological influences, the capacity of social forces to rapidly modify diverse aspects of human sexual psychology, behavior, and identity is indisputable. A historical review of sex, gender, sexuality, and orientation in the West makes it clear that human social conditions, including rapidly changing ideologies and cultural norms, have a lot to do with sexuality, gender, sexual practice, and our understanding of the wider subject.

Some of these recent changes have been so substantial they have earned the appellations of "movement" and "revolution." For instance, it was only about a century ago that the suffragette movement gave women the vote, and other incremental legal-political changes since then have gradually contributed to the slow conversion of women to full personhood. With increasing female involvement in the workforce, partly fuelled by the loss of male labor during World Wars I and II, the feminist movement achieved great momentum in the tumultuous 1960s. Concurrent with that, and no doubt partly fueled by it, was the start of the sexual revolution. In Canada, for instance, it was in 1969, shortly after Prime Minister Pierre Trudeau famously said that the state had no business in the bedrooms of the nation, when numerous changes to the law related to sex were made. Abortion, which is the medical termination of a pregnancy, became available, albeit with certain restrictions. Contraception, whose purpose is to prevent pregnancy (Section 10.4.2), became more widely available and more effective with the legalization of the "new" oral contraceptive known as "the pill." And homosexuality, which is sexual attraction to and activity with a member of the same sex, was decriminalized. Widespread societal acceptance of homosexuality as well as other orientations and identities has been more recent, with same-sex marriage being legalized first in the Netherlands in 2001 and in numerous other countries since then, including Canada in 2005 and the United States in 2015. So much has changed in one human lifetime (Figure 1.2)! While Western democracies protect people from discrimination based on sex, sexual practice, and sexual orientation, inequities remain, many of which are huge in other parts of the globe.

Homosexuality is one form of *sexual orientation*. (We use the term lesbian for female homosexuals; there is no real counterpart for this term among males.) There are other forms of sexual orientation, of course. Heterosexuality is sexual attraction to and activity with a member of the opposite sex, while bisexuality is an orientation in which attraction or activity includes members of both sexes All three orientations represent ideas and practices that have existed for a long time, although with varying amounts of acceptance. A more recent challenge to conventional thinking about sex and sexuality is the development of gender as a concept other than as a synonym for sex. Whereas *sex* is used as a descriptor of an individual as male or female and is based on biology, *gender* describes an individual's interior experience and outward presentation related to sexual identity. In fact, the biological definitions of *male* and *female* are based on the type of internal reproductive organs a person has, known as gonads, or, even more fundamentally, the types of sex cells produced by those gonads—cells known as gametes. Where the gonads are testes (singular testis, and referred to as testicles in humans) and the gametes are sperm, the individual is a male. Where the gonads are the ovary and the gametes

FIGURE 1.2 **CULTURE IS A POWERFUL MODIFIER OF BIOLOGICAL SEX DIFFERENCES** Although biology plays a huge role in the differences among men and women, the power of culture to influence values and behaviors and to liberate or constrain practices should not be underestimated. Whether it was the demand for political equality through "the vote" a century ago, or the more recent demands for safe and legal abortion and the acceptance and validation of homosexuality over the past few decades, culture is a significant force. [1.2a] Records of the National Woman's Party, Manuscript Division, Library of Congress, Washington, DC; [1.2b] iStock.com / jcarillet

are the eggs or *ova* (singular *ovum*), the individual is a female. An individual's exterior sex apparatus are the genitals (Section 2.3.1), which are what we naturally use to identify the sex of a human baby. Although it is logical for us to identify an individual's sex by reference to genitals, the fundamental male–female difference is determined by the gonads and gametes.

1.1.1 Assigned sex and gender dysphoria

Sex assignment is the determination of an infant's sex at birth, usually based on an immediate inspection of the genitals. With the widespread use of prenatal ultrasound, sex assignment now commonly occurs before birth, at least informally. In a small proportion of cases, the genitalia of newborns can be ambiguous, in which case sex assignment usually follows medical testing, sometimes coupled with surgical intervention or hormonal therapies.

Many gender scholars assert that the mainstream binary (male and female sexes, masculine and feminine genders) is a fiction. Accordingly, there are those who claim that sex assignment is a social or even arbitrary exercise, but this position fails to accept that in the great majority of cases the genitals are clearly male or female. As we will see, the binary nature of genitals in most cases correspond with other biological binaries at the genetic, chromosomal, developmental, gonadal, hormonal, morphological, and even behavioral levels. Rarely, some parents attempt to raise their children without assigning sex, but this remains very challenging in a world where the binary nature of sex prevails, reinforced by biology.

Most people's gender identity matches their biological sex; such individuals are described as cisgender. Sometimes, however, sex assignment based on genital inspection does not align with an individual's developing identity, and this mismatch is termed gender dysphoria. Transgender individuals experience a substantial degree of dysphoria between their biological sex and their gender identity, sometimes presenting and identifying as the sex opposite to their biological sex or even undertaking medical "corrective" surgery to align their biological sex with their gender identity.

As we review patterns in the natural world, both human and non-human, we will find that there are indeed individuals who are not clearly male or female and that at some life stages, many organisms are effectively sex-neutral. In addition, what constitutes masculine or feminine traits is highly variable. But we will also find that the binary model runs very deep in biology, and it is abundantly clear that the male–female binary is hardly a fiction in nature.

1.1.2 Sex as a biological construct and gender as a social construct

It would be naïve to argue that sex and gender are unrelated, but it would be negligent to ignore their deep differences, and it is important to understand these differences as we embark on our examination of the biology of sex. Although the words *sex* and *gender* are used interchangeably in some circles, academic treatment of the two establishes them as distinctly different, although not unrelated. In the biology context, for instance, we might ask, "How do genes influence what types of genitals develop?" or "How do differences between the production of eggs and sperm influence the different mating strategies of females and males?" or "What role does sexual compatibility play in the evolution of species?" In the gender context we might ask, "How does the degree of state support for maternity leave and paternity leave influence male and female access to wealth?" or "Do school efforts to regulate sexting behavior enforce gender norms?" or "To what extent does sex hormone therapy correct gender dysphoria?" If you examine these questions, you might fairly suggest the distinction is not 100 per cent. For instance, sex hormone therapy referred to in the final question has a biological basis, but its application in that example focuses on gender.

Accordingly, academics who ask and answer questions about the biology of sex and those who ask and answer questions about gender largely operate in separate realms: the biology questions are mostly asked by biologists in biology departments in science faculties and the gender questions are mostly asked by social scientists in gender studies departments in arts faculties. This is obviously generalized; enquiries made by psychologists, anthropologists, and health professionals, for instance, contribute to our understanding of either sex or gender or both.

1.2 A CAVEAT ABOUT BIOLOGICAL DETERMINISM AND IDEOLOGY

An ideology is a set of opinions or beliefs used by an adherent to make a system intelligible—usually a human system such as a political, economic, or social one. Ideologies can be defensible, especially if based on evidence; after all, they are assembled to be explanatory.

However, they are often not evidence-based, they are prone to rigidity through the selective rejection of valid evidence or arguments to the contrary, they are exploited by groups to further their own interests, and they tend to polarize exchanges about complex social issues and public policy.

Biological determinism, the position that behavior is solely controlled by genes and the physiological and anatomical components of the bodies that such genetic information encodes, is an ideology. Usually, the term is applied to human behavior, although it is a more general concept than that. For many years, the dominance of men in society—in politics, religion, the workplace, and most other public spheres—was justified by many means, among them biology. It was alleged that anatomy was destiny, with male anatomy fating men for the corridors of power and female anatomy relegating women to what was left over. Feminism has managed to change much of that.

In contrast, social determinism is the position that human behaviors are solely determined by social interactions and cultural forces. It, too, is an ideology. In fact, the old dichotomy of *nature versus nurture,* which roughly translates to *biology versus environment,* is almost universally rejected by researchers because the preponderance of evidence is great that both are influential; development is strongly regulated by genes, but development simply cannot proceed in an environmental vacuum. An early demonstration of this came from the research of Nobel-winner Konrad Lorenz. He demonstrated that greylag geese (*Anser anser*) goslings have a genetically determined sensitive period during which they imprint on the larger organism in their immediate vicinity (biology). Usually that larger organism is the mother goose, but if it is a human, that human thereafter is treated as a mother goose, in preference to a goose (environment) (Figure 1.3). This is an elegant example of genes and environments interacting to produce a behavior. Humans are sexual animals. Our complex brains (which are biological) and patterns of development (which are also largely biological) interact with the highly diverse social structures and cultural norms of our environments (which are not primarily biological) to generate diverse outcomes.

FIGURE 1.3 **MOST BEHAVIORS REFLECT BOTH BIOLOGICAL AND ENVIRONMENTAL INFLUENCES** The young of many birds have an instinctive (biological) tendency to follow shortly after hatching, but they will follow whomever they were imprinted on (as in exposed to) during a sensitive period—obviously an environmental influence. In the case illustrated, the young birds that had been imprinted on a human when hatched continue to follow that human in the light aircraft.

When E.O. Wilson wrote *Sociobiology: The New Synthesis* in 1975, it had tremendous influence, but it also triggered much controversy. In his book, Wilson introduced the concept of sociobiology: the idea that because our genes represent our evolutionary past, human behavior is largely determined by DNA (deoxyribonucleic acid), which is the molecule that comprises the genetic code. Shortly afterward came *evolutionary psychology*, a similar idea arguing that useful psychological traits that are related to a diversity of behaviors (e.g., perception and mate choice) are characteristics that have been sculpted by human evolutionary history and therefore have a genetic basis. The controversy with sociobiology and evolutionary psychology has revolved around the fear that they are a repackaging of a biological determinist ideology that fails to acknowledge cultural and other environmental influences. Challengers, including some biologists, have offered two major lines of opposition to these approaches: first, that humans are a special case where logic and culture can be used to counter biological impulses, and second, that not all behaviors are likely to be adaptations constructed by evolutionary means. Accordingly, trying to explain everything in evolutionary terms ignores other important factors.

Both of these arguments (that biology need not be deterministic and that not all behaviors need to be adaptive) have merit. However, in general, the intellectual legacy flowing from the idea of sociobiology in the intervening four decades has been rich in generating fruitful hypotheses and rich in evidence assembled in the defense of sociobiology, from both nature and experimentation. Yet in the interest of everyone, as we consider the biology of sex, especially as it pertains to humans, we are chastened to remember David Hume's *is-ought problem.* In his *A Treatise of Human Nature* (1739), Hume argued that in describing the way nature is, we don't necessarily conclude that nature therefore *should* form the model for the way humans ought to behave (Figure 1.4). To some extent, we can choose behaviors and appropriate codes of conduct. This is particularly relevant when we consider the subject of sexual coercion, which includes human rape. Conversely, we should be careful not to be tempted to apply human concepts like fairness and morality when we consider infanticide or other selfish behaviors that occur in non-humans when one sex gains the upper hand over the other.

Nonetheless, there is ample evidence that for both humans and other organisms, biology makes much about sex understandable; genes have a strong influence not only on bodies and their contents but also on behaviors. In this text, we take care not to treat biology itself as an ideology or to use it as a prescription of how things ought to be. But as we proceed to descriptively characterize sex in the context of biology, we show that biology yields a rich and illuminating perspective on the subject, including social systems and behaviors. As we will see, there are genetic influences on behavior and resultant social patterns, and behaviors that have yielded reproductive rewards for individuals in the past are those that have been most successful and that therefore remain with us today.

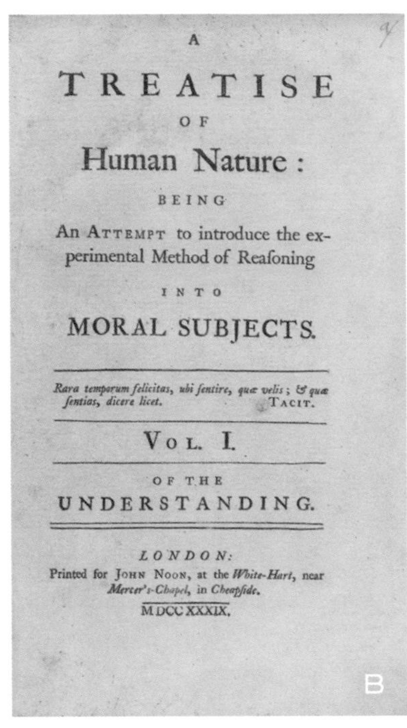
A

TREATISE

OF

Human Nature:

BEING

An ATTEMPT to introduce the experimental Method of Reaſoning

INTO

MORAL SUBJECTS.

Rara temporum felicitas, ubi ſentire, quæ velis; & quæ ſentias, dicere licet. TACIT.

VOL. I.

OF THE

UNDERSTANDING.

LONDON:

Printed for JOHN NOON, at the *White-Hart*, near *Mercer's-Chapel*, in *Cheapſide*.

MDCCXXXIX.

FIGURE 1.4 **DESCRIPTIONS AND PRESCRIPTIONS OF BEHAVIOR SHOULD BE DISTINGUISHED**

In 1739, Scottish philosopher David Hume (a) articulated in *A Treatise of Human Nature* (b) what has come to be known as the *is-ought problem*, arguing that it is not logical for humans to generate a *prescription* of behavior from a *description* of behavior. In 1975, Biologist E.O. Wilson (c) published *Sociobiology: The New Synthesis*, arguing convincingly that human behavior is made more understandable through the lens of evolutionary biology. Wilson did not promote *biological determinism* (that humans have no choice in their behavior), although critics feared sociobiology would be interpreted that way, and it sometimes has been.

(1.4c) Rick Friedman / Corbis Historical / Getty Images

1.3 THE COMPARATIVE APPROACH

Naturally, the most compelling perspective in examining the biology of sex will be its application to human beings. After all, we are sexual creatures and are curious about ourselves. One of the goals of post-secondary education is to understand ourselves better, and so this book focuses on humans more than any other species. But corollary goals in the tradition of liberal arts education in the West are to broaden perspectives, to challenge inherited knowledge that clashes with other knowledge systems, and to critically engage in the learning community. Without a doubt, the biology of human sex is far better understood when taught in a comparative manner by bringing in information from other biological systems; this deepens our understanding and illuminates the subject much more richly. Just as studying World War II while ignoring the political consequences of World War I, the Great Depression of the 1930s, and the legacies of imperialism and anti-Semitism would result in an incomplete and perhaps mystifying analysis, studying the biology of human sex without considering other systems—the fascinating and often surprising patterns we see in other organisms—gives the student a picture that is relatively bland and still leaves much of the subject poorly understood.

Accordingly, throughout this book there is a thread that revolves around the biology of sex as it is revealed in humans. But, there is also ample comparative material that sets the human material in a biological context, and this material deeply enhances the understanding of the human material (Figure 1.5). Some of this comparative material is built into the main narrative, and some of it is presented in stand-alone boxes that summarize other systems. In learning, it has commonly been found that using this comparative approach promotes a fuller explanation. For instance, one can make more sense of monogamy (the mating of one male and one female) if polygamy (mating involving more than one male or female) is presented in counterpoint.

To refer to other organisms, we need to apply a reliable naming system. Probably all students know what a chimpanzee is, but not all will know the Hanuman langur. Box 1.1 reviews the rudiments of biological classification so that we can place our periodic references to other species in a logical context.

FIGURE 1.5 **MAKING COMPARISONS AMONG SPECIES ENRICHES UNDERSTANDING**
The comparative approach takes the perspective that by comparing the characteristics and patterns of a diversity of species, our knowledge of all life, including humans, is enriched.
© Lev Frid

1.4 BIOLOGY AS A PART OF NATURAL SCIENCE

Having at least begun to flesh out what we mean by sex, we should also consider what is meant by biology, and by science for that matter. The natural sciences are divided into two main branches—life science, where biology as the study of life squarely sits, and physical science, which includes physics, chemistry, astronomy, and geology. Some sub-branches of physical science straddle the divide, such as biochemistry. Whether natural, life, or physical, the term common to all is *science,* a word that really means knowledge, but that in its modern use has come to have a more particular meaning.

Science, in fact, has a two-part meaning. It refers to the body of knowledge that deals with the material (or "natural" world), and it refers to a methodology that matches its goal of understanding that natural world. This body of knowledge changes over time, as new evidence is processed, making science a very dynamic discipline. Mathematics is not a natural science, but there is a natural link between the natural sciences and mathematics because math is a systematic and formulated kind of knowledge, and because it is the language that natural scientists often employ to describe the natural world. In our examination of the biology of sex, we rely little on math, although we will encounter it occasionally at the high school level. The social sciences such as psychology, economics, anthropology, political science, and many others, are distinguished from the natural sciences on the basis of the body of knowledge they investigate—broadly, human societies and relationships among its members in the one and the natural world in the other.

Biology itself has many sub-disciplines: evolution, genetics, cell biology, anatomy, behavior, and ecology, to name a handful. In our examination of the biology of sex, we will rely on material from many of these sub-disciplines because they collectively facilitate a more complete understanding of the subject.

1.5 THEORIES AND THE METHODOLOGY OF SCIENCE

A word that sometimes causes a conceptual challenge for students is theory, because of the mistaken notion that it is equivalent to hypothesis, or more pejoratively, a guess. In science, we use *theory* very differently. A theory is a well-supported explanation of a broad class of observations about the natural world that relies upon many sources of carefully collected data and experimental results. So, when scientists refer to a theory, they are referring to an organizing idea for which there is tremendous support. In biology, three major theories that we will encounter are *Cell Theory* (that the cell is the basic unit of structure in all organisms), *Gene Theory* (that genes are the units of inheritance), and the *Theory of Evolution* (that all species are related by patterns of ancestry and that they change over time).

A hypothesis is narrower. It is a proposed explanation for a phenomenon that usually includes predictions about observations or experimental results that will allow it to be refuted or supported.

BOX 1.1 WHO'S WHO IN THE MULTICELLULAR WORLD?

Biodiversity is a contraction of "biological diversity." Although it includes the diversity at genetic and ecological scales, the usual way we think about biodiversity is *species* diversity. We do not know how many species there are on Earth—there may be a half-million beetle species alone—but there is certainly a mesmerizing variety. During the Age of Exploration, the imagination of the European public became enchanted as descriptions and specimens of exotic species made their way back to Europe. In the 1700s, as specimen after specimen made it apparent that the number of species on Earth was almost overwhelming, natural scientists faced the challenge of categorizing species in a systematic way that was consistent with the developing views of scientific methodology.

That challenge was met by several, but it was an eighteenth-century Swedish naturalist, Carl Linnaeus, who formalized the system of arranging organisms in hierarchical categories according to similarity—called classification or taxonomy—and of naming species uniquely—called binomial nomenclature. Throughout the intervening two and a half centuries, this system has been modified. However, we retain much of what Linnaeus established, notwithstanding that the perspective over that period has switched from a concept of unchanging species to the view that species diversity is driven by evolution, the concept that species change over time and that current species are descended from extinct ones. Although it is now a simplification, organisms continue to be placed in a category at each of seven ranks. From the largest category to the smallest, these are Kingdom, Phylum, Class, Order, Family, Genus, and Species (Figure 1.6). The name of a species is a *binomial* ("two-name") made up of the last two levels.

We will periodically rely on a variety of species to compare humans with other organisms. It is helpful in these cases to be able to classify these organisms—not just to provide a name but also to give context to the species we compare. Consider the image in Figure 1.6, where we have a sample of five species. Note that each species is in a category at all seven ranks. All five are in the Animal kingdom and all are Chordates (the phylum that is basically all animals with backbones). There are several Chordate classes, including mammals, but of the species here, we have representatives from the Amphibian class and the Reptile class. It is at the class level that biologists decided long ago that of these species, the spotted turtle is the most distinctly different, so they placed it in a different class (the reptiles) from the others (the amphibians). As we proceed in this manner, you can see that the Amphibian class is here divided into two orders (there is a third tropical one not represented in the image) and that the Frogs and Toads order is divided into two families (and there are additional families too that are not shown).

There are two true frogs in our sample, and because these two are thought to be quite closely related, they are also placed in the same

	Green Frog	Leopard Frog	Gray Treefrog	Spotted Salamander	Spotted Turtle
Kingdom	Animals				
Phylum	Chordates				
Class	Amphibians				Reptiles
Order	Frogs and Toads			Salamanders	Turtles
Family	True Frogs		Treefrogs	Mole Salamanders	Pond Turtles
Genus	*Lithobates*		*Hyla*	*Ambystoma*	*Clemmys*
Specific epithet	*clamitans*	*pipiens*	*versicolor*	*maculatum*	*guttata*

FIGURE 1.6 **ONE CLASSIFICATION SYSTEM IS USED FOR ALL ORGANISMS**
All organisms are classified in the same manner by reference to seven taxonomic ranks (kingdom through specific epithet), yielding a unique *binomial name* or species name derived from the genus name and the specific epithet, such as *Clemmys guttata* for the spotted turtle.
Drawings by Peter Mills

genus, *Lithobates*. To distinguish the leopard frog and the green frog in a formal sense, they have different *specific epithets* (the second half of the binomial), and so the species name for the former is *Lithobates pipiens* and for the latter is *Lithobates clamitans*. Whenever two species have been placed in the same genus, it reflects their high degree of similarity. If two species have been placed in different genera (the plural of genus) but in the same family, they are still similar but less so. The gray tree frog (*Hyla versicolor*) and another common amphibian not shown, the spring peeper (*Pseudacris crucifer*), are examples; both are in the tree frog family (Hylidae), but as is obvious from their names, they are in different genera. Where biologists refer at one time to several species in a genus, such as the genus *Ursus* for bears, they use the form *Ursus* spp. Also, where mention is made in sequence of two species of the same genus, as in black bear and then grizzly bear, the genus is abbreviated to the letter in the second instance, as in black bear (*Ursus americanus*) and grizzly bear (*U. arctos*).

An epithet is a word or phrase attached to a name that describes a quality, like *Conan the Barbarian* or *Linnaeus, the father of biological classification*. Sometimes you can derive information from specific epithets. *Hyla versicolor* can change color, for instance. In any event, this system of classification is a universal one, covering everything from primates to starfish to water lilies to field mushrooms. By convention, all species names are written in italics. *Homo sapiens*—humans—is a familiar one, as is *Canis familiaris*, the domestic dog.

What constitutes support for a hypothesis, or, more broadly, for a theory? In the natural sciences, it is empirical or experimental evidence. *Empirical* evidence is data that comes from experience in the sense that it comes from the senses, although with modern technology we include data from all sorts of gadgets that can sense things that humans cannot. Empirical data are counted, measured, and characterized in diverse ways related to observation. Sometimes empirical and experimental are distinguished from each other, although experimental results are often a principal part of empirical information. When we distinguish the two, empirical evidence comes from non-experimental but careful and systematic analysis of data (discovery science) and experimental evidence comes from experimentation (experimental science). Both are used to test hypotheses.

In biology, there are countless hypotheses. For every pattern or mystery, there is never fewer than two, because a hypothesis is usually framed in a dual way, with the *research* or *alternative hypothesis* (that says there is a difference between two groups, for instance) and the *null hypothesis* (that says there is no support for the claim that there is a difference in the two groups). Not surprisingly, there are often several competing research hypotheses. Experimenters are careful to design their procedures so that differences that appear in the results can be attributed to a factor articulated in the hypothesis, leading to support for, or evidence against, the hypothesis.

For instance, if you were to hypothesize that being under the influence of alcohol increased a person's willingness to engage in risky sex, at the minimum your investigations would make sure that you had two groups, one whose behavior was assessed while under the influence of alcohol and one whose behavior was assessed while sober. We would call the group under the influence the *treatment group* and the sober group the *control group*, and we would ensure that otherwise there were no differences between the groups. For instance, if both men and women were involved in the study, both sexes would be evenly distributed between the two groups; if not, then it might be a person's sex that influenced their willingness to engage in risky sex, and you could not draw conclusions about alcohol. This is what is meant by a controlled experiment. This often-simple approach has proved to be perhaps the most powerful tool in science.

Science works with data sets. To analyze those data sets, scientists typically employ mathematical tools. Accordingly, the results of scientific investigations tend to be reported quantitatively, i.e., by using numbers and statistical analysis. In accordance with the phrase "a picture is worth a thousand words," those same quantitative results are usually presented graphically, meaning that most scientific papers have one or more graphs that present results visually (Box 1.2).

Finally, students may also struggle with the terms *deductive* and *inductive* reasoning and the relationship of the two terms to theories, hypotheses, and experiments. Theories are used deductively in the sense that they inform thinking that leads to the development of hypotheses to test particular phenomena, which are indeed then tested with experiments or other data sets. On the other hand, theories are formulated inductively in the sense that empirical and experimental data are collected to generate a concept that logically accommodates all the evidence. As Charles Darwin collected data on the reproductive lives of organisms, he used that inductively to formulate his theory of sexual selection (Chapter 5).

CHAPTER 1 SUMMARY

- Sex is the prevalent mode of reproduction in complex organisms.
- An individual organism's biological sex relies upon the definition of sex organs (ovaries or testes) and the types of gametes (reproductive cells) they produce—either eggs or sperm.
- Sex is a biological construct used as a binary descriptor of biological maleness or femaleness, and gender is a social construct used to describe an individual's interior experience and identity.
- In humans, sexual behavior is strongly influenced by both biology (genes, gonads, hormones) and cultural factors.
- Understanding the biology of sex benefits from the comparative approach that considers patterns in multiple species.
- Organisms are classified and named in accordance with evidence about their relatedness through evolution.

BOX 1.2 INTERPRETING SCIENTIFIC DATA USING GRAPHS

Depending upon the type of data, different types of graphs are called for, and so the first thing to be done when seeing a graph is to note the type it is. Unless a graph is looked at systematically, it is easy to derive an incorrect message. For instance, if it is a graph that has two axes, an x-axis on the horizontal and an y-axis on the vertical, it is always a good idea to first read the axis labels so that you are thinking about the two attributes whose relationship is being presented.

To consider graphical presentation of scientific data from either discovery science based on empirical data or experimental science based on experimentation, we'll consider examples pertinent to the biology of sex. In Figure 1.7, we see a *line graph* that shows the relationship between the age of onset of menstruation and year. In line graphs, we usually have a single type of data (e.g., age of onset) that is plotted on the vertical (y) axis against some independent variable on the horizontal (x) axis, most commonly the passage of time. Usually, the points are joined by a line that shows the pattern over the period covered. These data are from South Korea and Canada, but this pattern of progressively earlier onset during modernization exists throughout the developed world. This is an example of discovery science because although it tests the hypothesis that girls now reach sexual maturity younger, it is not based on a controlled experiment but is instead derived from careful use of historical observational data.

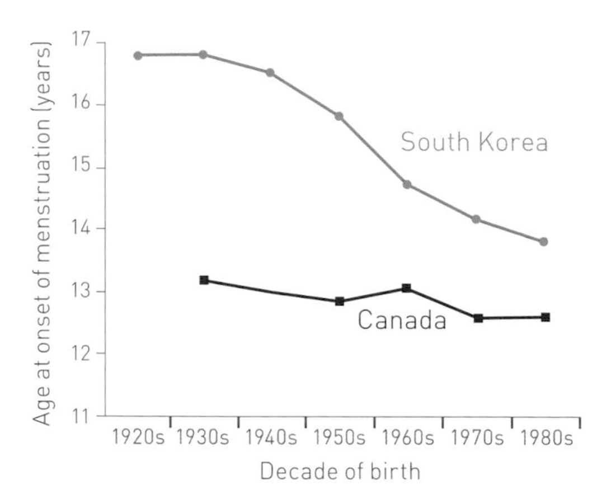

FIGURE 1.7 **LINE GRAPHS SHOW CHANGES IN A DATASET CONNECTED BY A LINE**

This line graph shows how the age of onset of menstruation has changed over time. These patterns are typical for developed nations. For many nations, the change began a century or more ago, and further changes have been modest (e.g., Canada). For others, the change has been more recent (e.g., South Korea).

In Figure 1.8, we see the results of a controlled experiment presented as a *column graph* (known as a bar graph if the orientation is horizontal instead of vertical). Column graphs are particularly appropriate where the data are used to compare categorical pairs, such as whether or not a person was on the pill, and whether or not a person made an advantageous mate preference. The difference in the column heights is the graph's message.

In the figure, a factor in the question being addressed was whether the women were taking the "pill," meaning the popular hormonal contraceptive pill, which triggers several changes in the woman's body through the influence of hormones. The question was whether taking the pill influences young

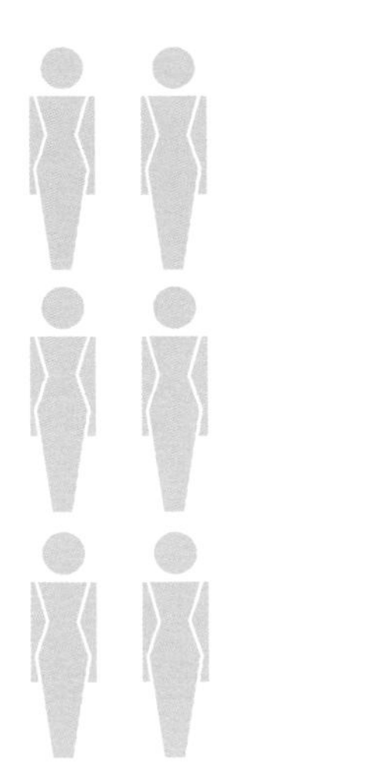

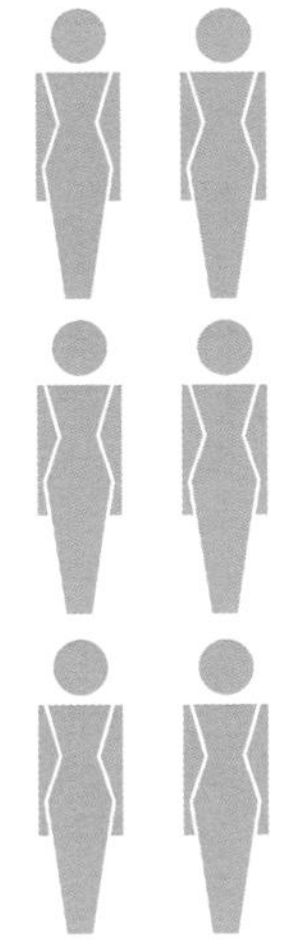

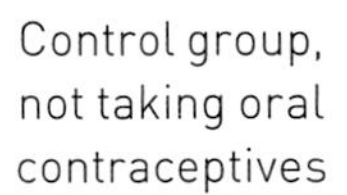

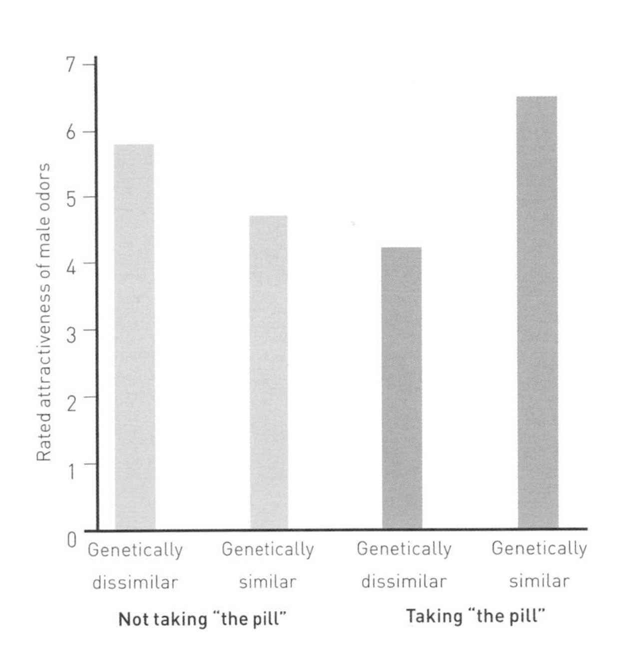

FIGURE 1.8 **COLUMN GRAPHS DISPLAY VALUES FOR DIFFERENT CATEGORIES**

On the right, a column graph indicates patterns of human female preference based on the subjective attractiveness of male odors. To the left of the graph is an image contrasting the treatment group (participants taking oral contraceptives) and the control group (participants not taking oral contraceptives). The responses of the women were assessed for genetic appropriateness, where a dissimilar male is a better choice genetically and a similar male is a worse choice, based on immunity genes.

From Wedekind, Seebeck, Bettens, & Paepke (1995). By permission of the Royal Society

women's assessment of the attractiveness of potential mates. Young heterosexual women were placed into one of two groups, those on the pill (the treatment group) and those who were not (the control group). They were then asked to use their sense of smell to rate the attractiveness of T-shirts that had been worn by young men for an extended period. Genetic tests were done on both male and female participants to assess mate suitability (based on dissimilarity of immunity genes). The young women tended to choose genetically dissimilar males (a good thing) when not on the pill, but they chose inappropriately by preferring genetically similar males when on the pill, suggesting the pill interferes with their ability to make advantageous mate choices.

A third type of graph is the *pie chart*. It is conceptually simple: some category represented by the whole circle is divided into its component parts (i.e., pie pieces) whose relative sizes indicate the proportions of those components. For instance, Figure 1.9 shows the proportions of species in the various vertebrate classes.

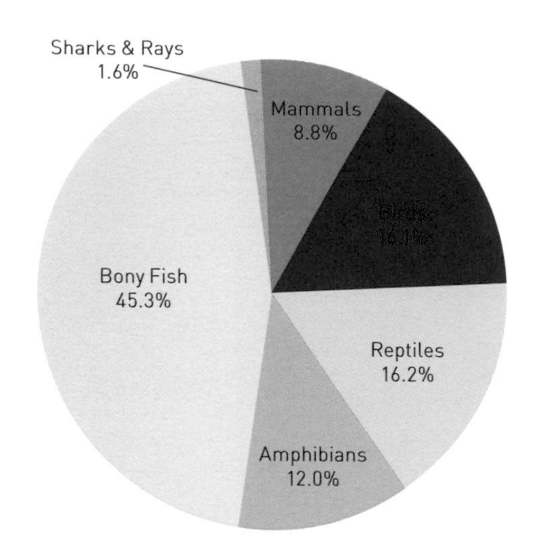

FIGURE 1.9 **PIE CHARTS ARE USED TO INDICATE NUMERICAL PROPORTIONS**

This pie chart shows the distribution of the approximately 62,000 known vertebrate species among their component classes. When pie charts provide percentages, they must add up to 100.

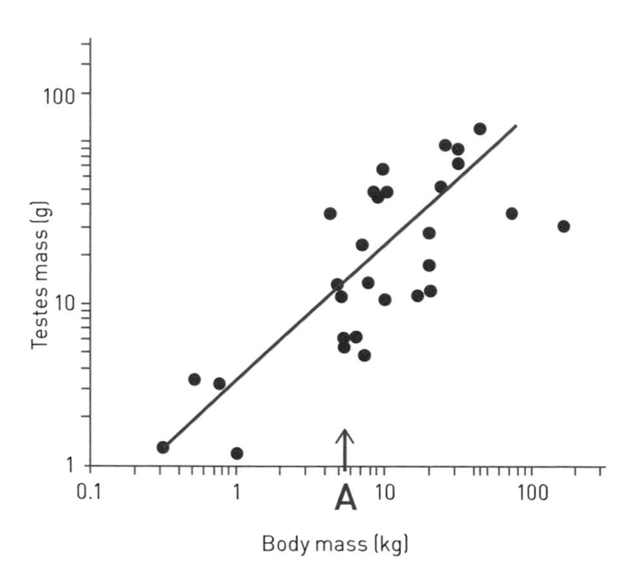

FIGURE 1.10 **SCATTERPLOTS ARE LIKE LINE GRAPHS BUT HAVE KEY DIFFERENCES**

Whereas a line graph has one value for every equally spaced value on the x-axis, a scatterplot can have multiple values for each, and the x-values are commonly not evenly spaced. This scatterplot shows the relationship in primates between testes mass (i.e., gonad size in males) and body mass. Each dot is a species whose position on the graph is determined by body mass (the x-axis value) and testes mass (the y-axis value). As is typical for a scatterplot, for a given body size (e.g., at point A on the x-axis), there is a variety of species that have different relative gonad size (the various y-values indicated by the dots above A). The line is known as a *regression line* (or *line-of-best-fit* or *trend line*), and it represents the relationship between the two axes.

We'll consider one more very common graph type in science: the *scatterplot*, shown in Figure 1.10. It resembles the line graph, but it is used when the researcher wishes to present the relationship between two measures from samples that have been taken. So, each point on a scatterplot is a sample, and the points collectively show the relationship between whatever traits are measured on the x-axis and on the y-axis, known as a *correlation*.

In Figure 1.10, note that we have two anatomical size measures. Along the x-axis, we have body mass (increasing rightward) and along the y-axis we have testes mass (increasing upward). Recall that the testes are the male gonads, referred to in humans as testicles, and sperm are produced in the testes. So, testes mass on the y-axis is a measure of the size of the male gonads, which in turn reflects their capacity to generate sperm. Because this is a scatterplot, each point is a sample, and in this

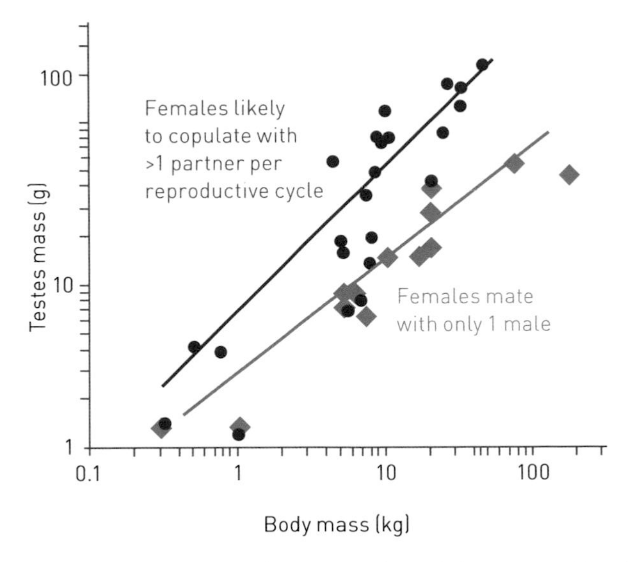

FIGURE 1.11 **MORE THAN ONE SET OF DATA CAN BE SHOWN IN A SINGLE GRAPH**

This is the same scatterplot as in Figure 1.10, but here the sample of species is divided between those primates that are non-monogamous (red circles and red line-of-best-fit) and those that are monogamous (blue squares and blue line-of-best-fit). In species where females are non-monogamous, males are equipped to produce relatively more sperm.

case, each point is a primate species. (Primates include monkeys, gorillas, and humans, among others). Note that there is diversity in the relationship; not all points fall exactly on the line. If you put your finger on point A on the x-axis (representing one body size) and drag your finger upward, you will encounter several species. The first one you hit (below the line) has relatively small testes for body size A, and the last one you hit (above the line) has relatively large ones for body size A.

What is the line on the scatterplot? This is a line generated mathematically that is known as a *regression line* (often called a *trend line* in business or, more generally, a *line-of-best-fit*). It is the line that is the best representation of the data points. Scatterplots and the regression lines associated with them are sometimes misinterpreted; they show the relationship between two measures (body mass and testes mass in this case), but they don't show that one measure *causes* another. That is why we say they represent *correlation*, not *causation*.

The researchers who generated the graph in Figure 1.10 were not simply interested in gonad size relative to body weight. The primates in this sample can be divided into two groups, depending upon their mating systems (Chapter 6). If you switch your attention to Figure 1.11, you will find these two groups marked separately. The red circles represents species with a non-monogamous mating system, where females mate with more than one male during each reproductive cycle. In the group represented by blue squares, females mate monogamously during each cycle. Treating them separately, we now generate separate regression lines, and see that the non-monogamous line is higher and more leftward than the monogamous line, indicating that the non-monogamous species are characterized by relatively larger testes, and hence a relatively greater capacity to generate sperm.

- Scientific theories are explanations of broad classes of observations supported by a wealth of evidence, especially experimental and empirical evidence.
- Scientific data are often best presented in one or more graphic forms.

FURTHER READING

Buss, D.M. (2009). The great struggles of life: Darwin and the emergence of evolutionary psychology. *American Psychologist, 64*(2), 140–148. https://doi.org/10.1037/a0013207

Cleveland, W.S., & McGill, R. (1985). Graphical perceptions and graphical methods for analyzing scientific data. *Science, 229*(4716), 828–833. https://doi.org/10.1126/science.229.4716.828

Gould, S.J. (1996). *The mismeasure of man.* New York: Norton.

Lane, D.M., & Sándor, A. (2009). Designing better graphs by including distributional information and integrating words, numbers, and images. *Psychological Methods, 14*(3), 239–257. https://doi.org/10.1037/a0016620

Lewontin, R.C. (2001). *Biology as ideology: the doctrine of DNA.* London: Penguin.

Ngun, T.C., & Vilain, E. (2014). The biological basis of human sexual orientation: Is there a role for epigenetics? *Advances in Genetics, 86,* 167–184. https://doi.org/10.1016/B978-0-12-800222-3.00008-5

Sanford, G.M., Lutterschmidt, W.I., & Hutchison, V.H. (2002). The comparative method revisited. *Bioscience, 52*(9), 830–836. https://doi.org/10.1641/0006-3568(2002)052[0830:TCMR]2.0.CO;2

Wedekind, C., Seebeck, T., Bettens, F., & Paepke, A. (1995). MHC-dependent mate preferences in humans. *Proceedings: Biological Sciences 260*(1359), 245–249.

Wilson, E.O. (2000). *Sociobiology: The new synthesis.* Cambridge, MA: Belknap Press of Harvard University Press.

Zimmer, C. (2008). What is a species? *Scientific American, 298*(6), 72–79. https://doi.org/10.1038/scientificamerican0608-72

2 Sex and Reproduction

KEY THEMES

Sexual motivation ultimately has an evolutionary explanation, being procreation that passes on genes, but there are immediate motivations too, including pleasure. Many relatively simple organisms reproduce asexually, but there are also such cases among some complex organisms. In addition to sex organs, males and females differ in a diversity of secondary characteristics that include body features and behaviors. Most plants reproduce sexually, and sexually reproducing animals can have internal or external fertilization.

Most of us have seen old black-and-white family photos where the mother and father are surrounded by numerous children, sometimes as many as ten or more (Figure 2.1). But in twenty-first century households in the developed world, this is now a very rare thing. A replacement rate that maintains a steady population size is two babies reaching adulthood per woman—one to replace her and one to replace the father(s). Yet the current rate (2015 data) in most of the developed world is below this—for instance not quite 1.9 in the United States and the United Kingdom, about 1.8 in Australia, and as low as 1.6 in Canada. What has changed? Part of it is technological change, in particular improved methods of birth control, especially the contraceptive pill. But such technological change reaches much farther when it's coupled with cultural change. With a waning of religious authority over sexual relations, or at least a softening of position, the old view that sex was only legitimately for procreation has continued to erode. A dramatic expansion of the socially acceptable roles of women outside the home has also contributed to the low birth rate, as has the broadening of what constitutes a "family" to include childless ones.

Yet there is no evidence that the decline in birth rates is matched by declines in sexual activity. In fact, we can say with confidence that the birth rate decline is solely due to people, mostly women, taking control of their reproductive lives. It seems likely there may be more sexual activity now than in the times of large families, given current liberal views about sexual freedom that contrast with the older prohibitions about premarital sex.

So if we ask why people have sex, we expect only a small minority will answer, "for producing children." Most will provide an answer in which that is only part of the explanation. After all, women engage in sex after menopause (when pregnancy is no longer possible) and during pregnancy (where further pregnancy is not possible), and both sexes engage in creative sexual activity (whether heterosexual or homosexual) that does not include sexual intercourse. Some religious doctrines have asserted, or continue to assert, that sex is only for procreation, and that any other indulgence in it should be treated as illegitimate. But because people's experience aligns so poorly with this

FIGURE 2.1 **FAMILIES TEND TO BECOME SMALLER AS SOCIETIES MODERNIZE**
Large families like this are mostly a thing of the past in the developed world. There is no doubt that sex in humans is rooted in reproduction, just as it is in other species. But in the liberalized cultures of the West, most sexual activity is clearly pursued for non-procreative purposes, especially given that birth control (various methods used to prevent full-term pregnancy following sexual intercourse) is easily available.

view, sexual activity has long been a common battleground between people and religious authority. Most people in the relatively sexually liberated West will answer the question of why people have sex by referring to several complexly related ideas that propose love, intimacy, pleasure, power, and other possibilities as alternatives to the baby-making function of sex.

2.1 DOES SEXUAL MOTIVATION REVEAL ITS "PURPOSE"?

To answer this question, we must pay attention to the level at which we form our answer. *Proximate answers* resort to the most immediate levels of causation, while *ultimate answers* resort to more removed, but deeper, levels. The explanation for sex at a proximate level would focus on the pleasure rewards. This could be done by reference to nerve networks that transfer impulses experienced by the brain as pleasure, or by the rush of endorphins associated with sexual arousal and fulfillment. In short, a good proximate answer is because it is pleasurable. We will consider these proximate causes in more detail later, when we discuss anatomy and sexual function in humans (Chapter 9).

The explanation for sex at an ultimate level is evolutionary, and we spend time considering the evolutionary explanations for sex patterns in Chapter 4 and beyond. But we should articulate a brief ultimate explanation here: sex persists because it has reproductive rewards. Individuals through history who have engaged in sex (specifically intercourse) are those who have produced descendants. Those whose patterns of sexual behavior (partner choice, frequency, etc.) led to more offspring have contributed more to subsequent generations than those whose patterns yielded fewer offspring. Looked at from the perspective of the current generation, everyone alive today has ancestors who were reproductively successful.

Admittedly, this is not how sex is experienced. It is only a relatively small proportion of the time that people engage in sexual activity with the goal of pregnancy in mind. In fact, it is much more common that the goal includes a non-pregnant result. If people were to rely simply on the goal of pregnancy as a reason for sexual intercourse, without the intervention of the rewards of pleasure, there would doubtless be far fewer offspring produced. This is what Jared Diamond was getting at when he produced his popular 1997 book *Why is Sex Fun?* The pleasure rewards of sex have a genetic basis: the greater the pleasure, the more likely it is that individuals will pursue sexual activity, and the more likely it is that pregnancy will result (in the days before birth control, at least). The genes that contribute to that pleasure and that therefore result in more reproduction will inevitably contribute disproportionately to subsequent generations (ultimate causation), thereby genetically entrenching a high interest in sexual activity.

FIGURE 2.2 EXAPTATION OCCURS WHEN AN OLDER FEATURE IS CO-OPTED FOR A NEW FUNCTION

Feathers (a) likely evolved for insulation and were later co-opted for flight. Similarly, there is no doubt sex arose as a mode of reproduction, but in social species, such as common chimpanzees (*Pan troglodytes*) (b) and humans, it is an exaptation that has been co-opted for additional social purposes.

(2.2b) iStock.com / Windzepher

2.1.1 Sex has diverse purposes through co-option

Sex is rooted biologically in the reproductive function. This is patently obvious, because wherever we observe sexual relationships in the natural world, we see reproductive consequences. In most species, we can be reasonably certain that the sole purpose of sex is for reproduction. While the basic logic of that is evident, you may find yourself resisting with respect to human sexuality, having the sense that sex has other legitimate purposes, even perhaps biologically based ones. In this, your resistance has a strong footing. For a more limited number of social species, the most interesting of which are the primates (including humans), we also see sex filling biological functions that are independent of reproduction.

To explore this more fully, we need to introduce the term exaptation (Figure 2.2). Exaptation was a term coined by evolutionary biologists in the 1980s to describe the concept where features adapted for one original purpose are co-opted for a new purpose. Sometimes authors use the term pre-adaptation for this concept instead, but that is a poor term because it suggests there was some anticipation of the new purpose.

Before we apply the concept of exaptation to sexual behavior, consider a simpler example to assist in clarifying the concept: feathers. There is good fossil evidence that the theropod dinosaurs that are the ancestors to modern birds sported feathers without being able to fly. Although a number of explanations for the evolution of feathers have been proposed, the most plausible is that they provided insulation. That is, feathers yielded evolutionary rewards because heat retention led to better survival, which in turn led to better reproduction and better-adapted offspring. As an advantageous device, the feather therefore spread in the bird world, making it the most distinctive feature of the class. Once feathers existed, they evolved to serve other purposes as well—most notably flight—which provided an evolutionary advantage. Note that the original purpose, thermoregulation, did not disappear, even though further evolutionary change meant feathers made flight possible. Feathers continue to provide birds with thermoregulatory survival rewards today. This is an exaptation because the original function for the feather was insulation, but this was later exploited for flight. Both are consistent with evolutionary explanations, and hence both are biological in nature.

Similarly, although a review of all sexual organisms makes it clear that the original purpose of sex is reproduction, an examination of the social lives of socially complex vertebrates, especially humans, also makes it clear that sex has non-procreative purposes. And because both purposes have biological rewards (the production of offspring in the former case, and better health, survival, and supportive relationships in the latter case), both are aspects of the biology of sex. We will expect to be able to explain much about sexual behavior in organisms by considering the deep relationship between sex and reproduction, but we will also expect to find some patterns that are better explained by this co-opted view, especially in humans.

Research that has focused on why people engage in sex other than for procreation has produced many specific reasons, but they can be generalized to four categories: (a) physical rewards, including pleasure, exercise, and stress management, (b) emotional rewards, including love, bonding, intimacy, and connectedness, (c) goal attainment, including revenge, the exercise of power, the forging of alliances, and access to resources and social status, and (d) the reduction of insecurity, including self-esteem boosting, conflict resolution, fulfillment of duty, and guarding against infidelity in one's mate. We are complicated organisms, and it is no surprise that our sexuality matches this level of complexity. As organisms, humans seek physical and emotional rewards, the attainment of goals, and ways of coping with insecurity. These certainly have biological components, since our continued existence depends upon them, but they are highly influenced by social circumstances, so we describe them as *biocultural* or *biosocial* features.

2.1.2 Homosexuality could be one manifestation of co-option

Co-option allows for many non-reproductive functions of sex, meaning there can be multiple functions of sex that remain biological yet are non-procreative. As soon as sex is liberated from its procreative function in this way, it can also be liberated from an exclusive heterosexual orientation. Accordingly, given the multiple possible functions of sex through co-option in complex organisms like humans, it would be quite incorrect to claim, as once was done, that homosexual behavior is unnatural. In retrospect, free from many of the prejudices of the past, this should not be surprising, since homosexuality and bisexuality are widespread across different cultures.

In addition, homosexual behaviors including same-sex courtship, pair-bonding, and copulation have been witnessed in hundreds of species, including mammals, birds, reptiles, and insects, among others. Careful examination of many of these species, however, indicates that long-term or exclusive homosexual preference in individuals is rare. In other words, although homosexual activity is widespread in nature, an exclusive homosexual orientation appears to be infrequent, except among humans.

Accordingly, the human situation remains something of a puzzle. This is because although non-exclusive homosexual behavior is widespread and can have all sorts of biological yet non-procreative rewards, a biological basis for strict homosexuality is theoretically unlikely. The argument is as simple as this: a hypothetical individual with a set of genes that results in a homosexual orientation is not going to breed, meaning those genes are not going to make it to the next generation. Contrast this with a hypothetical individual with a set of genes that results in a heterosexual orientation. This individual will likely breed and is therefore likely to produce progeny, meaning that those genes do make it into the next generation. This is the genetic paradox of exclusive homosexuality.

Although there are some genetic indicators that are statistically associated with homosexual orientation, suggesting a biological basis, persistent searching has shown there is hardly a gene or set of genes "for" homosexual behavior. Yet there may be sets of genes that on the one hand tend to orient an individual to homosexuality—which reduces the chances of those genes being passed on—but which in other related individuals, such as siblings of the opposite sex, for example, increase the likelihood of successful reproduction—which increases the chances of those same genes being passed on. This is very challenging to demonstrate genetically. Even though we know without doubt that the information encoded in genes represents the foundation of behavior, allowing us to say that all behaviors have a biological basis, humans are among the most malleable creatures behaviorally, especially under the power of culture and experience.

2.1.3 Sexual features can also be co-opted for non-social purposes

In exceptional circumstances, sexual traits can be co-opted for non-social purposes. You may be aware that bats find insects by using sonar. They emit high-pitched sounds (beyond the sensitivity of the human ear) whose echoes bounce off objects, including airborne insects. Bat ears are highly sensitive to these echoes, and bats are able to generate a picture in their brains using the echoes, just as most humans can generate a picture in their brains based on information received from the eyes. Many moths can hear the bats. Some simply take evasive action, but some species are able to "jam" the sonar of the bat by distorting or over-riding the echoes. In several Malaysian moth species, each individual can make these high-frequency sounds by rubbing parts of its genitals together—a surprising exaptation!

2.2 REPRODUCTION: SEXUAL, ASEXUAL, AND NON-SEXUAL

When people apply the word *reproduction* to living things, they are almost certainly using it in the sense of procreation: the biological process that produces new individuals. In biology, we think of reproduction a bit more broadly. When we apply the concept to cells, biological reproduction is like the copy-paste function on your word processor (not the cut-paste function). Where there was one cell (the copied text), there are now two identical cells (the original and the pasted text). This is true whether we are thinking about "simple" single-celled organisms like bacteria splitting into two, or about the trillion-plus cells in our own bodies dividing during the process of development, maintenance, or repair.

Even when reproduction is more limited to the idea of producing new individuals, as opposed to cell proliferation in complex organisms, not all reproduction is sexual. Here, we briefly compare reproduction in these different senses, so that we can then focus on the reproductive elements of sex in subsequent sections.

2.2.1 Not all organisms reproduce sexually

Evolutionary biologists have concluded that life arose from complex non-living molecules about 3.7 billion years ago, which was about 700 million years after the formation of the Earth. This conclusion is based on fossil evidence of "simple" single-celled organisms, on the presence of life-influenced chemical composition of rocks of that age, and on the shared molecular apparatus of all organisms, especially *amino acids*, which are the building blocks of proteins, and *nucleic acids*—DNA (deoxyribonucleic acid) and a related molecule, RNA (ribonucleic acid)—which store inherited information. The DNA molecule is packaged in association with certain specific proteins into very thin molecules called chromosomes, which are made up of sequential coded sections known as genes. Accordingly, all current organisms are connected to this almost incomprehensibly long-ago point in time through an unbroken reproductive linkage. Without reproduction, life would be limited to always starting over, and the basic building blocks of living systems could never accumulate to

BOX 2.1 DO SINGLE-CELLED ORGANISMS REPRODUCE USING SEX?

For bacteria, the answer is no. But they do something akin to sex that is rather interesting and worth briefly considering. And they may have been doing this long before sex was "invented" by more recently evolved organisms. Our working definition of sex is the joining of two gametes (egg and sperm) to form a new individual that becomes independent of the adults and shows traits of both parents. Bacteria clearly do not fit this pattern; in fact, they usually just clone themselves through binary fission (Section 2.2.1). Despite the fact that daughter bacteria are genetically identical to their parent, they can gain different genes later on through a process called conjugation. This may remind you of a term you've heard before, that marriage in humans is a "conjugal" union where sex is referred euphemistically as "conjugal relations."

Conjugation in bacteria is shown in Figure 2.3. As we've noted elsewhere, bacterial genes reside on their one chromosome (of which they may have multiple exact copies). But a few genes can be located on smaller stretches of DNA called *plasmids*, of which there can be many copies. The genes on these plasmids are often encoded for making proteins that provide immunity for the bacteria in their constant battle against foreign cells. During conjugation, two bacteria become aligned in close proximity to one another. The cell containing the plasmids, which we'll call the donor cell, grows a tube out from its cell wall called a *pilus* (sometimes even sex pilus). Upon the attachment of the pilus to the recipient cell, the two cells are drawn close together, allowing the movement of a plasmid from the donor cell to the recipient cell. The donor cell preserves copies of the plasmids, retaining the ability to make the immunity proteins and to be a donor another time.

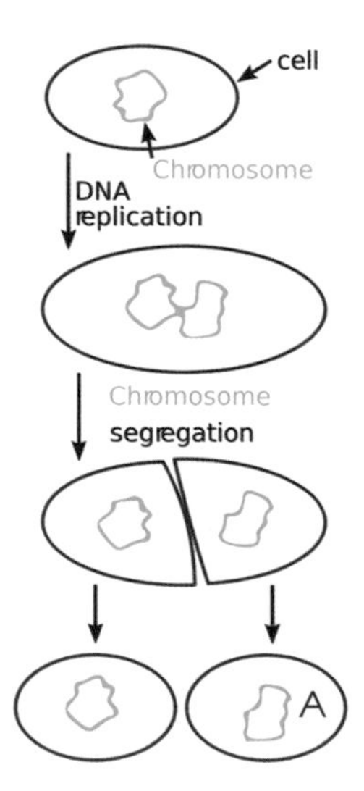

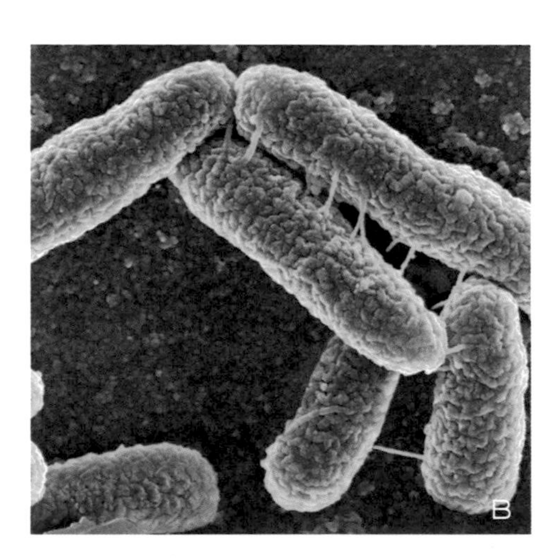

FIGURE 2.3 BACTERIAL CONJUGATION INVOLVES A GENE TRANSFER BUT IS NOT SEX

Most bacterial reproduction is by binary fission (a), but they do have ways of transferring genetic elements among individuals, known as conjugation. During conjugation (b), a sex pilus grows from a donor cell and temporarily joins its interior with the interior of a second cell. Genes located on small pieces of DNA known as plasmids move from the donor to the recipient, endowing the recipient with novel genes.

[2.3a] JWSchmidt / CC-BY-SA 3.0
[2.3b] Science Photo Library

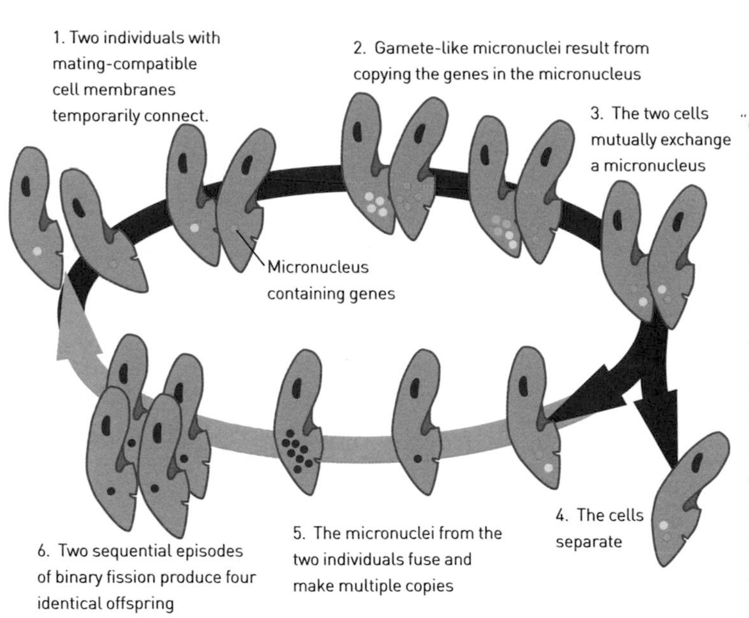

FIGURE 2.4 CONJUGATION IN THE *PARAMECIUM* IS MUCH MORE LIKE SEX
For one thing, it occurs between "opposite types," but this is based on cell membrane proteins so we cannot truly call them male and female. For another, before the exchange of genetic material, each individual makes a special nucleus that contains a single copy of its genes, similar to the way the human testis or ovary packages a set of genes into sperm or eggs. Then, the two individuals join cell membranes and make a mutual exchange of these nuclei. Unlike animal sex, there is not a donor and a recipient of sperm; instead, there is a balanced and mutual exchange. Nor are they making new individuals concurrent with the exchange.

Meanwhile, the recipient cell has the internal machinery to make multiple copies of the received plasmid, and it now benefits by having the immunity function associated with the plasmid genes. The recipient cell can now also be a donor cell. One can see parallels between bacterial conjugation and sex as we've defined it, but it is really quite different—no sperm and egg, no male or female, no new individual. But, we do have the mixing of genes from two individuals in the recipient, and this is reminiscent of sex.

If bacteria don't reproduce using sex, how about other single-celled organisms? You may at some point have watched a shoe-shaped *Paramecium* (a protist) moving about under a microscope. It is a single-celled organism, but it is many times larger than a bacterium, and its one cell is much more like one of your own cells in terms of complexity, in contrast to the bacterial cell. Genetic exchange in *Paramecium* is much closer to our working definition of sex, but it will still seem far from what is familiar. This quasi-sex is also called conjugation (Figure 2.4).

With *Paramecia*, the genetic exchange has to be between two compatible individuals. While this does not mean male and female, it does refer to certain opposing characteristics that are detectable on their cell membranes. Prior to the exchange, both individuals produce extra cell nuclei that contain copies of their genes, just as a human's testes or ovaries produce gametes with copies of their genes. Then, the two *Paramecia* sidle up to each other, their cell membranes temporarily merge, and each donates to the other one of these gamete-like nuclei. That done, they separate and swim away. Neither is a new individual. Neither is solely a donor or a recipient, but each now has new genes from another "opposite-type" individual.

yield the complicated interacting networks of cells that are characteristic of multicellular organisms.

The simpler method of reproduction is asexual, which simply means without sex. It is certainly the original form of reproduction. Sexual reproduction, which entails the merging of genetic material from two individuals to make a new individual, is thought to have first arisen about 1,200 million years ago, and is inherently more complicated. Despite the development of this new form of reproduction, asexual reproduction has remained commonplace. Bacteria, for instance, reproduce asexually (but see Box 2.1 for a bacterial system that involves the mixing of genes from two individuals within one individual). Bacterial reproduction is known as binary fission—*binary* referring to the "two-ness" of the end result, known as daughter cells, and *fission* referring to the splitting of the "parent" cell. Note that these offspring cells are not female, despite our use of the term *daughter*.

Prior to the splitting of the cell in two, the bacterium duplicates its genetic material, which is embodied in one or more circular chromosomes of DNA. The duplicated chromosomes are distributed between the two daughter cells such that each daughter ends up genetically identical to the parent. In this way bacteria are clones of the parent (and are identical to each other), unless they are subject to the infrequent intervention of mutation.

Although we don't usually think of cell reproduction within our own bodies as asexual, it technically is since it is the reproduction (of cells) that doesn't involve sex, although we are more likely to describe it as non-sexual, to distinguish it from organisms that reproduce new individuals asexually. In the human body, as in other multicellular organisms, cell reproduction is superficially like bacterial binary fission, generating two cloned daughter cells from one parent cell. However, it has certain technical differences and is known as mitotic cell division. Mitotic cell division, which is occurring in your body in countless places as you read this, has some similarities to another cell process known as meiotic cell division (Section 3.7). Because meiotic cell division is of utmost importance in sexual reproduction, we will consider it in more detail in Chapter 3, but at this point it is worth spending a few moments to understand mitosis, so that the two may be distinguished later.

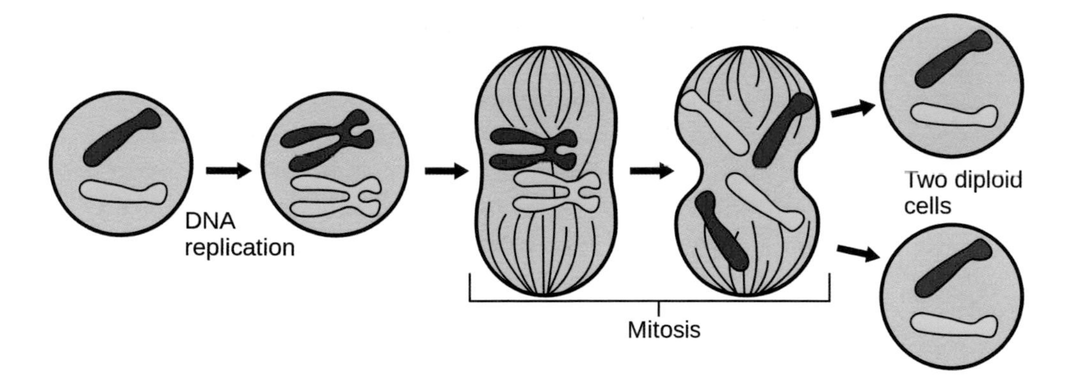

FIGURE 2.5 MITOTIC CELL DIVISION RESULTS IN THE PRODUCTION OF TWO IDENTICAL CELLS

Mitosis is the fundamental method used by complex organisms, such as humans, to create new cells from old cells, whether that is during growth and development, during maintenance, or during repair. Mitosis starts with a parent cell. In the simple species shown here, there is only one pair of chromosomes, whereas in humans there are 23 pairs. The chromosomes are duplicated, so that each chromosome becomes a pair of identical chromatids that are attached, resulting in an X-shape. These chromosomes line up along the cell's equatorial plate, and the cell's spindle machinery pulls them apart. To complete the process, the cell then splits in two (cytokinesis), and each daughter cell is a replica of the parent cell.

2.2.2 Non-sexual reproduction of human cells is through mitosis and cytokinesis

Mitosis and cytokinesis are illustrated in Figure 2.5. We need a few technical terms to describe the process and end result. We don't need to consider genes any further yet, but keep in mind that genes are made of DNA and they encode information for the building of molecules (mostly proteins) that the cell (and hence the body) needs. Recall also that genes are located sequentially on chromosomes, so chromosomes are also largely made of DNA. The human has perhaps 20,000 genes, and they are located on 23 pairs of chromosomes (so that there are two copies of most genes). Except in bacteria, which have no nucleus, chromosomes reside in the cell nucleus, a membrane-bound structure within the cell.

Prior to mitotic cell division in humans, there are 46 (2 × 23) chromosomes. In the early stages of mitosis, the chromosomes are copied within the nucleus, so that instead of two copies of each chromosome, there are effectively four. However, we don't say there are 92 (2 × 46) chromosomes, since the duplicated ones are still attached to each other, so at this point there are still 46 chromosomes, but each is made of a pair of identical chromatids. In a non-mitotic cell, the chromosomes are very long and thin, and although there is order to them, they evoke the image of a bowl of spaghetti. As the cell moves through the stages of mitosis, however, the chromosomes take on a coiled form that gives them a much shorter, thickened appearance, and they are visible under the microscope. In addition, because each is made of a pair of chromatids that are attached by a structure known as a *centromere*, they have an X-like appearance. (We will spend considerable time speaking about *the* X-chromosome later, but you should be aware that all the chromosomes have an X-like appearance at this stage).

As mitosis proceeds, the chromosomes "line up" along the cell's equatorial plate, and the chromatids are pulled apart by cell machinery known as the spindle apparatus. The result? Each half of the cell now has 46 chromosomes, and they are identical copies of the starting chromosomes (again, unless there has been an intervening mutation). The cell finishes the process with a pinching-off of the cell membrane along the equator, known as *cytokinesis* (cell movement), yielding two daughter cells that are clones of the parent cell.

2.2.3 Asexual reproduction in complex organisms

We've seen that bacteria reproduce asexually through binary fission, and that cell reproduction through mitotic cell division in your own body is fundamentally similar because both processes produce cloned daughter cells (i.e., "offspring" cells that have the same genetic content as the parent cell).

Bacteria are hardly simple, but when compared to some other groups that appear in the fossil record much more recently, they do indeed have a simpler plan: they have no nucleus, although they have a chromosome, and their internal cell structure is not nearly so populated with sub-structures known as *organelles*. Bacterial cells are known as *prokaryotic cells*, meaning "before the nucleus."

More familiar multicellular organisms such as animals, plants, and fungi have *eukaryotic cells*, meaning "true nucleus." The cells of these organisms are characterized by having a diversity of internal specialist organelles, and the non-circular (i.e., linear) chromosomes reside in the nuclei. There are many other less familiar groups of generally smaller eukaryotic organisms, both multicellular and unicellular, one example of the latter being the *Paramecium* featured in Box 2.1. In animals, plants, fungi, and many other multicellular eukaryotic groups, we can find examples of both sexual and asexual reproduction, including organisms that use both. Naturally, we will be examining sexual reproduction in greater detail, but a brief comparative review of several asexual methods will enrich our understanding.

When unicellular protists like the *Paramecium* reproduce asexually, it is much like mitotic cell division—a duplicating of chromosomal material, followed by a splitting-up of the duplicate chromatids and a splitting of the cell in two.

Multicellular organisms, which are the ones we usually associate with sexual reproduction, also engage in a variety of asexual methods (Figure 2.6). In plants, the most familiar type of reproduction is by seeds, which are produced sexually. But many plants have the capacity to reproduce asexually by *vegetative* means. This is achieved when a part of the plant other than a seed, such as a node on a creeping stem or even a fragment that manages to get rooted, forms a new individual. Again, because the cells come from the parent plant through mitosis, the offspring plant is a genetic clone.

Although most animals are sexual reproducers, there are a few that are asexual, as well as some that can switch between the two approaches. Reproduction in animals that does not involve fertilization of an egg is known as parthenogenesis. This means "virgin birth," since *parthenos* is Greek for virgin and *genesis* means "to create." One animal of interest is the relatively simple animal known as *Hydra,* named after the many-headed snake of Greek mythology whose heads grew again when cut off. In *Hydra*, an elongated "bud" growing off the main body breaks and proceeds to grow and live an independent life. It is a genetic replica, because the cells of the bud were produced mitotically. Single-celled yeasts, which are fungi and not animals, reproduce asexually in a similar manner.

Asexual vertebrates are relatively rare, and known cases have arisen through the past hybridization of two different, but related, species. Some mole salamanders (Class Amphibia) of North America's Great

FIGURE 2.6 **ASEXUAL REPRODUCTION IS PROCREATION OTHER THAN BY THE FUSION OF GAMETES**

This is a very widespread method of reproduction. There are multiple means, three of which are shown here. In water hydra (*Hydra* spp.) (a), a "bud" of cells identical to the individual's other cells breaks off but continues living, now as an independent individual. In some whiptail lizard species like this New Mexico whiptail (*Cnemidophorus neomexicanus*) (b), all individuals are females, which produce eggs that are genetically the same, so that all daughters are clones of each other (and of their mother, their aunts, their cousins, etc.). Plants have many ways of asexually reproducing; this seashore dropseed (*Sporobolus virginicus*) (c) is generating "offspring" by the use of *stolons*, which are above-ground creeping stems from which new plants (genetic clones) grow at junctions called *nodes*.

[2.6a] Biophoto Associates / Science Source / Getty Images [2.6b] Elliotte Rusty Harold / Shutterstock.com [2.6c] F. Neidl / Shutterstock.com

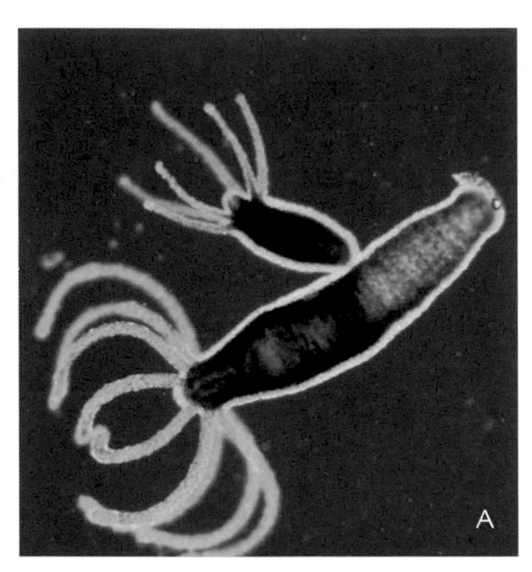
A

B

C

Lakes region, for instance, reproduce by cloning themselves, as do some whiptail lizards (Class Reptilia) of the southwestern United States and adjoining Mexico. Since they do not reproduce through sex, these populations are not only unisexual (i.e., all female) but they also are made of clones, where offspring are genetically identical to their mothers (and to their aunts, cousins, and even more distant relatives). The mole salamanders actually cheat a little: they mate with the males of a similar sexual species, "stealing" their sperm, but they do not incorporate the genes from the sperm into their eggs. Oddly, offspring production seems to require that the females participate in an ancestral mating ritual with males. Rather than being simply called parthenogenesis, the system used by these salamanders is called *kleptogenesis*, meaning the generation of offspring through theft (of sperm).

2.2.4 Switching between parthenogenesis and sexual reproduction

Before we move on to sexual reproduction, we might briefly consider those species that can and do engage in both sexual and asexual reproduction. We'll consider two species.

Water fleas (*Daphnia* spp.), which exhibit a pattern of switching between sexual and asexual reproduction, have not surprisingly drawn attention (Figure 2.7). During good times, meaning when the environmental conditions are not too challenging and when there is enough food, *Daphnia* reproduce asexually. These are female individuals whose eggs develop without having been fertilized by sperm, and so these eggs develop into clones of their mother. They mature fast, and so these daughters also produce eggs asexually. But when environmental circumstances become challenging, such as during drought or when population density is high, the *Daphnia* switch methods. Now some of the eggs hatch as males; however, these are still clones of their mother! How? Some genes are no longer active while others become active, making these sons cloned male versions of their mothers. These males produce sperm and they mate with other female *Daphnia* (that had been parthenogenetic to this point). This coupling produces eggs that are genetically unique, since they are produced by sexual reproduction. These eggs wait out the poor conditions, only hatching when good

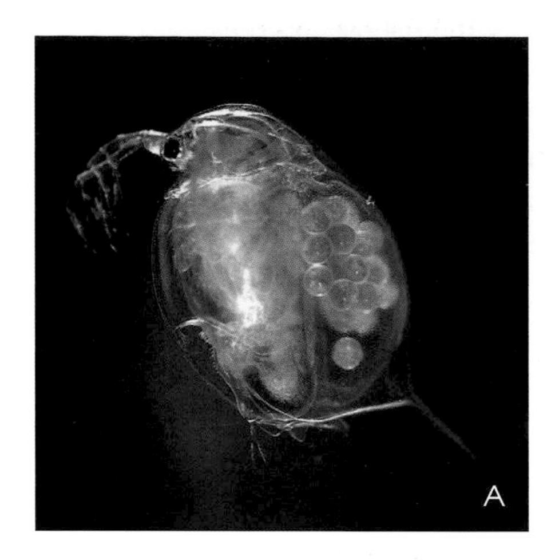

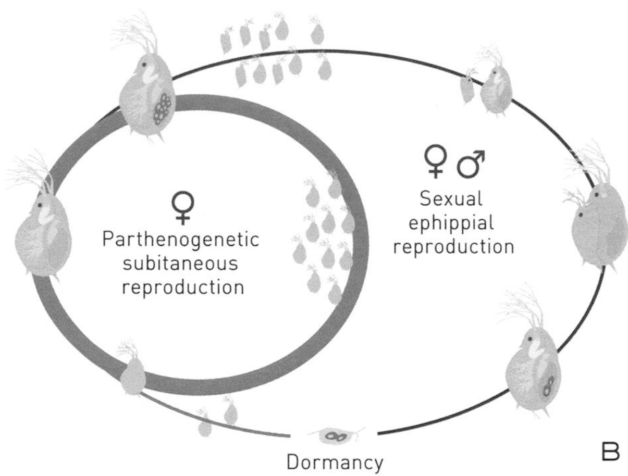

FIGURE 2.7 **CYCLIC PARTHENOGENESIS ALTERNATES BETWEEN SEXUAL AND ASEXUAL REPRODUCTION**

During good conditions, all individual water fleas (*Daphnia* spp.) are females that produce eggs. These eggs develop into clones of the mother without fertilization (asexual reproduction), often for several generations. When conditions deteriorate, some of these eggs hatch as male, but are otherwise cloned versions of their mother. Males and females now produce fertilized eggs through sexual reproduction, and these eggs can survive harsh conditions for long periods before "hatching." When they do hatch, they produce parthenogenetic females.

times return. This is called *cyclic parthenogenesis,* since the mode of reproduction cycles between sex and parthenogenesis.

The second example is not exactly an asexual one. Transparent roundworms (*Caenorhabditis elegans*) exhibit an added twist. Most roundworms of this species are hermaphrodites, meaning they have both male and female anatomy and that they can produce both eggs and sperm. Often, they reproduce without partners, mating with themselves to produce more hermaphrodites. Technically, this is sexual reproduction because it results from the merger of egg and sperm (from one individual), although it is like asexual reproduction in that there is only one parent. This is known as selfing. But a few of the offspring are males (there are no females), and these males only reproduce sexually because they can only produce sperm. When they mate sexually with a hermaphrodite, sperm from the male couples with eggs from the hermaphrodite, producing genetically unique offspring. Approximately half the offspring produced sexually are males, and half are hermaphrodites.

What is common to the many species that switch between sexual and asexual reproduction? It is the pattern already alluded to: during good times there appear to be advantages to asexual reproduction and during poor times there appear to be advantages to sexual reproduction. We will consider what the respective advantages might be when we consider the costs and benefits of sexual reproduction in Chapter 4.

2.3 SEXUAL REPRODUCTION FUNDAMENTALS

University-level students know the basic facts of life. In many modern Western education systems, sex education is taught even at the elementary level with the goal that developing young people will have reliable information as they encounter sex and sexuality as it is presented in a complex society that often distorts the truth. But most of the rudiments taught at earlier academic levels refer to body parts, and the goal is less about the biology of sex and more about arming the young person with a vocabulary and a level of comfort for meaningful communication about sex, when necessary. So, some of the following material students may already know, although research has shown that many of the biological fundamentals are in fact poorly understood by most people.

2.3.1 Female and male: Eggs and sperm

In sexually reproducing animals, there are two sexes, the male and the female. There may be different ways to express maleness or femaleness, as we will see, but the divide is a real one. Whereas a social analysis of sexuality and gender might construct a very different view, the message from biological analysis is that sexually reproducing species constitute a binary system of male and female (although individuals in some species can manifest both sexes as hermaphrodites). We have already defined male and female based on gonads and on the type of gametes produced in those gonads, either eggs or sperm. Which is the caring sex, the nurturing sex, the smaller sex, the aggressive sex tells us nothing about male and female in the animal kingdom, although we can often characterize sex-influenced differences in these traits in humans and other species. Simply put, an individual animal is a male if the sex

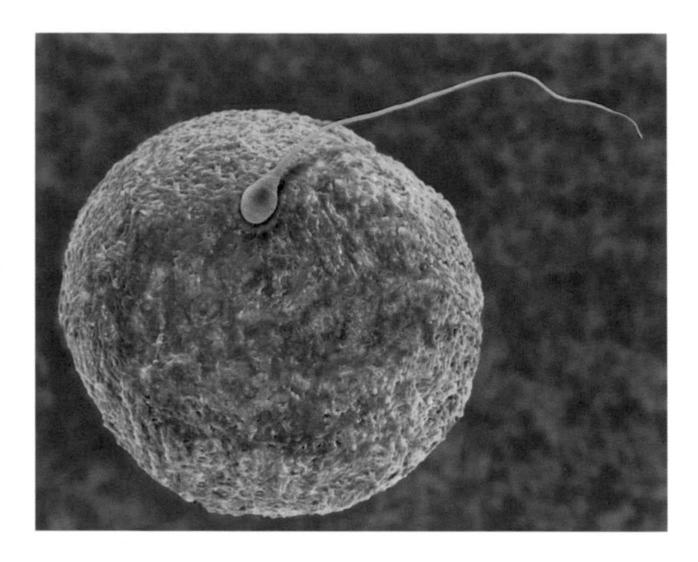

FIGURE 2.8 **SPERM ARE THE SMALLER MOTILE GAMETES AND EGGS ARE THE LARGER IMMOTILE ONES**

The motile sperm is much smaller than the immotile egg, but sperm contribute about the same amount of genetic information to the offspring that develops. The merger of egg and sperm is the essence of sexual reproduction, and it is by reference to these two packages from the parents that we define male (sperm producer) and female (egg producer), regardless of other traits exhibited by the individuals.

Dennis Kunkel Microscopy / Science Photo Library

organs are testes, which produce sperm. Sperm are known as *spermatozoa* (singular *spermatozoon*) if they have a whip-like *flagellum* (plural *flagella*) and therefore "swim." An individual animal is a female if the sex organs are ovaries, which produce eggs (Figure 2.8).

It is worth noting that a minority of species have *isogamous* gametes, which means the gametes are similar in size and shape, meaning the individuals that produce them are not classified as male or female. Instead, it is common for a "+" and a "–" to be used. Yeast is one such example. Far more common are *anisogamous* gametes like those of humans, which are not the same and which usually differ in motility (meaning ability to move) and size.

In this way, we have defined male and female without reference to accessory body parts, whether the external genitalia or the other parts that differ in degree between males and females, such as shape and hairiness. Because we are humans (and hence mammals), we most readily associate biological sex with mammalian genitalia. In females, the genitals are the opening of the vagina, the clitoris, and the labia majora (outer lips) and labia minora (inner lips), collectively referred to as the vulva. In males, the genitals are the penis and scrotum, as well as the "internal" gonads, which are the testicles that are technically outside the body cavity to maintain them at a slightly lower temperature. These attributes of mammalian genitalia are not universal attributes of all males or females in the natural world, however, and therefore sperm and egg form the definitive foundation of the difference between the two sexes.

Recall that sexual reproduction is the creation of new individuals from the merger of genetic material from two parents, through the fertilization of the female's egg with the male's sperm. Both eggs and sperm are single cells. The sperm is a particularly modified cell. It contains chromosomes that contribute genetic material to the offspring, but it is small and contains relatively little else other than machinery related to movement, since the sperm of most organisms are equipped with flagella—the whip-like tails—so that they are motile. Both egg and sperm contain a full set of chromosomes, meaning one of each kind. In humans, who have 46 chromosomes (23 pairs) in each cell nucleus, the egg contains 23 chromosomes, as does the sperm. When the sperm fertilizes the egg, the sperm contents can penetrate the egg's cell membrane, so that the chromosomes from the egg and those from the sperm constitute the 23 pairs of the fertilized egg. Once fertilized, the egg is known as a zygote.

2.3.2 Primary versus secondary sex characteristics

Modern humans have been able to achieve a certain amount of liberation from the sex stereotypes of the past, even those of a mere lifetime ago. In the mid-twentieth century, it was feared that if women, as the weaker sex, ran too fast—as they might do if the Olympics were to allow such racing—their health would suffer severely. Men were valued for manly traits like physical toughness and bravery, and women for feminine traits like caring and, alas, subordinate personalities. The successes of feminism have been great, such as equal access to education, highly paid employment, and social and sexual freedom. Indeed, women can now be "manly" and men can be effeminate, and technically at least, society no longer discriminates along these lines.

Yet, we know this is not entirely true. That sexual stereotypes of male and female remain largely entrenched is clear when a critical eye is turned to social motifs, especially in popular culture, advertising, and various media. There is an idealized female form, highly exaggerated by unusually tall and unnaturally slender models in the fashion industry, or by exceptional voluptuousness in popular media, including pornography, that set the standards of the flawless female. While not accepted overtly, these pervasive images have tremendous unconscious impact on what we think is beautiful or desirable. Images of the idealized male form are also pervasive, although both men and women, and society in general, seem to be more forgiving of deviations from the ideal among men. Nonetheless, research shows that males are more favorably considered when they manifest certain physical characteristics such as height, square shoulders, lean musculature, and deep voices; these individuals are much more likely to command positive attention than their opposites.

We have defined the male–female binary biologically based on sperm-and-testes and eggs-and-ovaries. We have also referred to the genitals, which are the external components of the sexual and reproductive system. These are known as primary sex characteristics—those features that are directly related to the creation and delivery of sperm or the creation of eggs and the receiving of sperm.

Most species also exhibit secondary sex characteristics. These are biological differences that manifest as differences in the body or in behavior that are predictably associated with one sex or the other. In animals, while all males have sperm and testes and all females have eggs and ovaries, we can list no such universals for secondary sex characteristics. They are simply too varied as we survey the animal kingdom. For instance, you may remember the clownfish that were introduced at the start of Chapter 1. There, the female in the "family" is the largest individual, she is the more aggressive of the two breeders, and she has the highest status, since males turn into females when opportunity presents, but not the other way around.

Can we list secondary sex characteristics in humans? Here we must be somewhat less definitive, as there is substantial diversity and some overlap in the bodily and behavioral manifestation of each sex within our species. Yet it is easy to do statistically for many features. In humans, males tend to be taller, larger, and more hairy, with greater muscle mass and deeper voices. Behaviorally, males tend to be more aggressive and more likely to partake in risky behaviors, and research shows males are relatively more interested in inanimate things and relatively less interested in relationships. This last characterization is a contentious one, with arguments on both sides regarding the extent to which it is biological or cultural. Certainly, if we look at non-human mammals, numerous human male secondary sexual behavioral characteristics line up with those of males in other species. Remember! This does not mean human males have to be aggressive, in compliance with their biological sex, as a form of biological determinism. But documenting these patterns can assist us in understanding male–female differences in humans and in trying to minimize costs associated with aggression.

Most animal species show sexual dimorphism to some extent (Figure 2.9). In other words, the sexes in most species can be characterized by having distinctly different bodies, over and above differences in gonads and genitals. Something that is dimorphic is something that has two distinct appearances—literally two distinct shapes—but the difference may be in size, shape, or even color or pattern. Humans are sexually dimorphic. Even if you were to remove cultural information like clothing and hairstyle but you covered the genitals, most such naked individuals would be easily (and correctly, by reference to gonads) classified.

Sexual dimorphism is manifested in diverse ways in the animal world. We cannot call it near universal, but it is an extremely common feature. California's northern elephant seal (*Mirounga angustirostris*) is a commonly cited example of extreme sexual size dimorphism. At over 2,000 kg, adult males in their prime weigh about four times as much as females. Not only is the male bigger, but it also has a giant inflatable snout used to make sounds connected to territorial and breeding behavior. The antlers of moose (*Alces alces*), which are grown and shed annually and can reach almost two meters from tip to tip, are only grown by males, and again, they are used in male–male combat.

In most migratory songbirds, males are only about three or four per cent larger than females, a rather modest dimorphism. But for many of those species, the sexes have dramatically different feather patterns, with males sporting bright reds, yellows, and blues, while females are more likely to be a subdued brown, gray, or green. Male and female

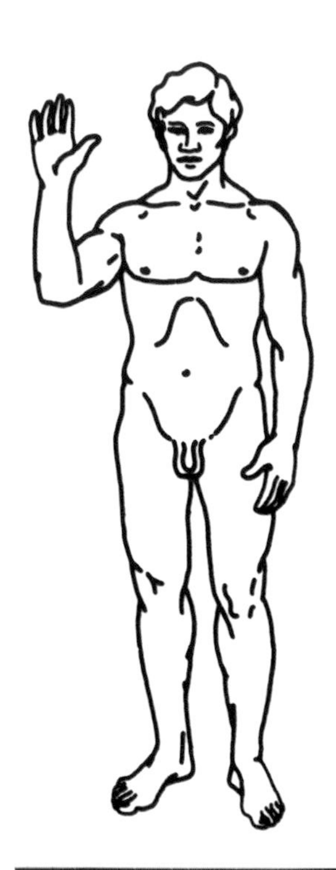

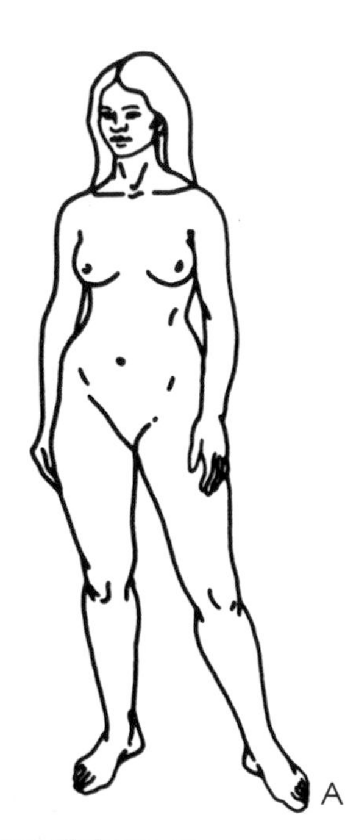

FIGURE 2.9 **SEXUAL DIMORPHISM IS A PERVASIVE PATTERN IN SEXUALLY REPRODUCING SPECIES**

There are many manifestations. Males are more often the larger sex, as in humans (a), but there are many opposite examples, such as the anglerfish (b), where the male is a relatively tiny, parasitic individual, and in many species of spiders, like this female golden silk spider (*Nephila clavipes*) with two attending males (c). Males more often than females have "embellishments," as in the antlers of the moose (*Alces alces*) (d), and they more often have more dramatic patterns and colors, as in many birds like these mandarin ducks (*Aix galericulata*) (e). Sexual selection (Chapter 5), a significant part of evolutionary theory, is thought to usually drive these patterns.

birds of prey tend to have similar feather patterns, but here again we see differences in size, with females almost always being the larger sex. In the goshawk (*Accipiter gentilis*), the female can be close to twice the size of her partner.

One of the most dramatic cases of sexual size dimorphism is found in the deep-sea anglerfish (Family Ceratiidae, Figure 2.9). Here, the female is many times larger than the male. It may be difficult in the deep dark depths of the oceans where they reside for individuals to find mating partners. When they do, in this case, the male latches on to the female by biting her and holding on. Soon, their flesh merges, and he becomes a permanent fixture, not even maintaining his own digestive system. He provides sperm for her when such are needed, but otherwise he is little more than a small appendage attached to her body.

When we consider sexual selection in Chapter 5, we will consider some of the ideas about what has generated and maintained some of these interesting secondary sexual characteristic differences in the bodies of males and females. When we do this, it will not be just bodily differences, but dimorphic behaviors too, since behavior differences between the sexes also fall under the umbrella of secondary sexual characteristics.

2.4 FROM FERTILIZATION TO THE PRODUCTION OF OFFSPRING

We've established that females have ovaries that produce eggs and males have testes that produce sperm. These are the definitive and primary sex differences, but we've briefly examined secondary differences that are common too: that males and females have accessory distinctions, both bodily and behavioral. And we've also established that eggs and sperm are gametes, each with genetic material, which fuse to form a fertilized egg known as a zygote.

How broadly does this summary apply to all organisms? It is true for all vertebrate animals (fish, amphibians, reptiles, birds, and mammals) that reproduce sexually. With modifications, however, we can stretch this definition much farther. Other non-vertebrate animals that are not hermaphrodites (most insects, starfish, lobsters, and scorpions, to name just a few) also have similar patterns of primary and secondary sex characteristics. The sexes have ovaries or testes, for instance, although sometimes the sperm-delivering organ is sufficiently different to be given a different label. We don't even have to qualify this summary much for flowering plants, although in the case of plants, both sexes are usually found in the same individual and even in the same flower (Box 2.2). But there are female flower parts that contain ovaries that produce eggs, and male parts (not testes) that produce pollen that carries sperm. Flower sperm is usually not mobile once the pollen reaches the female flower part, but the sperm creates a tube that achieves the same thing—fertilization of the egg to produce a zygote.

This is the fundamental scaffold of the essentials of biological sex differences and sexual reproduction—ovaries, testes, eggs, sperm, secondary sexual characteristics, and fertilization. Yet it is only a small part of the story.

2.4.1 External versus internal fertilization

A key consideration with significant ramifications is where the sperm meets the egg, resulting in a fertilized zygote. If it is inside the female, the process is called internal fertilization, and if it is outside, it is called external fertilization (Figure 2.11). The most common means by which internal fertilization occurs is by delivering sperm directly inside the female through the use of the male's intromittent organ. The usual manner of such delivery is through copulation, where the intromittent organ is inserted into the female reproductive tract and sexual excitement—in the male at least—results in the release of sperm in a fluid known as semen or seminal fluid. In humans, copulation is also referred to as sexual intercourse, the intromittent organ is the penis, and the reproductive tract is the vagina. The terms *penis* and *vagina* are used universally for mammals (and are used for some non-mammals too), but *intromittent organ* and *female reproductive tract* have more general, broader usage. Commonly, the intromittent organs of different groups are given special terms since they develop from different tissues. In insects, for example, the intromittent organ is called the *aedeagus*. *Phallus* is a term related to penis but it has a wider variety of meanings; it can include the clitoris, since the mammalian penis and clitoris have

BOX 2.2 SEXUAL REPRODUCTION IN PLANTS

Animals and flowering plants are dramatically different. Plants are immobile, unlike most animals, although their pollen and seeds often travel great distances. Plants don't have nervous systems, so their "behavior" is very rudimentary compared to animals. Animals are unable to convert sunlight into chemical energy, so that, unlike plants, they have to consume other organisms for energy. But both are complicated, multicellular organisms whose representative species mostly reproduce using sex. That means that both reproduce by generating eggs and sperm and that the union of egg and sperm creates a zygote that becomes a new individual. How egg and sperm get together is very different between the two, obviously, but the principle is the same. In most plants, the flower (Figure 2.10) is the sex organ, so here we find all the botanical equivalents of testes, ovaries, sperm, and egg.

The fundamental definition for distinguishing males and females that we've already established—that an individual is male if it produces sperm and female if it produces eggs—applies to plants. Yet in most flowering plants, we find that the flowers contain both male and female parts, so in this way plants are commonly like the animal hermaphrodite that contains both testes and ovaries. When a flower contains both male and female parts, which is the more common situation, it is a *perfect flower*. Otherwise it is an *imperfect flower*, in which case it has either male or female flowers but not both.

The typical perfect flower is made of four parts that are usually arranged radially around the tip of the stem. The outside part is the *calyx*. It is usually green and is made of leaf-like structures called *sepals*. The next part is the *corolla*, whose component parts are the *petals*. It is commonly brightly patterned and colored, at least when the plant depends on animal pollinators like bees and must have means of attracting such pollinators. The third ring of elements is the male component, collectively known as the *stamens*. Stamens have longish stalks (the *filaments*, as in a light bulb) that are terminated by sacs called *anthers*. It is within the anthers that pollen is made, containing sperm cells. Finally, the fourth part is the *pistil*, centrally located in the heart of the flower. This is the female part, and it contains the ovary in its base, as well as a stalk (the *style*) that is terminated by a usually sticky tip called the *stigma*.

Plants reproduce sexually when pollen from an anther gets "captured" by the sticky stigma of a pistil; the means by which this happens is known as pollination. In some cases, especially when pollination has otherwise failed, the pollen comes from the same flower, which is a form of *selfing* (Section 2.2.4). More often and more preferred is crossing, where the pollen comes from a flower on a different plant. To achieve successful pollination, a few species use

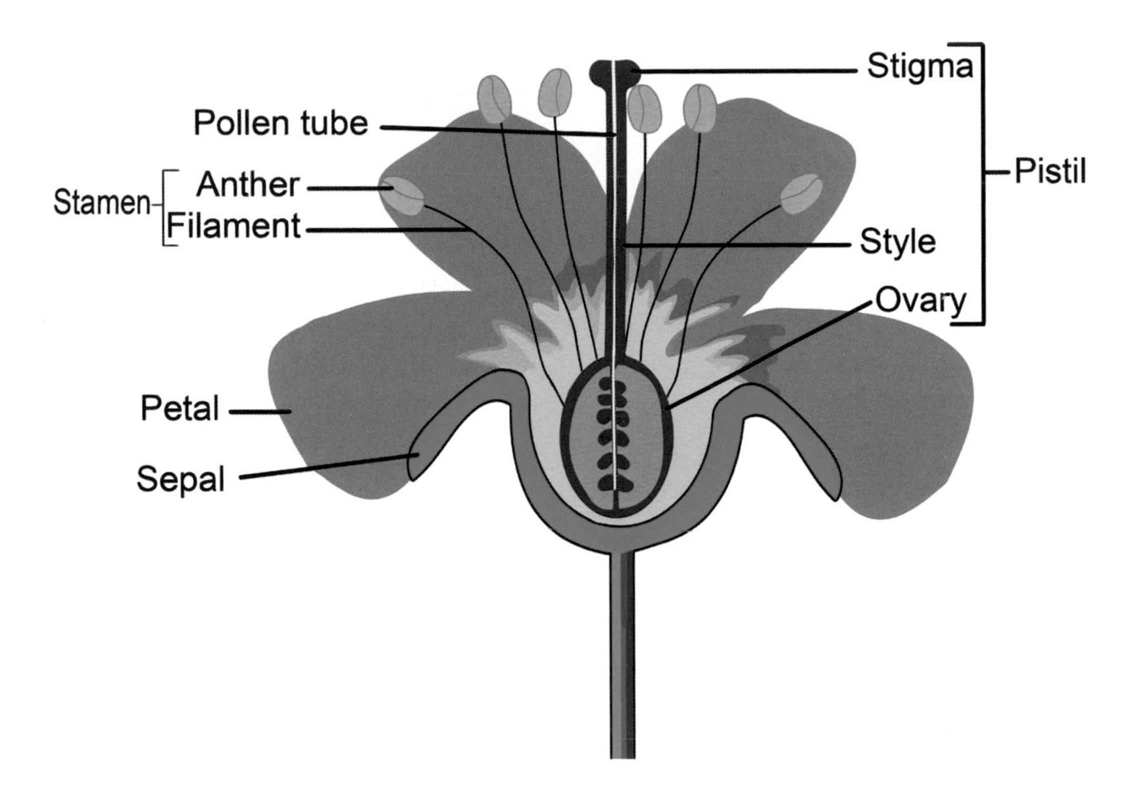

water and many use wind, but the majority use animals, whether it is bees, butterflies, moths, bats, hummingbirds, or even lizards. But people can also purposefully move pollen from one flower and apply it to the stigma of another, to control the parentage of the resulting seeds. This is the approach Gregor Mendel put to good use more than 150 years ago, as he set out to understand patterns of inheritance in sexually reproducing organisms (Chapter 3).

FIGURE 2.10 **THE FLOWER IS THE SEX ORGAN OF MOST PLANTS**
Some are quite elaborate, but the basic pattern is of four radially arranged components, with the outer two being the calyx (made of sepals) and then the corolla (made of petals). The male part is a series of stamens, the anthers of which produce sperm cells inside pollen grains. The female part is in the center, with the ovary at its base and the stigma at its tip. For successful sexual reproduction to occur, the pollen from one flower must get to the stigma of another, usually on another plant, and the sperm released from the pollen grain make their way to the ovary.
ducu59us / Shutterstock.com

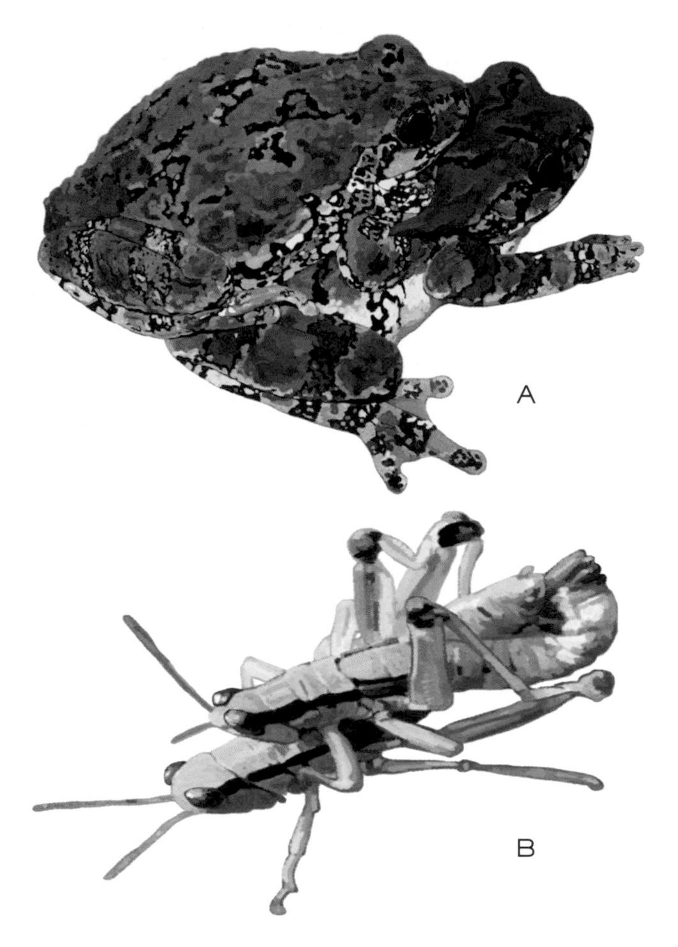

FIGURE 2.11 **FERTILIZATION CAN BE INTERNAL OR EXTERNAL**
The tree frogs (a) and grasshoppers (b) are similarly engaged in sexual reproduction. In the grasshoppers, however, fertilization is internal, with sperm being received internally by the female. Internal fertilization occurs in many vertebrate animals, including humans. In external fertilization, as in these tree frogs, sperm meets egg outside the body in a synchronized release.
Drawings by Peter Mills

the same embryonic origin, and it can also refer to cultural representations of the penis. There is tremendous diversity in morphology of intromittent organs (see Box 9.1).

Internal fertilization can be achieved without copulation. In several species, males package sperm in spermatophores, along with a complex of other materials that are likely to either benefit or manipulate the female. In some insects, these spermatophores are transferred to the female through copulation. In many aquatic salamanders, however, spermatophores are deposited on the substrate of the pond bottom, and they are only taken up by the female through her cloaca (the common opening for the reproductive and digestive tracts) if she is convinced to do so during the courtship performance of the male.

We don't apply the terms internal or external fertilization to plants, even though most plants reproduce sexually: the sperm is found in the pollen grain, which reaches the ovaries and eggs of another flower through wind or water action or by hitchhiking on a bee or other animal.

External fertilization is common in aquatic vertebrates, such as fish and amphibians, and in many non-vertebrate animals too, such as sea urchins. Here, because fertilization occurs outside the body, males and females are on a more equal basis once zygotes are formed, since the zygote is created outside the body and there is therefore no difference that automatically burdens the female with greater responsibility, unlike the case with internal fertilization. Although there can be parental care in some species that employ external fertilization, a common feature of this system is that once the eggs are fertilized, both parents are done, and the hatching young are on their own. For instance, when it is the season for frogs to breed, most species gather in water bodies that will be suited for the development of their aquatic larvae, known as tadpoles. Males and females engage in an embrace, known as *amplexus* (Figure 2.11), and as the female extrudes her unfertilized eggs through her cloaca into the water, the male extrudes semen through his cloaca. The close proximity of eggs and sperm in this watery environment allow the swimming sperm to fertilize the eggs, changing them to zygotes that will then hatch into tadpoles.

In most cases, especially in external fertilization, embryonic development begins immediately following fertilization. In humans, for instance, the single-celled zygote has already become a collection of several hundred cells within the first week. But in some species like bears, development is delayed by one of several methods, such that development occurs long after copulation.

As we will see in later chapters, that humans have internal fertilization has huge implications. This is because, as in most mammals, it leads to dramatic disparities in the commitment to parenting for males and females and also because males can't be certain whether a mate's offspring are his own or those of another male.

2.4.2 Other variations in patterns of sexual reproduction

There are many variants when it comes to reproductive characteristics. We might not wonder about why these characteristics exist, supposing it is "just the way things are." However, when we compare some of them to characteristics in other animals, including other mammals, we are more likely to realize that they did not develop randomly.

We've already acknowledged that humans have internal fertilization and that this will have significant—but different—repercussions for the two sexes. And, we also touched on another feature: that in many species that depend on internal fertilization, including almost all mammals, the fertilized egg develops inside the mother over an extended period, culminating in the birth of live young. We say live-bearing species like humans are viviparous (literally, to bring forth the living), while egg-bearing species like birds are oviparous (to bring forth ova, or eggs). Again, this can have consequences. Vivipary often entails much further commitments of the female, as is certainly the case in humans.

Parenting is such a highly variable aspect of sexual reproduction that we can make no generalizations. Some organisms, like humans, invest tremendously in parenting; human children simply would not survive without it. Others, such as most pond-breeding frogs, invest nothing beyond the efforts made in courtship and production of (small) eggs and sperm. We will look at the implications of parenting investment when we consider mating systems in Chapter 6.

We'll conclude this chapter by making two further observations about reproduction in humans. The first is that the female human has multiple serial pregnancies. The second is that each pregnancy usually involves one fetus, although twins are not particularly rare. Again, because we are people, we may take these features for granted, but they too, have implications for motherhood, fatherhood, mating systems, and behavior in general. Species that have multiple, sequential breeding events like humans are iteroparous breeders (literally, a repeated bringing forth). Other species that breed only once are semelparous breeders (literally, bringing forth once, which is what happened to the mythological *Semele* after she slept with the God Zeus). Sockeye salmon (*Oncorhynchus nerka*) are semelparous. For such species, sexual reproduction comes all at once at the end of their lives. Not surprisingly, most mammals are iteroparous, although male *Antechinus* marsupials in Australia never live past their one and only breeding season, so exhausted and damaged are they from male–male combat and the demands of trying to fertilize females (see Figure 4.10). Most adult female *Antechinus* manage to live at least two breeding seasons.

Humans and salmon not only exemplify the iteroparous-semelparous distinction, but they also represent another dichotomy, and that is reproductive investment per offspring. Most human pregnancies produce one infant, and that infant demands the attention of the mother for a further extended period, essentially until weaned. Accordingly, it is difficult and rare for a woman to produce more than a dozen offspring in her life. But the sockeye salmon produces tens of thousands of eggs, investing little in each one. After she lays her eggs in a riverbed and the male fertilizes them (much in the manner of frogs, discussed above), she abandons them. We will see that this idea of investment in offspring, particularly in the sense of disparities in investment between males and females, also has tremendous implications for males, for females, and for relationships.

CHAPTER 2 SUMMARY

- Humans are motivated to engage in sex for numerous reasons other than for procreation.
- Sexual motivation in organisms can be explained by *proximate* mechanisms (such as pleasure) and *ultimate* mechanisms (especially procreation).
- The original, procreative purpose of sex has been co-opted in many species, especially humans, for other purposes, such as bonding and the exercise of power.
- There is a diversity of asexual reproduction methods found among both unicellular and multicellular organisms.
- Males and females differ in their primary sex characteristics—referring to the gonads and genitals—and their secondary sex characteristics—referring to any other physical and behavioral differences between the sexes.

- Flowering plants reproduce sexually, although many can also reproduce asexually.
- In animals, the event of fertilization, which is the joining of sperm and egg, can be internal or external.
- Some sexually reproducing organisms produce live young and others produce eggs, and some reproduce one time, whereas others reproduce multiple times.

FURTHER READING

Avise, J. (2008). *Clonality: The genetics, ecology, and evolution of sexual abstinence in vertebrate animals*. New York: Oxford University Press. https://doi.org/10.1093/acprof:oso/9780195369670.001.0001

Diamond, J.M. (1997). *Why is sex fun?: The evolution of human sexuality*. New York: Basic Books.

Geary, D.C. (2010). *Male, female: The evolution of human sex differences* (2nd ed.). Washington, DC: American Psychological Association. https://doi.org/10.1037/12072-000

Gould, S.J., & Vrba, E.S. (1982). Exaptation: A missing term in the science of form. *Paleobiology, 8*(1), 4–15. https://doi.org/10.1017/S0094837300004310

Meston, C.M., & Buss, D.M. (2007). Why humans have sex. *Archives of Sexual Behavior, 36*(4), 477–507. https://doi.org/10.1007/s10508-007-9175-2

Moore, D.S. (2002). *The dependent gene: The fallacy of "nature vs. nurture"*. New York: Times Books.

Narra, H.P., & Ochman, H. (2006). Of what use is sex to bacteria? *Current Biology, 16*(17), 705–710. https://doi.org/10.1016/j.cub.2006.08.024

Pievani, T., & Serrelli, E. (2011). Exaptation in human evolution: How to test adaptive vs exaptive evolutionary hypotheses. *Journal of Anthropological Sciences, 89*, 9–23.

Schon, I., Martens, K., & van Dijk, P., (Eds.). (2009). Cyclical parthenegenesis in *Daphnia*: Sexual versus asexual reproduction. In *Lost sex: The evolutionary biology of parthenogenesis*. New York: Springer.

Tinbergen, N. (1963). On aims and methods of ethology. *Zeitschrift für Tierpsychologie, 20*(4), 410–433. https://doi.org/10.1111/j.1439-0310.1963.tb01161.x

3 Sex and Inheritance

KEY THEMES

Gregor Mendel discovered the basic rules of inheritance in sexual reproduction. Later work showed that genetic information is encoded in genes, which come in different versions called alleles. Genes are made of coded DNA found on chromosomes, and they give cells the information needed to make complex proteins. In most organisms, chromosomes come in matching pairs called *homologous* pairs. Eggs and sperm do not contain pairs, but when egg and sperm fuse upon fertilization, the cell again has homologous pairs.

Whether you like it or not, chances are good that you've been told you look like one of your parents, or perhaps a sibling. This is not surprising. Given that your appearance largely reflects your genetic makeup and given that your genes come from your mother and father, you are bound to show physical resemblances to them and your siblings more than to anyone else.

Until the mid-1800s, the details of how this worked were unknown. It was a classic black box: we knew that the inputs were two people of opposite sex who had engaged in sexual intercourse and that the outputs were offspring that resemble the parents, but the intervening details were mysterious and opaque. This did not prevent humans from exploiting the principles of inheritance in dealing with the animals and plants they depended upon. Farmers, for instance, would carefully consider which individuals they wanted to produce offspring, based on their observed traits. So, if you wanted sturdy horses for farm work, males and females exhibiting the desired traits—perhaps particularly powerful muscles, calm temperament, and resilience—would be used for breeding purposes, and others would be prevented from mating. On the other hand, if you wanted horses for riding and speed, the selected parents would look quite different indeed. Over many generations, separate working and racing breeds were developed, as were sweet corn, duck-retrieving dogs, and many other domesticated organisms.

Successfully attending to inputs and outputs in this process of applying the principles of inheritance was one thing, but once the scientific enterprise had sufficient momentum, people wanted to understand what was going on in the black box too. By the 1860s, with the increasing acceptance within the scientific world that evolution was the concept that best explained patterns of diversity and inheritance, thinkers about these things faced a conceptual obstacle. The obstacle was a prevailing idea that seemed to be true for parents and children, but was evidently not true when the bigger picture was considered. The idea was called blending inheritance (Figure 3.1). This was the intuitive notion that the offspring resulting from the mating of two individuals manifests a mix of traits from the parents—and of course, this is a pattern that is generally borne out. But such blending would always

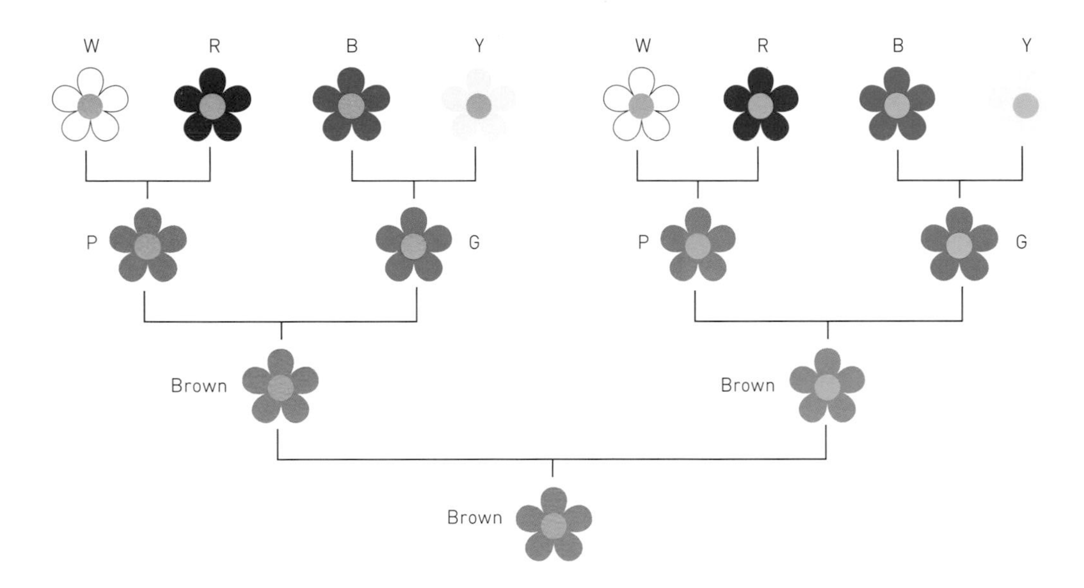

make averages, and after many generations, everyone would be the same height, with the same body proportions, and the same hair, etc. Figure 3.1 shows how this principle would affect flower color, if it were true. Remember that flowers reproduce sexually, although not using intercourse. We have a pair of grandparents that are white and red, and they have a pink (blended) offspring. We have another pair of grandparents that are blue and yellow, and they have a green (blended) offspring. When the pink and the green reproduce together, they have an olive-brown (blended) offspring. Brown, a mixing of all colors, is the inevitable result, and all variety would be lost. Yet populations are full of variety, so we know blending inheritance doesn't hold true.

FIGURE 3.1 **THE BLENDING INHERITANCE HYPOTHESIS WAS REJECTED BECAUSE GENETIC DIVERSITY PERSISTS**
The idea of blending inheritance is shown here for sexually reproducing flowers. The red and white flowers have pink offspring, and the blue and yellow ones have green offspring. When the green and pink mate, they would have brown offspring, as will all individuals thereafter—an inevitable result if blending inheritance were correct. Yet we know populations of sexually reproducing organisms retain lots of variety in colors, shapes, and other attributes. Blending inheritance was a popular notion, but was clearly not a workable theory, because it did not match reality.

Unlike today, when most scientific advances are made in universities and other high tech institutions, much scientific advance in the eighteenth and nineteenth centuries was the enterprise of a mostly private, privileged class of intellectual entrepreneurs. Some were independently wealthy, like Carl Linnaeus (Chapter 1) and Charles Darwin (Chapter 4). Others were less wealthy, but they still belonged to an intellectual class and sometimes had civic or religious positions of significance. One such person was Gregor Mendel. Born into a land-holding family in Moravia (now the Czech Republic), he studied physics and philosophy before becoming an Augustinian friar within the Roman Catholic Church. However, it was none of physics, philosophy, or theology that he is famous for. Instead, it is genetics, particularly the "rules of inheritance" in sexual reproduction.

3.1 HOW AN AUGUSTINIAN FRIAR DISCOVERED THE RULES OF SEXUAL INHERITANCE

Rather than take over the family farm, Gregor Mendel pursued what we would now think of as an academic life. He was drawn to the monastery in Brno. This was a regional cultural hub, with a library, experimental facilities, and a culture of research and teaching in science. He lived the remainder of his life there, although he also spent time in his twenties at the University of Vienna where, in addition to physics and math, he studied botany. While at the monastery, he settled on the common garden pea (*Pisum sativum*) as a study system to experiment with patterns of inheritance.

Although the beauty of plants is widely appreciated, when it comes to biology, students are usually more interested in animals. Accordingly, we mostly focus on the animal kingdom in examining the biology of sex. But at this juncture, we need to consider plants and plant sex so that we can follow Mendel's experiments and conclusions. Although he focused almost exclusively on peas, we know that the principles he articulated after his eight or so years of experimentation apply equally well to plant sex and animal sex. For a primer of the sexual characteristics of flowering plants, refer back to Box 2.2.

3.1.1 Mendel's peas were a "friendly" study system

While it is possible that Mendel could have elected to study an animal species to investigate inheritance in sexually reproducing species, there is no doubt he saw the advantages of using plants instead. Certainly, plants are less needy than animals, and given a big enough garden, very large sample sizes can be assembled. This is attractive because larger samples usually yield clearer patterns. Not only were plants a logical study system, peas were a particularly good choice. For one thing, they grow quickly, allowing for the reasonably rapid analysis of multiple generations. For another, peas existed in many *true-breeding varieties* that allowed Mendel to start with some distinctly different strains. True-breeding varieties are those where the parents look the same for a particular characteristic and the offspring reliably also show that trait, generation after generation. Mendel assembled numerous true-breeding pea varieties for several characteristics related to flower color, pea shape, pod characteristics, and plant height, among others.

Mendel's experimental set-up was not particularly complicated. Peas are seeds, and he planted not only the peas from the stock he obtained (the true-breeding varieties), but also the peas that his experimental plants produced. He had a large garden, so he was able to grow many plants in each successive generation. The key to his work was to control parentage. First, he chose which individuals would provide the pollen, and in the flowers of those plants he removed female parts (i.e., the pistil; see Box 2.2), leaving the stamens, and thereby rendering it a "male" flower. Similarly, he decided which individuals would produce the eggs (and ultimately, therefore, the pea pods), and in the flowers of those plants he removed the stamens, rendering them "female." Having manipulated his flowers to be imperfect (meaning only having one sex rather than both), his job then was to pollinate the "female" flowers. To do this, he became the pollinating agent, rather than letting insects like bees do the task. He used a small brush to pick up pollen from the flower that had been rendered "male" and then applied that pollen to the stigma of the one whose eggs would be used to produce pea pods (Figure 3.2).

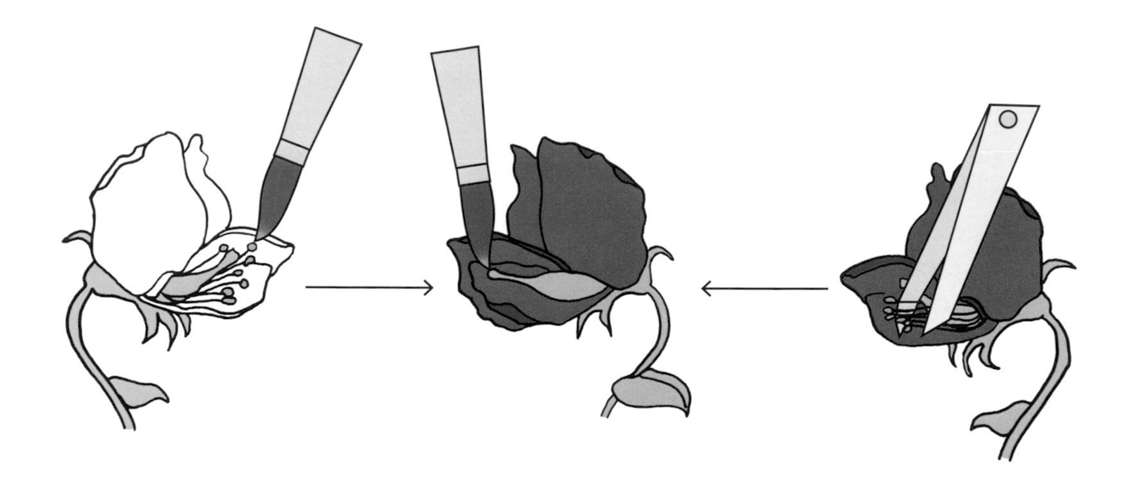

FIGURE 3.2 **MENDEL MANIPULATED POLLINATION IN HIS PLOTS OF EXPERIMENTAL PEAS**
When Mendel pollinated a purple flower with pollen from a white flower, effectively fertilizing the purple flower, and then planted the pea seeds that developed, all the offspring were purple-flowered.
[3.2b] © Alex Mills

Although Mendel considered seven different plant characters, we will refer to only a few as we discuss his two laws, the Law of Segregation and the Law of Independent Assortment. Note, however, that he found similar patterns of inheritance for all the characteristics he investigated, allowing him to inductively generate these two laws.

3.2 DEVELOPING MENDEL'S LAW OF SEGREGATION

Mendel started with a true-breeding purple-flowered variety and a true-breeding white-flowered variety. When Mendel used his brush to apply pollen from the stamens of one purple flower to the stigma of another purple flower and then planted the seeds that developed, all those seeds produced purple flowers. When he did the same for the white-flowered individuals, they too produced 100 per cent white-flowered offspring. This is in fact the definition of true breeding.

Naturally, Mendel wished to mate purple- and white-flowered individuals. When he took pollen from a purple flower and "painted" it on the stigma of a white flower and then planted the peas that developed, he found that all the offspring were purple-flowered individuals (Figure 3.2). This was certainly not blending inheritance, since one parent was white-flowered and one was purple-flowered, but the white flower trait seemed to have disappeared. When he switched the parents—taking pollen from a white flower and painting on the stigma of a purple flower—the results were the same, with all offspring being purple-flowered. These were called *reciprocal crosses*: pollen from the purple flower onto the stigma of the white in one case and pollen from the white flower onto the stigma of the purple in the other. He called the parental generation the P generation, and the first offspring generation the F1 generation. He used "F" for the Latin word *filial*, meaning "of or related to a child," and the "1" denoted the first generation.

3.2.1 The second-generation offspring showed a surprising pattern of inheritance

The next thing that Mendel did was to mate F1 (purple-flowered) individuals with other F1 (purple-flowered) individuals, again by taking pollen from one flower and applying it to the stigma of another, and then planted the seeds that were produced—the F2 generation. Here the results were more surprising (Figure 3.3). Although the F1 was entirely purple-flowered, in the F2 some of the individuals were white-flowered, just as half of their "grandparents" had been. In fact, in the 929 F2 plants, there were 705 purple-flowered offspring and 224 white-flowered offspring. What struck Mendel was that he found that

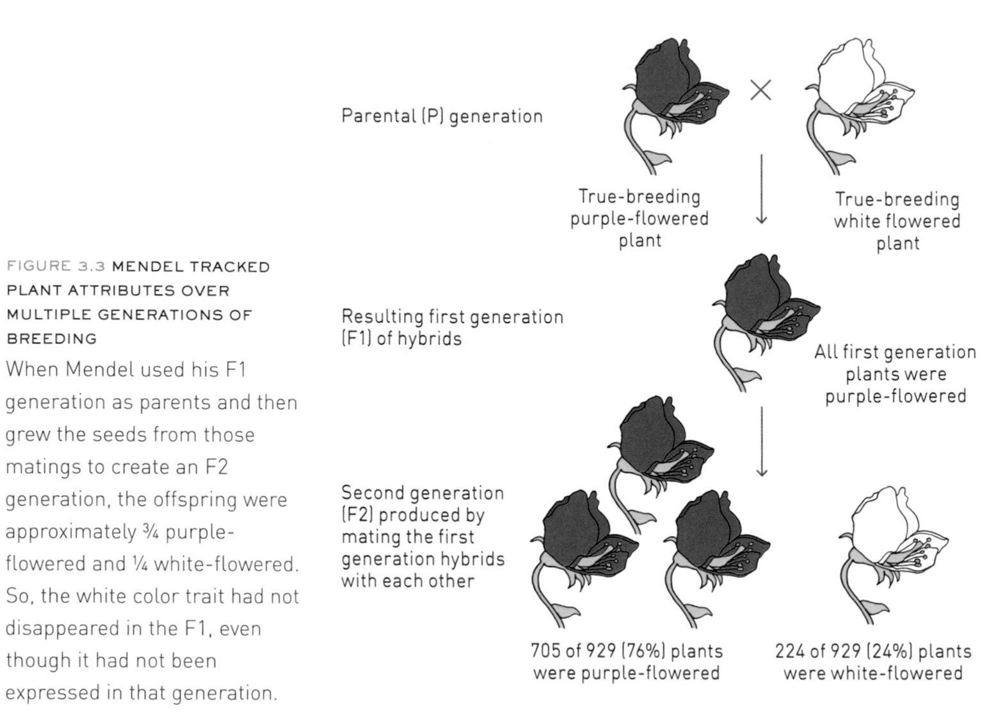

FIGURE 3.3 **MENDEL TRACKED PLANT ATTRIBUTES OVER MULTIPLE GENERATIONS OF BREEDING**
When Mendel used his F1 generation as parents and then grew the seeds from those matings to create an F2 generation, the offspring were approximately ¾ purple-flowered and ¼ white-flowered. So, the white color trait had not disappeared in the F1, even though it had not been expressed in that generation.

other characters, such as pea shape, followed a pattern similar to the one he observed for flower color: one trait would disappear in the F1 but it would re-appear in the F2. Furthermore, regardless of the character he was investigating, he noted a recurring ratio in the F2. Whenever a trait disappeared in the F1 but reappeared in the F2, that re-appearing trait was found in about one-quarter of the individuals. Or, expressing it as a ratio, in the F2 the ratio of the trait that completely dominated the F1 generation to the trait that disappeared in the F1 generation is 3:1. (For the flower color sample, for example, it was 3.15:1.)

3.2.2 The meaning of segregation in the making of gametes

Mendel was thoughtful about his experimental results. By interpreting the patterns that looked at a single character like flower color, Mendel inductively derived his Law of Segregation. He imagined that each parent carried two *factors* for a particular character. True-breeding purple-flowered peas had two purple factors, just as true-breeding white-flowered peas had two white factors. Each parent invested its eggs or pollen with one factor. A true-breeding purple-flowered pea plant had to invest its pollen or its eggs with a single purple factor since, after all, it only had purple factors. Similarly, the pollen and eggs of true-breeding white-flowered plants had to contain a white factor. The seeds that were planted to produce the F1 generation therefore had to have one purple factor and one white factor. Mendel reasoned here that one factor is dominant to the other. This is not necessary to the Law of Segregation, as we shall see, but it was the case with his pea plants. So, for pea plant flower color, the F1 individuals had a dominant purple factor and a second white factor that was recessive in the sense that it is not expressed; after all, the flowers of all these individuals were purple. But the factor had not disappeared as genetic information that can be passed on. Instead, when the F1 individuals make pollen or eggs, about half will get a purple factor, and about half will get a white factor. So the sperm and eggs generated by the F1 individuals are 50 per cent white-factor and 50 per cent purple-factor, and these are the gametes that create the F2 generation.

In the early 1900s, the Punnett square was developed to illustrate the proportions of factors in gametes and the traits of the offspring that develop (Box 3.1). Being able to understand and use a Punnett square is crucial for understanding the patterns of inheritance in sexually reproducing species.

We can express the Law of Segregation a little more formally:

1. For a given character (such as flower color) there can be different alternative states called traits (such as white or purple).
2. For each character, an individual inherits two factors, one from each parent. These factors can be identical, or they can be different.
3. If an individual has two different factors, one *may* be dominant to the other, in which case it is expressed in the individual regardless of the second, recessive factor.
4. When making gametes (eggs or sperm), the two factors separate (segregate), and single factors end up in different gametes in a 50:50 ratio.

BOX 3.1 WE USE PUNNETT SQUARES TO UNDERSTAND PATTERNS OF INHERITANCE

The Punnett square is a working diagram for illustrating the process and predicting the outcome of a particular breeding event between two individuals of a sexually reproducing species. Reginald Punnett was an early geneticist who devised the device in the early 1900s. The general pattern is shown in Figure 3.4 (a).

When Mendel began his pea breeding experiments, he started with a parental generation (P generation) that was true-breeding for the character of interest, such as flower color. He crossed a purple-flowered variety with a white-flowered variety, shown in Figure 3.4 (b) and (c). In such cases where the parents differ in a single trait, this mating is called a monohybrid cross. Note that the resulting offspring have the same pairs of factors in (b) and (c) because it does not matter in most cases, including for flower color, whether the purple factor traveled in the sperm and fertilized a white-factored egg or the white factor traveled in the sperm and fertilized a purple-factored egg. All the F1 offspring were purple-flowered, notwithstanding the fact that they had one purple-flowered parent and one white-flowered parent.

It was when Mendel crossed the F1 individuals that a surprising pattern emerged, shown in Figure 3.4 (d), that allowed him to hypothesize about inheritance factors. He drew the conclusion that the white factor was not gone from the F1 individuals but that the white factor was not expressed when in the presence of the purple factor.

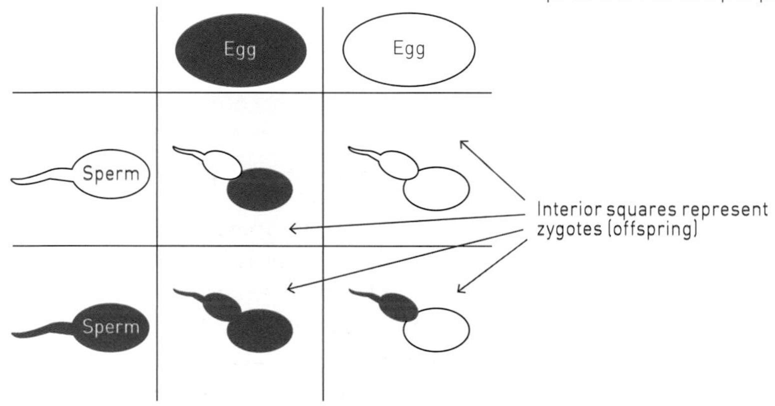

FIGURE 3.4A **PUNNETT SQUARES ARE USED TO ANALYZE OFFSPRING POSSIBILITIES FOR GIVEN PARENTS**
Information about the gametes of one parent is included in the row titles on the left; in this case it is the male, so the gametes are sperm. Information about the gametes of the other parent is included in the column titles above; in this case it is the female, so the gametes are eggs. Mendel hypothesized that each parent has two factors for a given characteristic, one of which is invested in each gamete. The resulting possible combinations in zygotes, through sexual fertilization, are shown in the interior boxes. Where the Punnett square presents a cross between two individuals that differ in the one character featured (such as flower color), it is called a monohybrid cross.

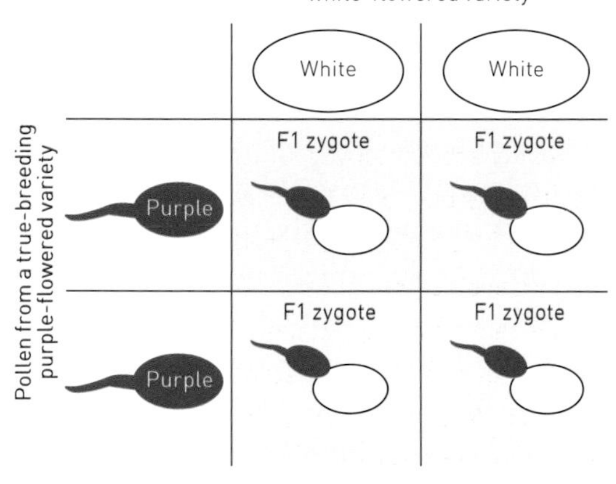

FIGURE 3.4B **MENDEL'S FIRST EXPERIMENT CROSSED WHITE-FLOWERED AND PURPLE-FLOWERED PEAS**

In his first experiment, Mendel crossed individuals from his true-breeding purple-flowered variety with those from his true-breeding white-flowered variety (a monohybrid cross). In the case shown here, he took sperm-carrying pollen from a purple-flowered individual and pollinated the flower of a white-flowered individual. All the offspring were purple, but by also observing the patterns of flower color in the offspring of the F2 generation (see the Punnett square in [d]), he correctly surmised the white factor was present in the F1 individuals, but it was recessive to the dominant purple factor.

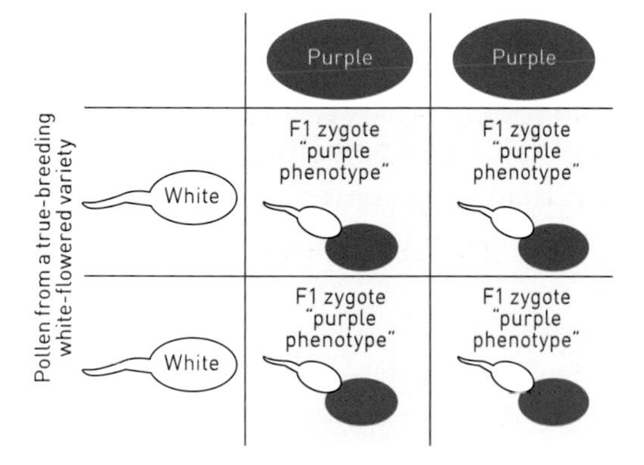

FIGURE 3.4C **MENDEL ALSO EXPERIMENTED WITH RECIPROCAL CROSSES**

The breeding experiment shown here was also part of Mendel's first experiment. He showed that for flower color, the offspring were the same whether he used pollen from a purple flower to fertilize a white flower (as in b) or whether he used pollen from a white flower to fertilize a purple flower (c).

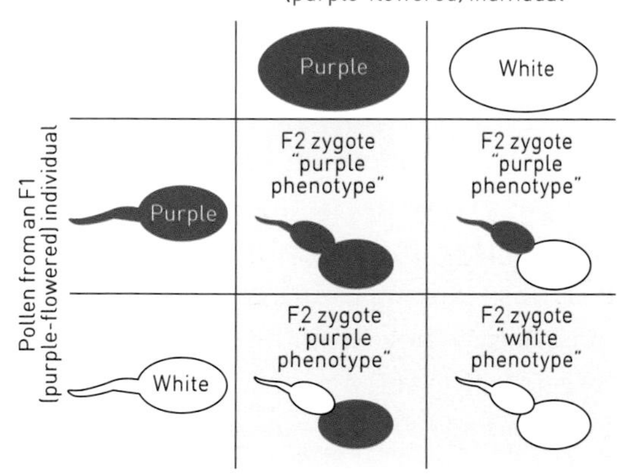

FIGURE 3.4D **MENDEL'S EXPERIMENTS DID NOT STOP AFTER JUST ONE GENERATION**

Mendel took the F1 generation and bred them together to observe patterns in the next generation, known as the F2. Here, white-flowered individuals reappeared in the F2 generation, and the ratio of purple-flowered to white-flowered individuals was approximately 3:1. Based on his idea of inheritance factors and that purple was dominant to white, Mendel reasoned correctly that one-quarter were purple because they had two purple factors, one-half were purple because they had one purple and one white factor, and one-quarter were white because they had two white factors. Although by flower color it is a 3:1 ratio, by considering the factor arrangement, it is a 1:2:1 ratio.

3.3 MENDEL'S LAW OF INDEPENDENT ASSORTMENT

Mendel's second law was generated when he looked at two characters at once. We'll leave flower color behind, and instead focus on two attributes of peas themselves: their color, which can be yellow or green, and their surface texture, which can be round or wrinkled. When Mendel looked at these two characters independently, he found the same pattern he had observed for flower color. When he crossed a true-breeding yellow-pea variety with a true-breeding green-pea variety, all the F1 were yellow. He therefore concluded yellow was dominant and green was recessive. Consistent with that conclusion, in the F2 generation, three-quarters were yellow and one-quarter was green. And, when he crossed a true-breeding round-pea variety with a true-breeding wrinkled-pea variety, all the F1 was round, as were three-quarters of the F2, but one-quarter of the F2 was wrinkled. So, he concluded that round was dominant and wrinkled was recessive.

To consider both characters at once, Mendel needed and obtained a true-breeding yellow/round variety and a true-breeding green/wrinkled variety. He surmised correctly that individuals of the first variety contained two yellow factors and two round factors, but no green or wrinkled factors and that the individuals of the second variety contained two green factors and two wrinkled factors but no yellow or round factors. When he crossed these two varieties (i.e., from the P generation) and grew the offspring (i.e., the F1 generation), all the peas were yellow and round. He was not surprised by this because when he had considered these two attributes independently, yellow and round were dominant traits. So, he figured the F1 individuals each had one yellow and one green factor for pea color and one round and one wrinkled factor for pea shape.

When he crossed the F1 individuals to produce an F2, there were two plausible hypotheses. In one case, given that in the P generation yellow and round were associated together and green and wrinkled were associated together, perhaps those pairings were fixed and could not be broken up. In the other possibility, they could be broken up, so that an offspring might have a green factor and a round factor, or another might have a yellow factor and a wrinkled factor.

	Yellow Round	Yellow Wrinkled	Green Round	Green Wrinkled
Yellow Round	YY RR	YY RW	YG RR	YG RW
Yellow Wrinkled	YY RW	YY WW	YG RW	YG WW
Green Round	YG RR	YG RW	GG RR	GG RW
Green Wrinkled	YG RW	YG WW	GG RW	GG WW

FIGURE 3.5 **THIS IS A PUNNETT SQUARE FOR A DIHYBRID CROSS**

In this case, the cross involves two characters (pea shape and pea color). Notice that there are 12 yellow-pea and 4 green-pea offspring here, which is a ratio of 3:1, as expected. Notice also that there are 12 round-pea and 4 wrinkled-pea offspring here, also a ratio of 3:1 as expected. But there are actually four different patterns: nine yellow-round offspring, three yellow-wrinkled offspring, three green-round offspring, and one green-wrinkled. In a dihybrid cross, this 9:3:3:1 ratio is typical. (Conventionally, capital letters are shown for dominant traits and lower case letters for recessive traits. Had we followed that, we would have used Y for yellow and y for green, and R for round and r for wrinkled. Instead, here we have simply used Y, G, R, and W for ease of understanding.)

When Mendel analyzed his F2 generation, he didn't only find yellow round peas and green wrinkled peas; he also found green round ones and yellow wrinkled ones. Because there were four types of peas instead of the two that he started with, he concluded that in the making of gametes, the factors for pea shape and the factors for pea color *assorted independently* of one another. So, just because a gamete contained a yellow factor didn't mean it also had to carry a round factor. We illustrate what Mendel found in Figure 3.5. Note that this Punnett square is bigger, because it tracks two independent traits (pea shape and pea color). Where the parents differ in both traits, it is called a dihybrid cross.

Mendel found a predictable ratio in his dihybrid crosses. He wasn't surprised to find that the round to wrinkled tally was 3:1 and that the yellow to green tally was also 3:1. The more interesting result related to the proportions of the four different combinations. This was a ratio of 9:3:3:1 for yellow-round to yellow-wrinkled to green-round to green-wrinkled, and this has proved to be a common pattern in dihybrid crosses in many systems. This told him that there was independent assortment—that whether a gamete carried a round or a wrinkled factor had no bearing on whether it also carried a green or yellow factor.

We can formalize Mendel's Law of Independent Assortment this way: when making a gamete, the single factor selected from one pair of factors that determine a particular character (like pea color) has no bearing on the factor selected from another pair of factors that determine a different character (like pea shape). We now know there are exceptions to this, although it is generally true.

3.4 UPDATING MENDELIAN GENETICS

One of the great successes of science is its ability to take a pattern based on results from investigating one system and to apply that pattern to another system. Mendel's pea experiments are just such an example. He studied patterns of inheritance in the sexually reproducing pea, and he was able to derive two laws that apply to all sexually reproducing species, including humans. Naturally, since the laws were developed in the mid-1800s, the principles in them are rudimentary. They remain essentially true, but they have been developed further and integrated with much more detailed knowledge at the level of the cell and of DNA. Having said that, Mendel's laws, which are often referred to as *Mendelian genetics* or "classical genetics," apply today, and they apply to humans as well.

We began this chapter by noting that the stages between sexual intercourse and childbirth were a black box for scientists of the 1800s. As part of this black box, we also noted the intellectual challenge that people experienced when trying to understand patterns of inheritance; this was especially so, since blending inheritance was not supported by the reality of so much persistent diversity of traits in populations. Mendel's work, published in German in 1866, shone considerable light into this black box. Surprisingly, however, his work went unrecognized for almost 35 years. It is certain that our understanding of genetics, including the genetic basis of evolution, might have advanced much sooner had his important work not languished in an obscure journal that relatively few had read and whose significance was not appreciated by those who had read it.

In 1900, several researchers independently rediscovered Mendel's work. At first it was controversial, but by repeating his experiments, sometimes with other organisms, his laws were found to be reliable and formed the basis of a new science, coined in 1905 as "genetics."

3.4.1 Using modern terminology

Science is about ideas, not about memorizing terms and diagrams. But in conveying ideas, there is commonly a lexicon of unfamiliar terms that can be intimidating to students. Yet learning essential terms is helpful for students, not a burden, since these terms have meanings that assist in understanding the narrative of the scientific ideas being considered. In reviewing Mendel's work, we've already made use of several terms that continue to be used in the twenty-first century: *segregation* and *independent assortment* are two, as are *selfing* and *crossing*, *recessive* and *dominant* traits, *monohybrid* and *dihybrid crosses*, and *reciprocal crosses*. From previous chapters we also know *gametes*, *gonads*, and *genitals*. Having read the foregoing sections, you should make sure you know what each of these means.

Now that we understand the genetics of sex to the extent that it was revealed by Mendel's work, there are a few more key terms that will be useful as we delve deeper into the subject of sex and inheritance. In most cases, these are paired terms.

Mendel's *factor* as a discrete unit of inheritance is a term that is no longer in use. Since 1905, this unit has been called the *gene*, defined as the physical and functional unit of heredity. As Mendel recognized, for many genes there are different variants. For instance, there is a gene in pea plants that determines flower color, but different versions of this gene generate different flower colors. These gene versions are called alleles. So, keeping with the pea example, white-flowered plants have two white alleles and purple-flowered plants have either two purple alleles or one purple allele and one white allele.

Recall that individuals have two copies of each gene, but the gametes they make contain only one copy of each gene. So, in humans, body cells such as muscle cells, heart cells, and nerve cells contain two copies of each gene and are therefore said to be diploid. But human sperm and eggs contain only one copy of each gene and are therefore said to be haploid. Always remember this distinction. It is apparent in the Punnett squares we have seen: for a given gene, the sperm and eggs (row and column titles) are haploid, containing one allele each, but the zygotes (interior boxes, following fertilization) are diploid, containing two copies of each gene. From Mendel's work and from reviewing the Punnett squares, you will recognize that although zygotes are diploid, for a given gene they can have two identical alleles or two different ones. For instance, in the pea plants, the white-flowered individuals have two white alleles. They are thus called homozygotes (zygotes having the same alleles). The purple-flowered individuals can have either two purple alleles (and are therefore also homozygotes) or one purple allele and one white allele, making them heterozygotes (zygotes having different alleles).

So, there are two ways to be a purple-flowered pea (homozygous purple or heterozygous) but only one way to be a white-flowered pea (homozygous white). This leads to yet another pair of essential terms, which are genotype and phenotype. Genotype is the collection of allele-pairs, and phenotype is the visible expression or the physical result of those genes—sometimes modified by environmental influences. So, white-flowered and purple-flowered pea plants are two different phenotypes. There is one genotype for the white-flowered phenotype: two white-flower alleles, making it a homozygous genotype. But there are two genotypes for the purple-flowered phenotype: either two purple-flower alleles (homozygous) or one purple-flower allele and one white-flower allele (heterozygous). Homozygous parents have gametes that all carry the same allele, because the parents only have one allele (in two copies, admittedly). Heterozygous parents, however, have gametes that carry different alleles (generally on a 50:50 basis), because for a given gene, half the sperm or eggs get one allele, and half get the other allele.

3.5 APPLYING MENDELIAN GENETICS TO HUMANS

Like most sexually reproducing animals, humans are diploid creatures. The only cells in your bodies that are haploid are the sperm or the eggs that you make in your gonads. So, all the understanding we were able to extract from Mendel's work on diploid peas applies to humans. Collectively, the genes in the nuclei of our cells make up our genotype. Those genes are both the units of inheritance that we package into eggs or sperm and the units of information that we use to make our own bodies—our phenotypes. For some of your 20,000 or so genes, you will be homozygous (having got the identical allele from each of your mother and your father). For others, you will be heterozygous, meaning the allele for that gene in the sperm that conceived you was different from the allele for that gene in the egg. Collectively, all the alleles for all your genes make up your total genotype, and since they are a unique combination (unless you have an identical twin), you have a unique phenotype that you recognize when you look in the mirror.

There are not really many characteristics of complex organisms like pea plants and humans that are governed by one gene of two alleles. Most traits, whether it is eye color, baldness, or height, are affected by multiple genes in the body and also sometimes by environmental influences during development as well. This does not mean that Mendel's laws do not generally apply; they apply simultaneously on numerous genes such that the phenotypic result is complicated! So, although it would be tempting to use earlobe attachment in humans as an example (as a number of textbooks have done), it would be an over-simplification to say such a phenotypic feature is determined by one gene of two alleles, one dominant and one recessive. Instead, we'll tackle the ABO blood system in humans to demonstrate the Mendelian inheritance of traits in a sexually reproducing couple.

3.5.1 Mendelian inheritance of blood groups in humans

There are two ways in which blood grouping is important for people. One is now mostly a thing of the past, but for most of the twentieth century blood grouping was used to test for paternity, sometimes being evidence used in family courts. That is, in some cases it could prove that a particular man could not be the father of a particular child. Now that DNA can be sequenced, more exact paternity tests are available, but the principle remains and it still can be an inexpensive first-attempt system in testing for paternity. Second, blood grouping continues to be important for the world of blood donations, as transfusions of blood from one person to another must take blood groups into account.

We'll restrict our examination to the most familiar blood group classification, the ABO system. This is a one-gene system. That is, there is a single gene for which all humans have two copies, and humans can be homozygous (two copies of the same allele) or heterozygous (two different alleles) for this gene. What is the gene for? It encodes information related to the building of a protein known as an *antigen* that is found on the surface of red blood cells. Where is this gene? Like all genes, it is found in the nuclei of all your nucleated cells. However, like many genes, it is not expressed in all cells. Instead, it is only expressed in those cells that themselves make red blood cells, because it is only on the surface of red blood cells that the antigen is phenotypically expressed.

What are the allele choices? Here we diverge a bit from Mendel's system. In the ABO system there are three alleles, the A, the B, and the O. The A allele codes for an A antigen, the B allele codes for a B antigen, and the O allele codes for neither. Don't forget a crucial point: even though there are these three alleles in human populations, in any individual there can only be two alleles (or two copies of one allele), since humans are diploid, not triploid.

There is a second way that the behavior of these alleles diverges from Mendel's system. If we were to compare the A and the O alleles, we can say the A allele is dominant (like Mendel's purple-flower allele) and the O allele is recessive (like Mendel's white-flower allele). Similarly for the B and O alleles, B is dominant to O. But when we compare the A and the B alleles, neither is dominant nor recessive. If you are type AB and you therefore have one each of the A and the B alleles, then your red blood cells express both antigens. We say A and B in this case are co-dominant, and it remains that O is recessive. The four blood group phenotypes are shown in Figure 3.6.

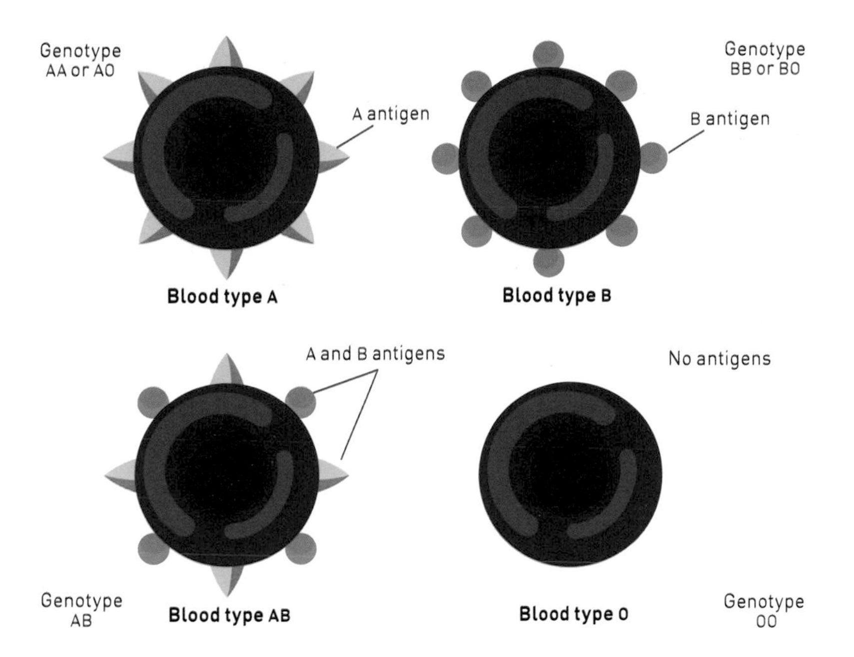

FIGURE 3.6 THERE ARE FOUR BLOOD GROUP PHENOTYPES

They are characterized by the antigens on the surface of the red blood cells. Two blood types have one genotype each (type O is OO and type AB is AB) and two blood types have two genotypes each (type A is AA or AO and type B is BB or BO).

As Figure 3.6 demonstrates, with three alleles (A, B, and O), there are six possible genotypes. You can see that the A phenotype can be produced from a homozygous AA genotype or a heterozygous AO genotype. The B phenotype is similar. But the AB phenotype can only be produced by the heterozygous AB genotype, and the O phenotype can only be produced by the homozygous OO genotype. The frequency of these three alleles in human populations varies ethnically and hence geographically. For instance, the B allele, which is the least common of the three, is found with the highest frequency among central Asians and the lowest frequency among Native Americans.

Why is the determination of a person's blood group important if he or she will be a donor, or a recipient, of blood? The proteins on the surfaces of cells, such as the A and B antigens, are not inconsequential. People with type O blood (about 45 per cent of North Americans, for example) have neither antigen, and so the immune response of a type O person when exposed to those antigens in the blood stream is significant.

BOX 3.2 WHO'S THE FATHER?

When a woman bears a child, she has no doubt that the child is indeed hers. In Chapter 2, we noted that one of the consequences of internal fertilization is that males, and their offspring, are not so certain about who the father is. That is, a particular male may assume, based on a variety of information, that the developing infant in his pregnant partner is his child. Yet we know that is not always the case, and one way this uncertainty shows up in human affairs is in disputes over paternity, with its related attributes of custody and support. It wasn't long after the ABO blood group system was figured out early in the twentieth century that it was realized it could be used to assist in cases of uncertain paternity.

To demonstrate, consider this scenario: a putative father is type O and he discovers at a routine doctor's visit that his daughter is type A. He wonders: is the daughter his? We could begin by considering the father's genotype. Since he is type O, he must have an OO genotype, since that is the only genotype that generates the type O phenotype. That means that all his sperm carry an O allele. That part is simple. The daughter is type A, but here we know there are two possible genotypes, AA or AO. He couldn't possibly be the father of an AA daughter, since his sperm could only contribute an O, but the routine blood test only tells us the daughter's type A phenotype, not which genotype she has. So, the child could be an AO and could therefore be his, providing it is possible that the she got her A allele from her mother. If the mother is type O or type B, there is a problem, because the mother has no A allele to invest in her eggs, and she could not produce a type A daughter with a type O partner. After all, there would be no parent to contribute the A allele. But, if the mother is a type A or a type AB, her eggs could contain an A allele, and she and a type O partner could easily produce a type A daughter. Using ABO blood groups can never confirm that a particular male is the father, but they can sometimes prove that a particular male is *not* the father.

Consider a second example, which we'll illustrate using a Punnett square. A woman and a man have sexual intercourse. Shortly afterwards, the woman becomes pregnant, and she chooses to not terminate the pregnancy. When she has the baby the man seeks access to the child,

Therefore, if a person has type O blood and he or she receives blood that has A antigens (from a type A donor) or B antigens (from a type B donor) or A and B antigens (from a type AB donor), the immune response causes red blood cell clumping that can result in death. So, type O individuals can only receive blood from other type O individuals. But since type O blood has neither antigen, type O blood can be donated to any individuals, whether O, A, B, or AB. If you are type AB, you can donate only to others who are also type AB. Where blood transfusions

arguing that he is the father, but the woman refuses. Prior to undertaking the considerably more expensive DNA testing, the couple opts for ABO blood group testing. The mother proves to be type B, the father is type AB, and the child is type O. Does the father get any support for his position with these results? The father has one possible genotype, being AB. The mother has two possible genotypes, being BB or BO. The child has one possible genotype, being OO. The two possible Punnett squares are shown in Figure 3.7, one for a BB mother and one for a BO mother. These analyses show that the putative father could not be the biological father, because he and the mother could never produce a type O child. He does not succeed in his claim for access; the mother must have had sexual relations with someone else around the same time.

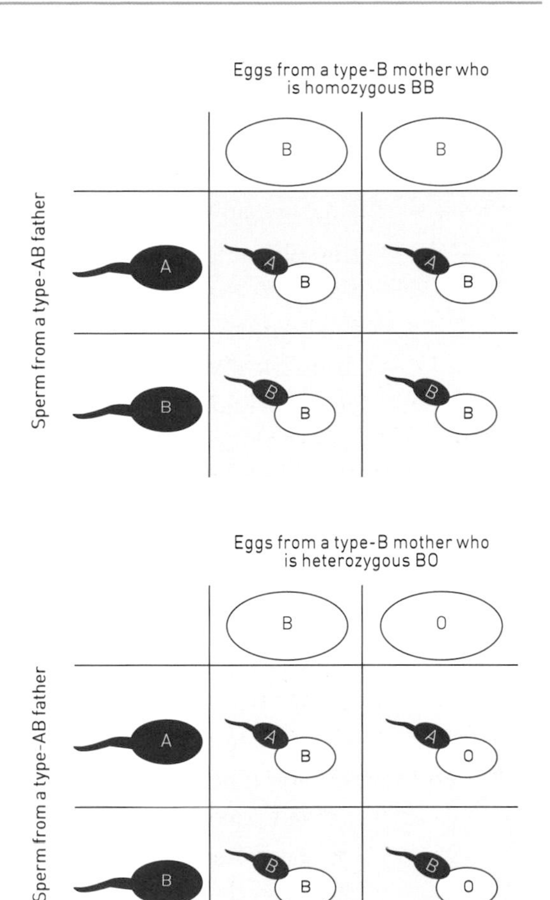

FIGURE 3.7 **THERE ARE TWO PUNNETT SQUARE POSSIBILITIES FOR A TYPE AB FATHER AND A TYPE B MOTHER**

This is because because there are two ways for the mother to be type B (homozygous BB or heterozygous BO). These show that an AB father and a type B mother can have offspring that are type AB, type B, or type A, but not type O.

are concerned, medical professionals pay particular attention to both the ABO system and the Rh system. This is a second system where a person is either Rh+ or Rh-; being Rh- means having no Rh antigen on red blood cell surfaces. This is why universal donors (who can donate to anyone) must have type O-blood (no antigens from either system) and why universal recipients (who can receive from anyone) are type AB+ since they naturally have all three antigens (A, B and Rh+). For the application of the ABO blood group system to paternity testing, see Box 3.2.

3.6 SEX, DNA, AND CHROMOSOMES

Once the discipline of genetics became established at the start of the twentieth century, human understanding of inheritance, as well as the cell machinery that governs it, increased rapidly. Through several experiments during the mid-twentieth century, DNA was confirmed as the molecule that encoded inheritance; genes were made of DNA, and DNA was the information component of chromosomes. By the end of the century, increasingly inexpensive methods of sequencing DNA were dramatically changing the field of genetics, and by 2003 the human genome had been mapped.

3.6.1 Chromosomes are the hereditary material

Improvements in microscopy during the 1800s, including advances in the technique of dyeing material that made cell internal structure more readily observable, led to many discoveries. One of these was the existence of chromosomes, located in the nuclei of cells. The term itself means "colored bodies," referring to the capacity of chromosomes to densely absorb stains (which are cellular dyes applied by investigators), thereby making them show in stark contrast to unstained or less stained parts of the cell. The idea that chromosomes were the vehicle of inheritance in the cell was promoted well before Mendel's work was "rediscovered" in 1900. Once Mendel's laws came to be accepted, the patterns they demonstrated could be logically combined with the behavior of chromosomes in cells, leading to the presentation in 1902 of the chromosome theory of inheritance.

Simply put, the chromosome theory of inheritance meant that chromosomes are the structures in cell nuclei that carry the genes. It was not immediately universally accepted. Key research that showed chromosomes assorting in cells creating gametes was that of Eleanor Carothers (1913), an American who had managed, against the odds imposed by a male-oriented establishment, to become a key player in the debate (Figure 3.8).

Using modern techniques, we can now characterize chromosomes this way. Chromosomes carry most of the genetic information con-

FIGURE 3.8 **ELEANOR CAROTHERS WAS A PIONEERING CHROMOSOME RESEARCHER**
In 1913, American geneticist Eleanor Carothers provided strong evidence, based on the movements of chromosomes in cells that make gametes, that chromosomes did indeed carry the "factors" or genes proposed by Mendel.
Smithsonian Institution Archives. Image #SIA2008-0358. Reproduced by permission

tained in organisms, including humans, and most of this information is in discrete units called genes. A gene is a coded section of chromosomal DNA, with each gene located at a particular place along a particular chromosome known as a locus (plural, *loci*). Most definitions of chromosomes focus on DNA, but they are actually a complex of proteins and DNA, with the former providing both structure and functional support when chromosomes are copied or read, and the latter providing the "information" content that is both passed on to offspring and used to make proteins. How the cells use genes to make proteins, or more broadly how the genotype is decoded to produce the phenotype, is summarized in Section 3.6.4.

Chromosomes exist in two different forms (Figure 3.9). When the cell is not involved in division—whether that is mitosis (Chapter 2) or in the making of gametes—chromosomes exist in the nucleus of the cell as chromatin. There is order to chromatin, but under the microscope it is likened to cooked spaghetti. This is a crude likeness of

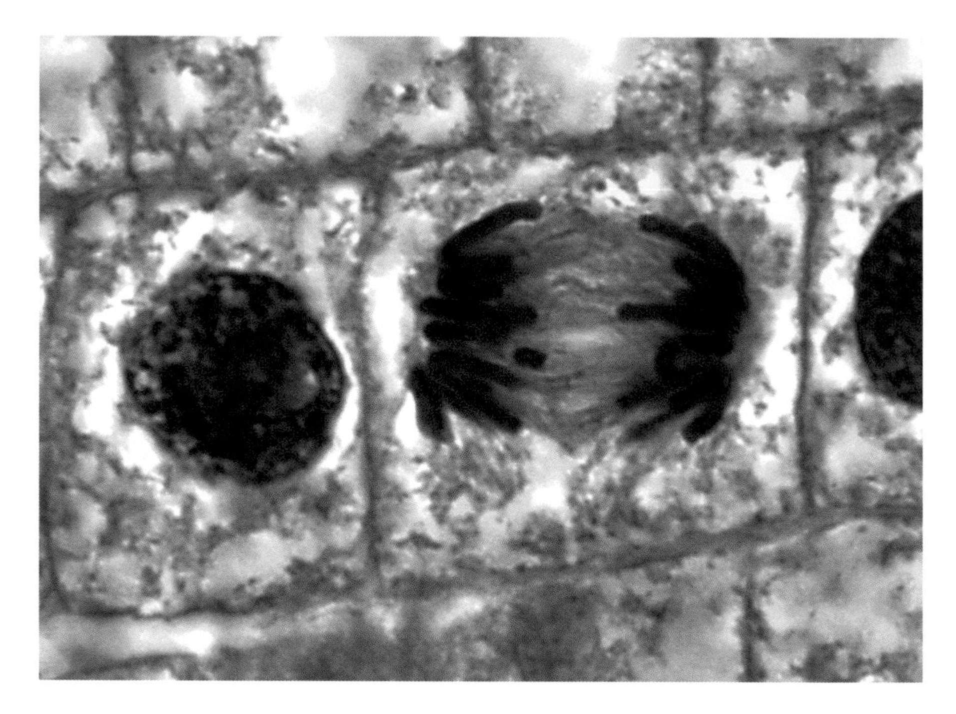

FIGURE 3.9 **CHROMOSOMES ARE MOST VISIBLE WHEN DIVIDING**
The cell on the left in this band of onion root cells appears to be in a non-dividing state, and accordingly the chromosomes are in their uncoiled state known as chromatin. The cell to its right is in a dividing state, with the duplicated chromosomes coiled up into a distinct form. At the time this image was taken, the chromosomes were being pulled to the cell's poles prior to cell division. The DNA in both cells contains the same information. It is the proteins associated with the DNA that are the agents that cause the dramatic change in appearance.
Jose Luis Calvo / Shutterstock.com

course; a mess of threads would be more accurate, as the thickness of the DNA is very small compared to the length. Even though we are referring to cells and their nuclei, which one needs a microscope to see, the chromosomes in this chromatin form are long. The longest human chromosome is chromosome 1, and it is about 8.5 cm. In contrast, when cells are engaged in duplication and division, the chromatin changes form, with the chromosomes coiling up into rod-shaped structures that so readily accept the microscopy stains that give them their name.

3.6.2 The human karyotype

A karyotype is the appearance, including the number, of chromosomes in the nucleus of the cell (Figure 3.10). It is applied to the cell in its dividing state, where the chromosomes are coiled into shorter rods,

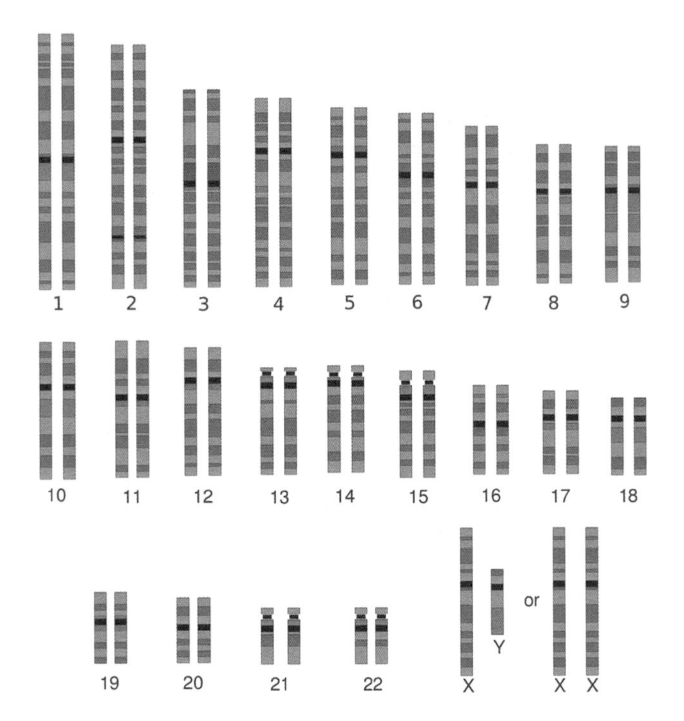

FIGURE 3.10 **THE HUMAN KARYOTYPE CONTAINS 23 PAIRS OF CHROMOSOMES**
This is a schematic drawing of the human karyotype where the chromosomes have been neatly arranged so that the members of homologous pairs are placed together. Here, the chromosomes have not yet replicated, so they are not X-shaped. Males have the XY pairing and females have the XX pairing.
Courtesy National Human Genome Research Institute

rather than in the thread-like chromatin. Because the karyotype is commonly of a cell in the process of dividing, the chromosomes commonly exist in replicated forms (not the case in Figure 3.10). These replicated forms of chromosomes are generally X-shaped because the chromosomes are attached by the *centromere* (section 2.2.2). Recall that at this stage, each replicate is called a chromatid. It is necessary to keep clear in your mind what is meant by chromosome, chromatin, and chromatid.

All humans have 23 pairs of (i.e., 46) chromosomes. These pairs are called *homologous pairs,* meaning that although they are different (since one came from the person's mother and the other from the person's father), they contain the same genes (but perhaps different alleles!) at the same loci along each chromosome. Twenty-two of these chromosome pairs are indistinguishable between males and females, but the 23rd pair constitutes the sex chromosomes. So, if you were presented with a karyotype of the 22 pairs (i.e., without the sex chromosomes), you would not be able to say whether it was from a cell in a male body or a female body. These 22 pairs are called autosomes, a term that distinguishes them from the sex chromosomes. The image in Figure 3.10 shows both possibilities for the sex chromosomes—two X-chromosomes for females and one X-chromosome and one Y-chromosome for males. Each person (and hence each karyotype) has one or the other, not both as shown in Figure 3.10.

We will encounter some anomalous karyotypes when we look more closely at the importance of the sex chromosomes in Chapter 8. Suffice to say at this point that there are rare individuals with karyotypes that don't match this pattern.

3.6.3 DNA is the genetic code

We've established that genes are the units of heredity, they are made of DNA, they exist at particular places (loci) on chromosomes, and there are different versions of these genes (alleles) that influence the phenotype of the individual. How does this genotype information (the genes made of chromosomal DNA) translate to the phenotype information (as in the organism that results)?

Proteins are the key macromolecules that determine phenotype. They are highly complex molecules that are diverse both in structure and in function. Major categories include antibodies, enzymes, structural proteins, transport systems, and cell messengers; in humans, there are many tens of thousands of proteins. DNA and proteins are intimately connected. The former represent information in the body, and the latter represent the expression of that information.

Much of the genetic work in the mid-twentieth century illuminated this relationship, resulting in the articulating of the central dogma of molecular biology (Figure 3.11) by Frances Crick, one of the Nobel prize-winning researchers who discovered the structure and function of DNA. This dogma captures the genotype-phenotype relationship by describing two universal processes: transcription and translation. Transcription is the copying of the encoded information stored in the DNA of the chromosomes to an intermediary molecule known as *messenger RNA* (mRNA). DNA is a complex molecule that exists in a double-helix form, which means that there

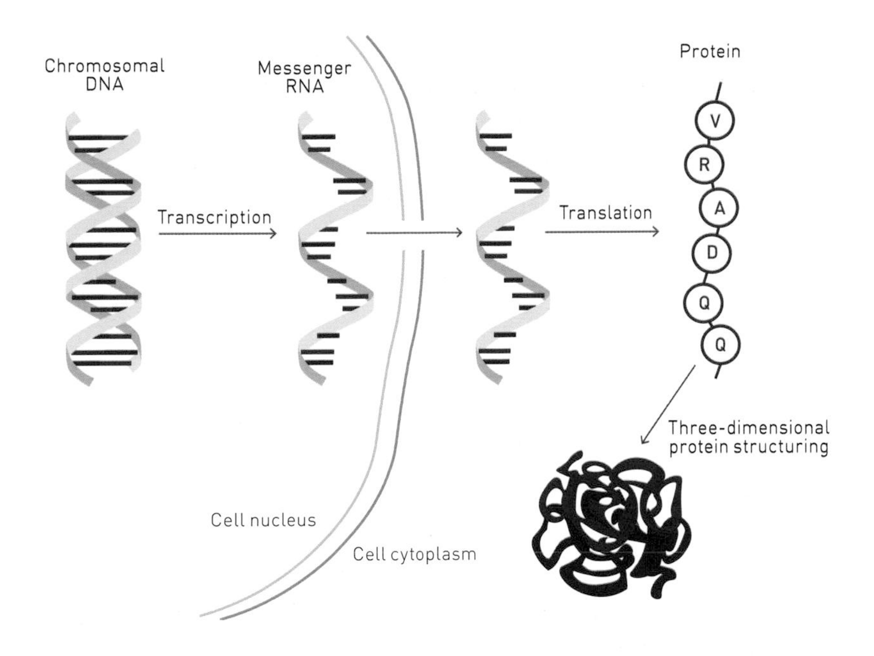

FIGURE 3.11 **TRANSCRIPTION AND TRANSLATION FORM THE CENTRAL DOGMA OF MOLECULAR BIOLOGY**

Inherited information encoded in the sequence of the components of chromosomal double-helix DNA in the cell nucleus is *transcribed* sequentially into single-helix messenger RNA (a process called *transcription*). The messenger RNA exits the nucleus and engages with cytoplasmic machinery to *translate* the information into proteins through the sequential assembly of protein building blocks known as amino acids, shown as "circles" with letters that signify which of 20 amino acids is called for (a process called *translation*).

are two connected strands that spiral together. The strands are essentially mirror images of each other, genetically speaking. Although DNA is this double-helix molecule and mRNA is a single-helix molecule, the mRNA copy has the same sequential information. The second step is known as translation. The mRNA molecule leaves the nucleus where it engages with cell machinery called ribosomes that assemble the protein that the mRNA code specifies.

In this manner, we are reminded to think of the information content of chromosomal DNA in two ways. In the central dogma sense, it is the information used by cells to create and maintain a phenotype from a genotype. In the heredity sense, it is the information from past generations that allows genetic information to survive from generation to generation.

3.6.4 How DNA encodes information

How does the genetic code work? Language encodes information, and it can be a useful metaphor for understanding the encoding that exists in the DNA of genes. The smallest division of a DNA molecule is called a nitrogenous base. There are four bases, and together they make up the four-letter alphabet of DNA. They are commonly designated by the first letter of their labels: A (*adenine*), C (*cytosine*), G (*guanine*), and T (*thymine*). The information in DNA is stored in 3-letter words called *triplets*; because there are only 4 letters, there are only 64 possible words (e.g., CCA, GTG, ACT, etc.).

Despite this small 4-letter alphabet and 64-word language, many words can be strung together, making for a great diversity of long sentences; each sentence is a gene. The chromosome is a string of sentences (as many as 1,000 or more), which might be thought of as a long paragraph. In humans, there are 46 such paragraphs, although some can have a thousand or more sentences. We might think of these 46 paragraphs as a book and the book as the individual's genotype. It's easy to see the imperfections of the alphabet-language metaphor though: there are only 64 words (i.e., triplets), the sentences (i.e., genes) are very long, there are as many as a thousand sentences per paragraph (i.e., per chromosome), and in humans at least, there are only 46 paragraphs. And, there are really 23 pairs of paragraphs, with the pair members being very similar except for some of the sentences that are slightly different (i.e., alleles). However, we can extend the metaphor to a breeding population, in which case the collection of books is a library, which represents the total gene pool—all the alleles for all the genes in all the individuals in the population. Remember that all the books would be highly similar though, with most of the diversity created by slight differences in some of the words of otherwise identical sentences.

The two helices of DNA are connected by pairs of nitrogenous bases such that the pairs are CG, GC, AT, or TA. That is, G and C always pair together, and A and T always pair together. We've noted that transcription produces an mRNA copy of the DNA. Structurally, mRNA uses the same bases as DNA, except that it uses *uracil* (U) wherever DNA uses thymine (T). So, when the DNA strand used for copying (known as the template strand) is used to make matching mRNA, a C in DNA calls for a G in mRNA, a G

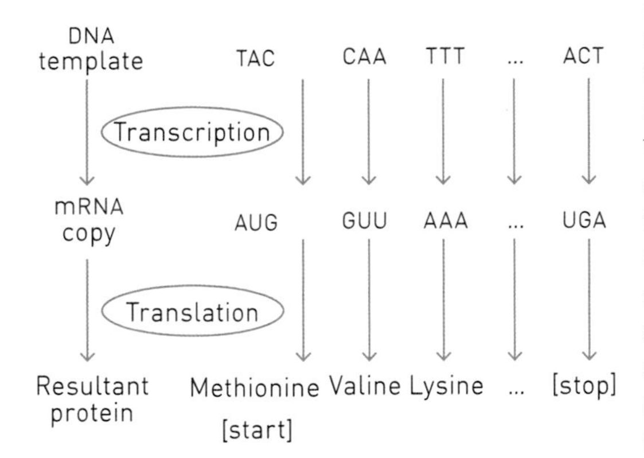

FIGURE 3.12 **CODED TRIPLETS IN MESSENGER RNA DICTATE PARTICULAR AMINO ACIDS IN PROTEINS** One half of the DNA double-helix molecule is known as the template strand, shown here as TACCAATTT, etc. Inside the nucleus, an mRNA strand is copied (transcribed) from the template, shown here as AUGGUUAAA, etc. Outside the nucleus in the cytoplasm, the mRNA is read (translated) to build a protein, shown here by its constituent amino acids. The amino acid methionine, which is translated from the mRNA triplet AUG, is universally used to start the protein-building. UGA is one of three mRNA triplets used to stop protein-building. The spaces between the triplets in the DNA and mRNA don't exist in reality.

in DNA calls for a C in mRNA, a T in DNA calls for an A in mRNA, and an A in DNA calls for a U in mRNA.

When this mRNA leaves the nucleus, it "docks" at a cell organelle called a *ribosome*. This is where translation into a protein occurs, which really means that this pairing of mRNA and ribosome is where proteins are assembled as the mRNA is read in sequence, one triplet at a time. Each triplet of RNA specifies one of 20 amino acids; because there are more RNA triplets (64) than amino acids (20), some amino acids are specified by more than one triplet. The code is illustrated in Figure 3.12.

Our understanding of the workings of the genetic code, and our more recent ability to manipulate it, has provided many applications in the fields of gene therapy, genetic engineering, conservation of threatened species, the study of human migrations, as well as much-improved paternity testing (Box 3.3).

3.7 WHAT HAPPENS TO CHROMOSOMES IN THE MAKING OF GAMETES?

Figure 2.5 illustrated mitotic cell division, the means by which multicellular organisms create new cells from older cells. That figure used only two pairs of homologous chromosomes to illustrate the process; if it had been for humans, it would have shown 23 homologous pairs, but

BOX 3.3 NOW WHO'S THE FATHER?

In Box 3.2, we saw how ABO blood types can be used in assessing paternity claims. This system dates back to the first part of the twentieth century, and although it never could determine that a particular person was indeed the father, it could often determine that a particular person was not. Our more recent capacity to take tissue samples and to sequence DNA has provided a very sophisticated tool to analyze ancestry, because a person's DNA is so unique.

One interesting and controversial application of these techniques was to consider the family tree of Thomas Jefferson, a founding father of the United States of America who is credited with being the primary author of the Declaration of Independence. Although he is associated with high principles and ideals, his legacy is a mixed one, especially when it came to slavery, since his considerable holdings remained dependent upon slave labor throughout his life.

Jefferson's wife Martha Wayles bore six children during their ten years of marriage, but only one daughter survived childhood. Martha herself died when Jefferson was only about 40, and he never remarried. However, Jefferson ran a complicated household and plantation. Later in life as a widower, he was accused of having had children through one of his slaves, Sally Hemings. Hemings herself was biracial, probably ¾ European and ¼ African, but she was a slave by heritage and by law. At the time, the allegation of having had children with his slave was scandalous, and it simmered without resolution for close to two centuries.

In the 1990s, as the technology to sequence DNA was becoming increasingly efficient, it was brought to bear on the question. To assemble evidence to test the idea, the focus was on the Y-chromosome for two reasons. First, unlike the X-chromosome and the 22 non-sex chromosomes, it is only found in males, and second, almost all of its DNA does not change from generation to generation, so for most of its length, a son's Y-chromosome DNA is identical to his father's.

Genetic material was needed from both the Jefferson family and the Hemings family. Jefferson had no legitimate male descendants, since he had no sons who had survived childhood through his wife Martha. But Jefferson's father's brother (i.e., his paternal uncle) had male-line descendants alive in the 1990s, five of whom provided samples. The researchers also found one male-line descendant of Eston Hemings, a son of Sally Hemings and allegedly the son of Jefferson, who also provided a sample. The family tree is shown in Figure 3.13.

When the modern day Jefferson and Hemings Y-chromosome sequences were compared, they were found to be identical. Does this confirm that Jefferson was indeed Eston Hemings father? No, because there were other Jefferson relatives (especially male-line cousins) who could have been the father, but based on opportunity it is likely that the father of Eston Hemings (and some of his siblings) was indeed Thomas Jefferson.

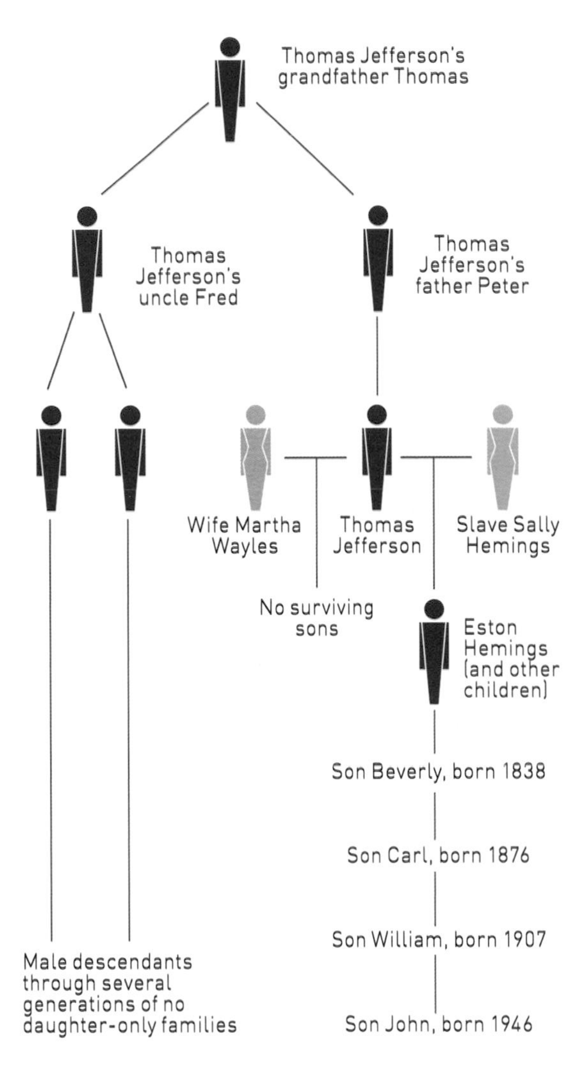

FIGURE 3.13 **BASED ON Y-CHROMOSOME ANALYSIS, THOMAS JEFFERSON PROBABLY HAD TWO FAMILIES**
This is the family tree associated with Thomas Jefferson. Jefferson had no male heirs through his wife Martha, but his Uncle Field, who would have had the same DNA sequences in his Y-chromosome, has a family tree in which every generation produced at least one male, preserving the Y-chromosome, so that modern-day samples could be obtained. Similarly, through his slave Sally Hemings, Jefferson's alleged son Eston Hemings also had a lineage that produced sons in every generation, so that Eston Hemings' Y-chromosome could be sequenced. The two Y-chromosomes matched, confirming that Hemings was indeed a Jefferson.

BOX 3.4 THERE ARE THREE LIFE CYCLE VARIANTS OF MEIOSIS

The *life cycles* of sexual organisms alternate between meiosis and fertilization, but there are three different modes that accomplish this (Figure 3.14). It is the *gametic* life cycle that we've previously described in detail that applies to humans and other animals. Here, a diploid zygote develops through mitosis into a diploid organism, in whose gonads meiosis occurs to produce haploid eggs and sperm. When these join through fertilization, a new diploid zygote is formed.

Most fungi, such as mushrooms, reproduce sexually by using the *zygotic* life cycle. The zygote never develops by mitosis into a diploid organism. Instead, it is the zygote that undergoes meiosis—not to produce haploid gametes, but to produce haploid spores. (You may have encountered fungal spores if you've ever discovered and disturbed the spore dust from a mouldy piece of bread). These minuscule haploid spores float through the air to land on suitable substrates where they then develop through mitosis into a haploid organism: the mushroom. The mushroom then undergoes mitosis to produce gametes (since the mushroom is haploid, it must use mitosis to produce haploid gametes). When these gametes unite, they form the diploid zygote, completing the cycle.

The *sporic* life cycle probably seems the oddest, especially the way it occurs in ferns and some fern relatives. As with animals, the zygote develops by mitosis into a diploid organism, which is the familiar fern plant. But, rather than undergo meiosis to produce gametes, this diploid fern uses meiosis to produce haploid spores, similar to fungal spores. These spores land in a suitable place, develop into a small fern plant that is the

---

the process remains the same. Using humans as an example, the key result in mitosis is that where there had been one cell with 23 homologous pairs of chromosomes, there are now two cells that each contain the exact same information—23 homologous pairs of chromosomes. A diploid *parent* cell has produced two diploid *daughter* cells. Mitotic cell division is the type of cell division that occurs in bone cells, nerve cells, and digestive tract cells—in fact in most cells other than the specialized set of cells in the gonads that produce eggs and sperm.

In ovaries and testes, a different type of cell division known as meiotic cell division occurs. Meiotic cell division starts in a similar way to mitotic cell division: before the chromatin coils up into chromosomes in preparation for cell division, it is replicated so that each of the 46 chromosomes is made of a pair of chromatids. But once

same species, obviously, but is structurally very different. This haploid plant then produces haploid male and female gametes by mitosis, which can unite through sexual fertilization to form a diploid zygote. In this way, ferns exhibit an alternation of generations, alternating between the familiar diploid plant and an obscure haploid one, generation after generation.

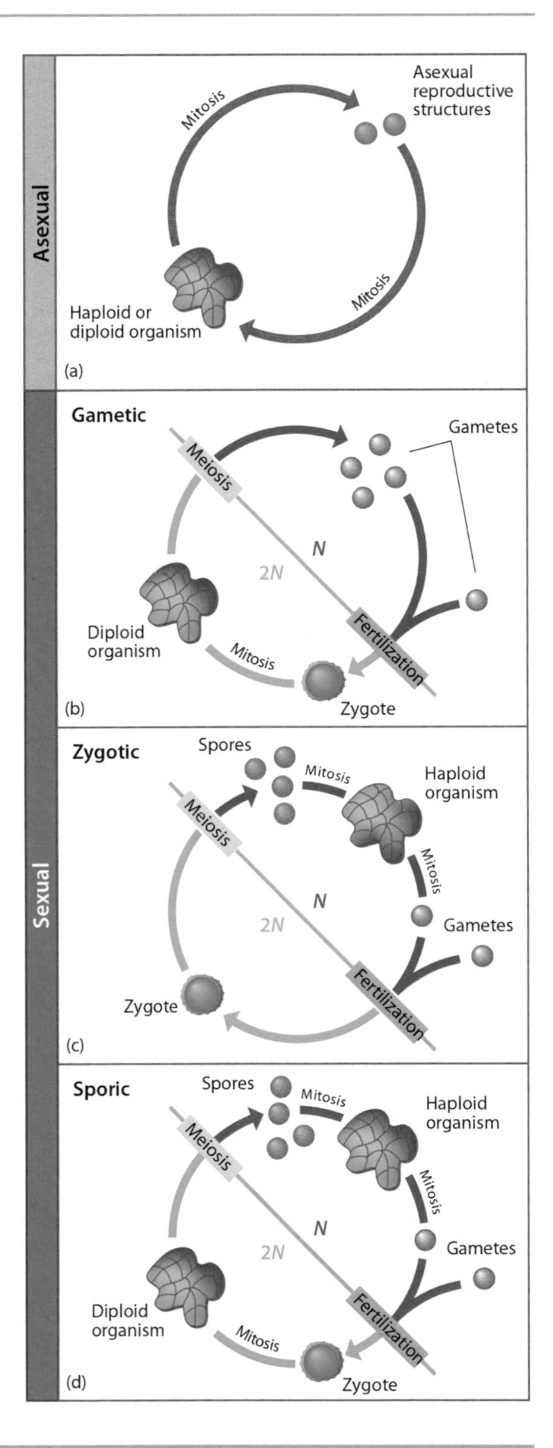

FIGURE 3.14 THERE ARE THREE ALTERNATIVE SEXUAL LIFE CYCLES IN NATURE

By way of contrast, asexual organisms like the *Amoeba* (panel a) reproduce using mitosis; there is no meiosis and no fertilization. The life cycles of sexual organisms always alternate meiosis and fertilization, but there are three different modes of doing this. In the *gametic life cycle* (e.g., mammals, panel b), the diploid zygote develops through mitosis into a diploid organism, in whose gonads meiosis occurs to produce haploid eggs and sperm. In the *zygotic life cycle* (e.g., fungi, panel c), it is the zygote that undergoes meiosis, producing haploid airborne spores that develop through mitosis into a haploid organism, such as a mushroom. These haploid mushrooms undergo mitosis to produce eggs and sperm. In the *sporic life cycle* (e.g., ferns, panel d) the zygote develops by mitosis into a diploid organism, as in humans. However, meiosis in the diploid organism produces haploid spores, similar to fungal spores. When these spores land in a suitable place, they develop by mitosis into a very small haploid fern plant, which then produces haploid male and female gametes by mitosis. The N half of each cycle denotes the haploid stage (following meiosis), and the 2N half denotes the diploid stage (following fertilization).
From Graham, Graham, & Wilcox (2016). Reprinted by permission of LJLM Press

the chromatin coils up into chromosomes, mitosis and meiosis diverge. In addition to taking place in a different location (the gonads, as opposed to anywhere else in the body), there are three other major differences. The first is a fundamental process in meiosis that is rare in mitosis, known as crossing-over. The second is that meiosis involves two cell divisions, not one. And the third is that the cells produced—the gametes—are haploid, not diploid.

This third difference—that meiosis produces haploid gametes—requires qualification, because it is not true in plants and fungi. Plant and fungus meiosis does include crossing-over and does involve two cell divisions. But in plant and fungal meiosis, the haploid cells are microscopic spores, not sperm and eggs. Plants do produce haploid eggs and sperm, but they do this through mitosis from haploid structures (Box 3.4).

3.7.1 What is crossing-over?

Recombination is the production in the gonads of haploid gametes in which each chromosome contains a mixture of genes: some that came from the individual's mother and some that came from the father. This mixture of maternally derived and paternally derived genes on a chromosome is accomplished through crossing-over, which occurs during the early stages of meiosis (Figure 3.15).

Crossing-over begins when the two chromosomes of each homologous pair associate side by side in a process known as synapsis, from the Greek word for conjunction. One chromosome of each pair came from the person's father (i.e., paternally derived), and the other came from the person's mother (i.e., maternally derived). Each chromosome at this stage is in the coiled-up, replicated form of two chromatids, and the two chromatids of each chromosome are called sister chromatids due to their common origin, but they are not inherently "female." Because there are four chromatids in this arrangement (two pairs of "sisters"), it is called a *tetrad.*

During synapsis, chromatids trade sections of DNA, resulting in the swapping of alleles. These trades are between a paternally derived chromatid and a maternally derived one. Because the trading chromatids involved are derived from different parents, they are called non-sister chromatids. When the chromatids are eventually pulled apart and distributed among separate gametes (Figure 3.16), they now merit the label "chromosome," and they will each be unique mixtures of the person's parents' genes. In contrast, in body cells the paternally derived and maternally derived members of each homologous pair do not mix; they retain their integrity although each independently influences the operation of the cells whose nucleus they occupy.

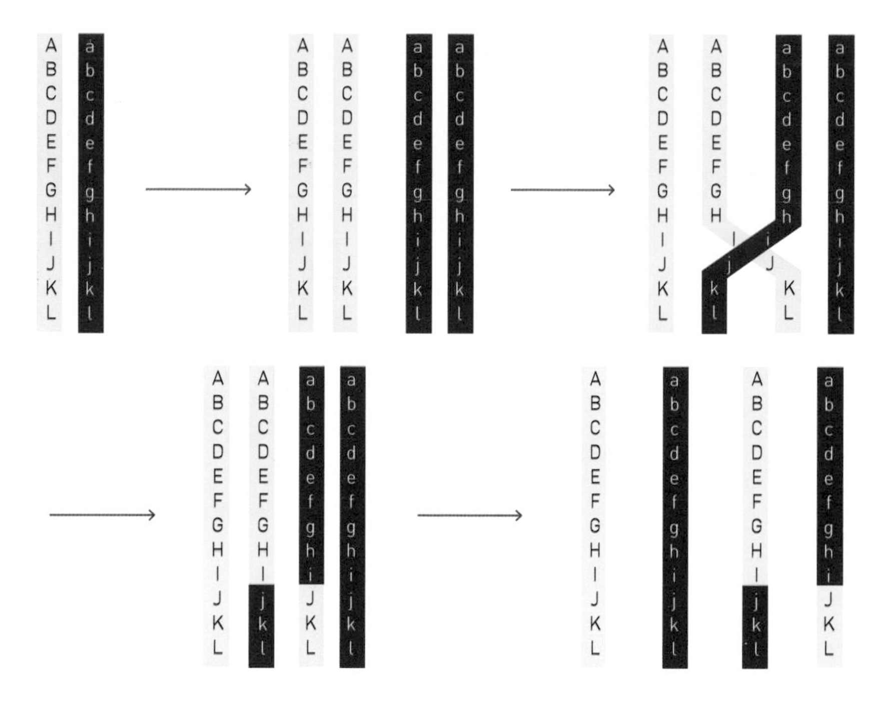

FIGURE 3.15 CROSSING-OVER IN GAMETOGENESIS CREATES UNIQUE SPERM AND EGGS

This sequence illustrates crossing-over during the prophase I stage of meiosis within a testis or ovary. The chromosomes are in the coiled-up, replicated form of two *sister chromatids* each, attached at a *centromere* (*centromere* not shown). The left chromosome (yellow) was inherited from this person's father, and the right (red) from this person's mother. The four chromatids are a *tetrad*. In a formation known as *synapsis* (second image), the DNA of *non-sister chromatids* breaks one or more times at a *chiasma* (third image). The pieces re-join after trading maternal and paternal sections (fourth image). The result is the fifth image: after separating into four separate chromosomes, the two middle chromosomes are true mixes of paternally derived (yellow) and maternally derived (red) genetic material. Crossing-over can involve all four chromatids, and there can be multiple points of exchange.

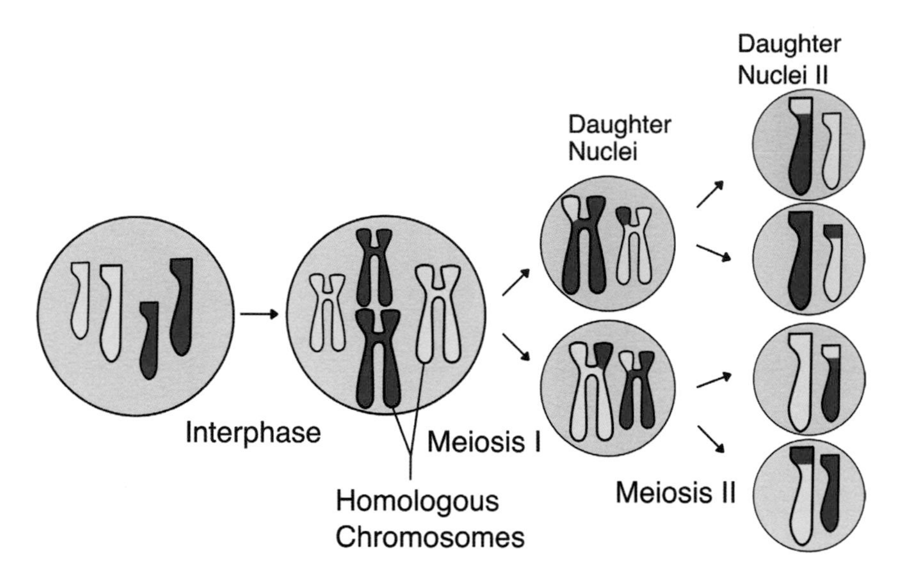

FIGURE 3.16 MEIOTIC CELL DIVISION IS THE PROCESS THAT CREATES HAPLOID GAMETES

Unlike *mitotic cell division* (Figure 2.5) where there is one cell division yielding two daughter cells identical to the parent cell, in meiosis there is crossing-over (Figure 3.15) and two divisions, yielding four non-identical daughter cells. During the first division, the chromosomes remain as paired chromatids, but the two members of each pair of homologous chromosomes are separated into daughter cells. In the second division, the chromatids are separated (as in mitosis), with one copy going into each of these daughter cells.
Rdbickel / CC-BY-SA 4.0)

Students sometimes have trouble "seeing the forest for the trees" when learning meiosis. Remember this. When you make eggs in your ovaries or sperm in your testes, the chromosomes you have to work with are the ones you got from your mother and father. In your body cells, these chromosomes remain unmixed, where you have a pair of each of the 23 chromosomes, with one member from each parent. It is only in your gamete-making gonads where genetic material sourced from both parents ends up on the same chromosome. This process yields haploid eggs or sperm that have one of each chromosome, but these chromosomes now contain sections from your mother's chromosomes and your father's chromosomes in a mixed form. So, your parents' genes never mix on chromosomes until you produce gametes. If your gametes produce offspring, the chromosomes they receive from you are combinations of your parents' genes. This means that a couple's genes do not mix on the same chromosome until their children make gametes, meaning that the mixed chromosomes appear in their grandchildren.

3.7.2 The two divisions of meiosis

Mitotic cell division involves one division, which yields two diploid daughter cells with identical copies and numbers of chromosomes as the starting cell. Meiosis, on the other hand, involves two cell divisions, known as Meiosis I and Meiosis II, and jointly these yield four haploid daughter cells (gametes) that have chromosomes that are new in the sense they mix paternal and maternal sources. The two divisions of meiosis are shown in Figure 3.16. For simplicity of illustration, this image is of a species that has two pairs of homologous chromosomes, in contrast to the 23 pairs found in humans.

Once crossing-over occurs in the first division of meiosis, the homologous pairs line up along the equatorial plate. Each chromatid of each chromosome member of each homologous pair is likely to be a unique set of alleles due to the crossing-over that has just finished. All four chromatids of each homologous pair will have the same genes but they will have unique allele combinations. The members of each homologous pair are separated, and once they move toward the poles of each forming daughter cell, cytokinesis follows.

At this stage—between the first and second divisions of meiosis—the two daughter cells are haploid, containing one member of each homologous pair. Of course, each member is in a 2-chromatid form, but that is still technically considered haploid. During the second division, the chromosomes are aligned in single file (i.e., not in pairs) along the equatorial plate through the cell, and the chromatids separate at the centromere, as they do in mitosis. Cytokinesis follows again, and there are now four haploid cells where each chromosome is made of one chromatid (but now again referred to as a chromosome). We have seen already that crossing-over, by trading alleles between non-sister chromatids, generates variation by making unique combinations of alleles. The cell divisions that randomly distribute the newly created chromosomes into gametes add an additional level of variation; this is Mendel's *independent assortment.*

3.7.3 Sex and chromosomes that don't crossover

The effect of crossing-over events, where the starting chromosomes trade sections, is that the new chromosomes that are packaged into gametes have genes from both paternally derived and maternally derived chromosomes. In your own case, when you make eggs or sperm, the chromosomes you invest them with are in this "shuffled" state, so that chromosome 15 in your gametes, for instance, has some alleles from your maternally derived chromosome 15 and some from your paternally derived one. In this way, generation by generation, the combination of alleles on a particular chromosome is constantly being shuffled.

There are three chromosomes that are exceptions to this pattern. The first case is the Y-chromosome (Box 3.3), because this chromosome is never paired with other Y-chromosomes (otherwise the individual would have no X-chromosome and would be missing hundreds of necessary genes). Y-chromosomes pair with X-chromosomes in male bodies, but there is only a very small amount of crossing-over between the two during male meiosis. The crossing over that does occur takes place at one end of the Y-chromosome, leaving most of the Y-chromosome untouched. Therefore, a male's Y-chromosome (with the exception of a few genes at one end) is the same as his father's and as his father's father and his father's father's father, etc. Because Y-chromosomes are only found in males, we say their inheritance is *patrilineal.* (See Box 3.3.)

The second case is the simplest exception, and it only applies in males. When X-chromosomes are in the paired state in a female, they cross over in the usual manner, with the result that the eggs generated have X-chromosomes that contain alleles from both the maternally derived and paternally derived X-chromosomes. In a male, however, with the exception of a small part of the X-chromosome at one end that exchanges with the Y-chromosome, the X-chromosome has no partner for crossing-over. So, when meiosis in a male invests a sperm with an X-chromosome, it is virtually unchanged from the X-chromosome that male received from his mother. X-chromosomes therefore engage in crossing-over about half as often as the autosomes, because it only happens in ovaries, not testes.

The third case requires us to introduce a small 47th chromosome that humans all have, known as the *mitochondrial chromosome.* The 23 pairs of homologous chromosomes on which we have so far focused (i.e., the 22 pairs of autosomes and the sex chromosomes) reside in the cell nucleus and collectively contain well over 99 per cent of all human genes. However, a handful of genes—37, in the case of humans—are found on a circular chromosome outside the nucleus, located in a cell organelle known as the mitochondrion (*mitochondria* in the plural). Mitochondria are essential to the cell for energy management. There are many mitochondria in each cell—thousands in the case of heart and liver cells—but the billions of copies of this chromosome in a person's body are all identical. In fact, they are identical to the mother's mitochondria and her mother's mitochondria and her mother's mother's mitochondria, etc. This is for two reasons. First, mitochondria exist in single copies, not in a diploid state, and they therefore reproduce in the cell by binary fission, as bacteria do, without crossing-over with other chromosomes. Second, although sperm have mitochondria (identical to the man's mother's mitochondria) to help them process energy, especially for swimming, the mitochondria in sperm do not get incorporated into the fertilized zygote. Although mitochondria are found in both sexes, we say their inheritance is *matrilineal* because they reach a dead-end in the male. As with Y-chromosomes, mitochondrial DNA experiences mutation very slowly over long stretches of time, but their DNA sequences do not otherwise change.

3.7.4 How meiosis relates to Mendel's laws

Return to Figures 3.15 and 3.16 as we consider this. In Figure 3.15, imagine that the factor (i.e., gene) for flower color is located on this chromosome and is in fact the gene at locus B/b. Imagine also that the allele B is the allele for the purple flower color and that the allele b is the allele for the white flower color. So, the chromosomes shown in the upper-left are like the ones you would find in an individual of Mendel's F1 cross, where a true-breeding purple-flowered variety and a true-breeding white-flowered variety produced this heterozygote. (Genes at A/a and C/c through L/l influence other non-color characters).

Now look at the bottom-right image in Figure 3.15, where we have the four resultant chromosomes in the gametes produced by this

F1 plant (whether they be packaged in the male pollen or in the female eggs). Fifty percent of the gametes carry the B allele (which we have imagined is the allele for purple color) and 50 percent carry the b allele (which is the allele for the white color). This is Mendel's law of segregation, where the two alleles separate on a 50:50 basis in the making of gametes. But notice what else has occurred. Although in this case two of the four resultant chromosomes match the parental chromosomes, two are novel combinations that mix paternally derived and maternally derived alleles. For these 12 alleles, one chromosome has alleles A through I and j, k, and l, and the other chromosome has alleles a through i and J, K, and L.

What about independent assortment? Consider Figure 3.16. Imagine that the pea color gene is located on the long chromosome and the pea shape gene is located on the short chromosome. Because they are on different chromosomes, you can see how they can assort independently. The number and extent of crossovers varies, so there can be many possibilities, but you can see here that whichever chromatid for the long chromosome ends up in a gamete is unrelated to whichever chromatid for the short chromosome ends up there.

CHAPTER 3 SUMMARY

- In the 1800s, most biologists believed characteristics were passed on through blending inheritance, where the parents' traits were simply averaged together.
- Mendel's Law of Segregation states that a parent has two inherited factors that influence a particular characteristic, but only one factor is incorporated in an egg or sperm.
- Mendel's Law of Independent Assortment states that the segregation that occurs for one physical characteristic occurs independently of the segregation that occurs for a second characteristic.
- Punnett squares are used to analyze patterns of inheritance, where column titles and row titles provide information about eggs and sperm and the interior boxes provide information about the resulting offspring.

- Genetic information is encoded in genes, genes are located on chromosomes, and different versions of genes are called alleles.
- Most sexual organisms are diploid, meaning their chromosomes exist in homologous pairs, one from each parent.
- In testes or ovaries, meiosis creates sperm and egg cells that are haploid, meaning chromosomes are not in pairs but are instead single.
- Among the 23 homologous pairs of chromosomes in humans there is a pair of sex chromosomes. These can be either two X-chromosomes (in the case of females) or one X- and one Y-chromosome (in the case of males). The other 22 pairs are called autosomes.
- The human ABO blood group system is useful for demonstrating patterns of inheritance in individuals.
- The central dogma of biology is that genes encode information in DNA that is used by cell machinery to make proteins; the DNA is first transcribed in the cell nucleus into messenger RNA, and that RNA is translated outside the nucleus into protein.
- During meiosis, crossing-over mixes sections of the paternally derived chromosome with sections from the matching maternally derived chromosome, making unique combinations of alleles.

FURTHER READING

Abbey, D.M. (1999). The Thomas Jefferson paternity case. *Nature, 397*(6714), 32. https://doi.org/10.1038/16177

Adams, J. (2008). Paternity testing: Blood types and DNA. *Nature Education, 1*(1), 146.

Brush, S.G. (2002). How theories became knowledge: Morgan's Chromosome Theory of Heredity in America and Britain. *Journal of the History of Biology, 35*(3), 471–535. https://doi.org/10.1023/A:1021175231599

Crick, F. (1970). Central dogma of molecular biology. *Nature, 227*(5258), 561–563. https://doi.org/10.1038/227561a0

Galton, D.J. (2012). Did Mendel falsify his data? *Quarterly Journal of Medicine, 105*(2), 215–216. https://doi.org/10.1093/qjmed/hcr195

Graham, L.E., Graham, J.M., & Wilcox, L.W. (2016). *Plant biology*, 3rd ed. Madison: LJLM Press.

Gunter, C. (2005). Genome biology: She moves in mysterious ways. *Nature, 434*(7031), 279–280. https://doi.org/10.1038/434279a

Lane, N. (2005). *Power, sex, suicide: Mitochondria and the meaning of life.* Oxford: Oxford University Press.

Lorenzano, P. (2011). What would have happened if Darwin had known Mendel (or Mendel's work)? *History and Philosophy of the Life Sciences, 33*(1), 3–49.
Nature News Feature (2001). What a long, strange trip it's been.... *Nature, 409*(6822), 756–757. https://doi.org/10.1038/35057286
Rose, S. (2003). *Lifelines: Life beyond the gene.* Oxford: Oxford University Press.
Singh, R.S. (2015). Limits of imagination: The 150th anniversary of Mendel's Laws, and why Mendel failed to see the importance of his discovery for Darwin's Theory of Evolution. *Genome, 58*(9), 415–421. https://doi.org/10.1139/gen-2015-0107
Wang, J., García-Bailo, B., Nielsen, D.E., & El-Sohemy, A. (2014). ABO genotype, "Blood-Type" diet and cardiometabolic risk factors. *PLOS One, 9*(1), 1–9.

4 Sex and Evolution

KEY THEMES

In Europe, the scientific worldview fostered during the Enlightenment paved the way for Charles Darwin to develop his ideas of evolution in the following century. In 1859, he published the dramatically influential *On the Origin of Species*, establishing that evolution explains diversity and that natural selection is the major mechanism. Allele frequencies in a population do not change unless there are evolutionary forces at work. Sex has costs, but the genetic diversity it generates has great benefits. Sex plays a role in the creation of species.

The central organizing concept in biology is evolution. This is true whether one is thinking about cell structure, predator-prey relations, or the physical and behavioral attributes associated with sexual reproduction—or any other sub-discipline of the life sciences. It is true that many biological attributes can be *described* without reference to evolution, but when *explanation* of attributes is the goal, evolution is always at the heart of that explanation. This has been true since the latter part of the nineteenth century, shortly after the publication of Darwin's *On the Origin of Species* in 1859 (Figure 4.1).

Yet if asked to characterize sex—even the biology of sex—and why sex is so significant, few people would include any evolutionary language in their explanation. This is probably for three reasons: because relations between the sexes are central to day-to-day life, because sexuality is such a subjective component of a person's identity, and because people inaccurately think of evolution only as something that has happened in the long-ago past.

We know that among humans, sex is not only about reproduction. In social species like ours, for instance, sex has been co-opted for purposes in addition to reproduction, such as bonding (Section 2.1). On the other hand, reproduction is *the* invariant feature associated with sex regardless of the species—spiders, pea plants, clownfish, or humans. In this chapter, we summarize the key concepts of evolution, and as we do so, we show that evolutionary change marches forward on the basis of reproductive success, generation by generation. Every person reading these words has inherited genes from his or her parents, and every person has a perfectly unbroken lineage of reproductively successful ancestors. This is the deep connection between sex and evolution.

4.1 A SHORT HISTORY OF EVOLUTIONARY THOUGHT

The study of remote or past societies by anthropologists and historians makes it clear that all societies have explanations about origins, whether for the Earth itself, the "heavens above," the diversity of living things, or humans. Such narratives become deeply culturally entrenched, with particular stories associated with particular peoples.

FIGURE 4.1 *ON THE ORIGIN OF SPECIES* WAS PUBLISHED IN 1859

This great work by Charles Darwin (1809–1882) was an immediate bestseller, has never been out of print, and is without doubt one of the five most influential books of all time. It presented evidence that evolution has occurred and continues to occur, making current organisms and their behaviors understandable, and it set out its major mechanism, *natural selection*.

As such, they become competing explanations when different groups encounter one another, often to the point of ideological or even aggressive conflict. Due to the power of their motifs, their associations with homelands, and their pervasiveness in their home cultures, many of these explanations of origins continue to have great currency and acceptance today, and continue to have the capacity to polarize people when explanations accepted by different groups conflict with each other.

Evolution is the concept used by science for explaining origins of biodiversity, including humans, and it is almost universally accepted by life scientists. Yet, even in societies that have embraced science, it still has to compete with opposing intransigent religious explanations; in the West, this is much more true in the United States than in Europe, Australia, and Canada. Unlike competing explanations, science relies uniquely upon an evidentiary foundation, both empirical and experimental. The principles of science summarized in Chapter 1—deductive and inductive logic, hypothesis testing, controlled experiments, careful measurement, and empiricism—are widely used in advancing scientific knowledge, and they constitute the tools used by evolutionary biologists.

4.1.1 Evolutionary theory is a natural result of the Enlightenment

It would be highly simplistic to suggest that before 1859 there was a universal acceptance that a divine entity created the Earth and its component species or that no one had ever conceived of evolution before. It is true that the idea of divine creation was widely held, and it is true that the publication of *On the Origin of Species* in that year constituted what we would now call a *paradigm shift,* but evolution was hardly a new idea. Even among some philosophers of ancient Greece, the evolution of species had been proposed. Anaximander (~600 BCE), for instance, suggested that animals (including humans) sprang from the sea. Still, such ancient evolutionary proposals were philosophical exercises, not experimental or empirical ones.

The eighteenth century in Europe is considered to represent the Enlightenment, a period where rationalism (i.e., reason) and empiricism (i.e., measurement) were deemed by philosophers and other intellectuals to be the key ingredients necessary for the generation of valid ideas and the determination of objective knowledge. This perspective supported a belief that the material world, including even its living components, could be logically and objectively deciphered. Naturally,

this attitude fostered the scientific enterprise, but at the same time it constituted a challenge to a religious authority that had enjoyed a monopoly for centuries on what was, or was not, valid knowledge.

FIGURE 4.2 PROGRESSIVE CHANGES IN EUROPE LED TO THE DISCOVERY OF EVOLUTION
The Enlightenment empowered Europeans to contemplate nature with an increasingly objective perspective, and fossils of what were eventually called dinosaurs (as well as fossils of other large exotic reptiles like the *plesiosaur* portrayed here) challenged people's entrenched views. Not only did these giants provoke awe, but their apparent *extinction* was almost as great a challenge to the idea that species were divinely created as was the idea of evolution.
iStock.com / estt

4.1.2 Exploration, fossils, and a very old Earth also led to evolutionary theory

As this young science legitimized by the Enlightenment forged ahead, biology was among the areas of knowledge that drew its attention. But it wasn't just the intellectual momentum of the Enlightenment at work. As the 1700s gave way to the 1800s and as Europeans traversed the globe, they encountered unusual plants and animals in faraway places, many of which were harvested to bolster ever-growing collections in Europe. People were struck by patterns of similarity and of dissimilarity among organisms that were curious and thought-provoking, and this in turn invited serious contemplation about the origins of these patterns.

Closer to home, as geologists studied rocks and landforms with growing sophistication, it became increasingly evident that the estimated time necessary for those landforms to develop was much more than the 6,000 years or so specified by biblical authority. With intensified quarrying, especially for the coal deposits and building materials that were driving the young Industrial Revolution, fossils were being discovered at an increasing pace. Some of these were giant, many were strange or even bizarre, and virtually all were of extinct organisms. Bones of giant extinct animals were found in England as early as the 1600s, but it would not be until 1842 that enough had been systematically assembled that the group famously known as dinosaurs would be described. Why, people began asking, do such creatures no longer exist? Extinction was almost as much a challenge to a religious explanation of the creation as was the idea of evolution (Figure 4.2).

4.1.3 Jean-Baptiste Lamarck proposed a mechanism for evolution in 1809

Jean-Baptiste Lamarck (1744–1829) was not the first evolutionist, but he gets credit for being the first to propose a mechanism by which evolution could proceed. His argument that life became more complex over time, as seen in the fossil record, and his idea that life forms transform gradually by accumulating small differences from generation to generation are two contributions that remain supported by the evidence today. However, his evolutionary mechanism, *inheritance of acquired characteristics*, has been long rejected as being without evidentiary support.

Lamarck's mechanism, published in 1809 (which was the year of Darwin's birth), basically said two things. First, an individual can change during its life so as to adapt to its environment, and it does so by either using, or not using, certain body parts. Those that are used become enhanced, and those that are not used degenerate. Second, these changes become part of the heritable information stored in the individual, and that information gets passed on to offspring, such that offspring exhibit the changes in body parts the parents developed. Interestingly, modern-day surveys show that many people, even those who have benefited from post-secondary courses in biology, wrongly believe that evolution is thought to proceed in this Lamarckian fashion. Individuals do not evolve; populations evolve from generation to generation.

4.1.4 Charles Darwin and the voyage of the *Beagle*

Most early European scientists were wealthy; they were the only ones who could afford to pursue their interests without the demands faced by ordinary folk. Charles Darwin (1809–1882) was no exception. His father, who was a wealthy physician, despaired that his son would not amount to anything, as Charles spent much of his youth hunting and beetle collecting and rather underperforming as a student. When he was 22, the opportunity arose for Charles to be part of a two-year voyage aboard the *HMS Beagle*. Mandated to chart South America's coastlines, this undertaking turned into a five-year venture (Figure 4.3). Darwin actually spent most of this time on land in South America, where he had the benefit of exploring tropical jungles, great grasslands, mountains, and remote island archipelagos, including the Galapagos Islands. Throughout, Darwin collected specimens that were periodically shipped back to England, and he documented his observations and thoughts.

We know that it was during the voyage of the *Beagle* that Darwin's ideas about evolution were developed, but it is also fair to say that his pre-voyage rambles in England, as well as the laborious studies and experiments that he conducted when he got home, also contributed. Although he returned to England in 1836, it was not until 1844 that Darwin produced an unpublished essay that set out his ideas about evolution. He was encouraged by friends to publish, but he is thought to have feared a backlash, and, as a thorough investigator, he wanted to refine his ideas and assemble more evidence. But 14 years later, in 1858, Darwin received an essay from Alfred Russell Wallace, another British

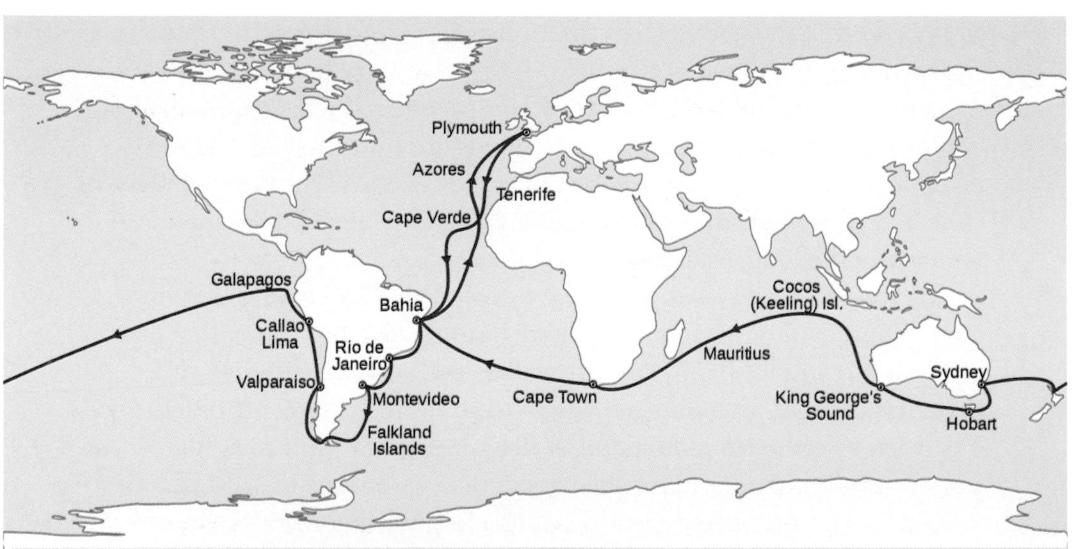
Plymouth
Azores
Tenerife
Cape Verde
Galapagos
Bahia
Callao
Lima
Rio de
Janeiro
Valparaiso
Montevideo
Falkland
Islands
Cape Town
Mauritius
Cocos
(Keeling) Isl.
King George's
Sound
Sydney
Hobart

naturalist who had spent an extended period in the tropics. Wallace was seeking feedback from Darwin, yet Darwin's main observation was that Wallace's theory matched his own. With some dejection, Darwin then expressed the view to acquaintances that credit for the idea should go to Wallace. Friends intervened. The views of both men were presented together in their absence at a meeting of the Linnaean Society in London in the summer of 1858, which was followed by publication later that year. Although this had some impact, it wasn't until Darwin's publication of *On the Origin of Species* later in 1859 that the idea broadly entered the public sphere, having a dramatic and irrevocable impact on people's thinking.

FIGURE 4.3 **CHARLES DARWIN TRAVELED THE WORLD FOR FIVE YEARS IN THE 1830S**
Charles Darwin is highly respected by modern biologists as a great observer, experimentalist, and thinker, and he indeed deserves the respect he receives. But he was also lucky, being born into a family with sufficient financial means and being given the opportunity at the age of 22 to travel around the world on the *HMS Beagle*—free to explore, to collect specimens, and to ruminate on his observations. He never had to work in the sense of paid occupation, but he worked diligently throughout his life, continuing to publish his findings.

Darwin's great book really made two contributions. The first was to provide considerable evidence that species do indeed evolve. This evidence has only increased since then (Box 4.1), and this is why evolution became the organizing principle of modern biology. The second was to articulate the mechanism by which evolution proceeds, which is natural selection (Section 4.2). Natural selection is not the sole driver of evolutionary change, but it remains the main driver. A special subset of natural selection is *sexual selection* (Chapter 5), another concept that was first set out by Darwin in an 1871 book.

Biological scientists and evolutionary psychologists, virtually all of whom accept the truth of evolution, sometimes despair that special interest groups methodically undermine its scientific merit by distorting its message in favor of scientifically unmerited alternatives, such as *creationism* and *intelligent design*. Evolution has tremendous power to explain patterns of biodiversity, inheritance, adaptation, and behavior and equally great power to generate hypotheses about how nature works. For our purposes, the biology of sex is only comprehensible in light of evolutionary theory, as we will see in this chapter and in Chapters 5 through 7.

BOX 4.1 CATEGORIES OF EVIDENCE FOR EVOLUTION

Scientists generally group the evidence supporting the Theory of Evolution into eight major categories.

1. *Paleontology*. The fossil record reveals many extinct forms. More complex forms and forms that are more similar to today's organisms are found in younger rock strata. Some evolutionary lineages contain multiple transitional forms that demonstrate how species change over time (Figure 4.4). Dating methods indicate that the fossil record is more than three billion years old.
2. *Comparative morphology* and *comparative development*. This line of inquiry compares structures (*morphology*), such as skeletons, with developmental patterns, such as embryonic and fetal stages, among different species, to draw conclusions about shared evolutionary heritage. For example, the bones in the bat wing, the bird wing, and the chimpanzee forearm look similar in early stages of development in the womb or the egg, but progressively change in terms of relative sizes and positions as the organism develops. As adults, these forelimb structures appear quite different in the three species, but they retain the same basic bones and basic structures.
3. *Comparative genetics*. DNA is the molecule used by all organisms for storing inherited information, and the DNA/amino acid code (Section 3.6.4) is the same in all organisms. Organisms share genes, and in most cases the more similar the organism, the more similar the gene sequences.
4. *Vestigial structures* and *imperfect structures*. Vestigial features show up in a phenotype but are of no use, with the interpretation being that they were of utility in ancestors. For example, the fossil record shows that snakes have lizard-like ancestors but progressively lost their limbs over time. Several groups of modern snakes still have remnants of the pelvic girdle, which are the torso bones into which the leg bones fit. Accordingly, the girdle is vestigial in these limbless snakes. Imperfect features result because evolutionary forces act on the structures that already exist, as opposed to "designing" something anew. For instance, the vertebrate eye has a blind spot on the retina, created because the optic nerves pass through the retina. Other unrelated organisms with complex eyes, such as octopus, have a different eye structure where the nerves pass behind the retina, avoiding the creation of a blind spot.
5. *Evolution in real time*. Many research programs actually document changes in species over relatively few generations. This is common for pathogens that evolve in response to antibiotic pressure, developing resistance. But, it also exists in complex organisms. For instance, guppy populations show genetic changes in growth rate, body size, reproductive traits, and body patterning in relation to predator exposure in as few as 30 or 40 generations.

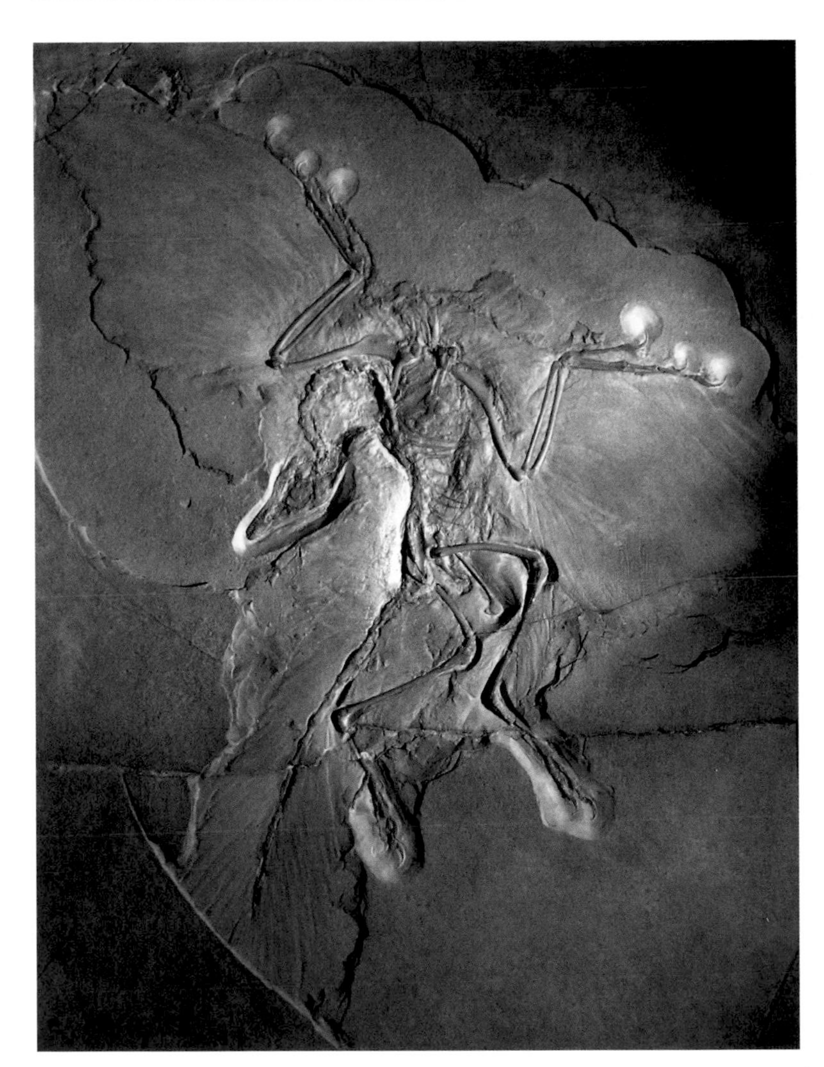

FIGURE 4.4 **PREHISTORIC *ARCHAEOPTERYX* HAD FEATHERS COUPLED WITH REPTILIAN TRAITS** Discovered in a German limestone quarry just two years after the publication of *On the Origin of Species*, *Archaeopteryx* dates from Jurassic strata laid down about 150 million years ago. *Archaeopteryx* shows features of typical (but small) dinosaurs, such as teeth, a bony tail, and claws on the forelimb, but it also shows features typical of birds, most notably feathers and reduced forelimb bones.

6. *Selective breeding*. Long before Darwin, animal and plant breeders changed organisms by selective breeding, also known as artificial selection. This was done by choosing parents for the next generation on the basis of desirable traits in the parents. In this way, humans created larger and sweeter corn, woollier sheep, powerful work horses, and so on.
7. *Biogeography*. This is the study of geographic distributions of groups of species. More similar species tend to be found in the same geographic regions. For instance, rattlesnakes (Crotalinae) are a highly

FIGURE 4.5 **BIOGEOGRAPHY PATTERNS REVEAL REGIONS WHERE DIFFERENT GROUPS HAVE EVOLVED**
All rattlesnakes (about 30 species) and all hummingbirds (about 350 species) are found only in the Americas, consistent with the idea that both groups evolved here. Indeed, all rattlesnake fossils are from the New World. Eastern massasauga (*Sistrurus catenatus*), above, and Mexican violetear (*Colibri thalassinus*), below.

diverse group whose species are only found in the Americas (Figure 4.5), with the interpretation that they evolved here and that the Atlantic and Pacific Oceans are significant enough barriers to have prevented their spread elsewhere.

8. *Internal consistency.* This describes the pattern where different branches of evidence (as in 1 through 7 above) support or are consistent with each other, without conflict or contradiction, and are consistent with evidence of concurrent geologic change. For example, it is well documented that the continents move (plate tectonics). Their relative positions, including their points of contact, have been reconstructed. The fossil record and biogeographical distributions are found to be consistent with these movements, including the timing attributed to them by dating methods. For instance, some groups of species that are shared between now-distant continents appear in the fossil record at times when those continents were in contact.

4.2 THE PRINCIPLES OF NATURAL SELECTION

Darwin showed not only that there was substantial evidence that evolution had occurred, but he also articulated the mechanism—natural selection. Because artificial selection, also known as selective breeding, (see Box 4.1) was a concept familiar to Victorians, Darwin relied on it to demonstrate the idea. Artificial selection is a process where humans select individual animals or plants on the basis of desirable traits and use those as breeding stock.

For instance, if one wishes to have a dog breed that has short legs so that one could have a dog that can access animal burrows, one would select as breeding stock those with the shortest legs. There are many genes that contribute to most traits, including leg length, and each of these genes is likely to have a series of alleles that make for shorter or longer legs. After doing this for many generations—always selecting the individuals with the shortest legs as the breeding stock—the breed will become one that reliably produces short legs. Generation after generation, the alleles that contribute to shortness will become more and more common in the breeding population.

Darwin was also influenced by Thomas Malthus, an influential social commentator. Malthus wrote about the difficult social conditions created as Britain became more urban and industrialized. He noted that humans, just like animals and plants in nature, produce more offspring than can survive, and he also observed that as a result, there is poverty, famine, and other human misery. Darwin read Malthus as he was formulating his ideas about nature, and the idea of over-production of offspring and consequent casualties contributed to his theory.

4.2.1 The argument for natural selection

Natural selection is similar to artificial selection, with a key difference: rather than humans being the agent that determines which individuals get to breed and therefore which individuals' alleles are preferentially contributing to the next generation, natural processes do the selecting.

Natural selection is not, therefore, that difficult a concept in its operation. Many authors have summarized it. Here, we will reduce it to five observations in nature, with three inferences that flow from those observations.

- Observation #1. For all species, populations have the capacity to increase exponentially, given that females can produce more offspring than are necessary for replacement.
- Observation #2. Resources relied upon by organisms (food, space, nest sites, etc.) are not infinite.
- Observation #3. Most populations remain relatively stable over extended periods of time.
    - Inference A. Production of more individuals than can be supported by the resource base in the environment means there is a struggle for survival among individuals of a population, resulting in the survival of only a fraction of those born.
- Observation #4. Members of a population vary extensively in their characteristics.
- Observation #5. Much of the variability in characteristics among individuals depends upon the alleles they have for genes, and so such variability is heritable.
    - Inference B. Individuals whose genetic traits yield a high probability of surviving and reproducing are likely to leave more offspring than others.
    - Inference C. The proportions of all the alleles for all the genes in a breeding population, known as the gene pool, will change accordingly.

These observations and inferences can be trimmed to the essential argument: evolution by natural selection results when populations change over generations because individuals that possess favorable heritable traits are more successful in leaving more offspring than other individuals. In short, evolution is the change in allele frequencies in a population over time.

4.2.2 The concepts of fitness and adaptation

Two important terms connected to evolution by natural selection are *fitness* and *adaptation*. Our everyday notion of fitness tends to bring to mind exercise and lean, healthy bodies, as in physical fitness (Figure 4.6). *Biological* or *Darwinian* fitness is different. Although it is likely to be associated with good health, biological fitness refers to the reproductive success of individuals as measured by how many offspring they produce. Fitness is relative because it measures an individual's reproductive success relative to other breeders in the population. Fitness is the outcome, and natural selection is the cause of the outcome.

Biological *adaptation* is also an outcome of natural selection. In this case, an adaptation is a phenotypic feature that enables the individual to better survive and reproduce in its environment. Any change in a population that occurs through natural selection and therefore improves the match between the organism and the environment in which it lives is an adaptation. Naturally, adaptations contribute to fitness, since adapted individuals are more likely to survive and reproduce.

Adaptation results because natural selection favors individuals with variants of traits that contribute to fitness and it disfavors those with inferior variants, meaning the former become more concentrated in the population (the "adaptation") and the latter tend to disappear. In this way, the process of adaptation is not one that happens within individuals. Instead, it happens within populations of breeding individuals over many generations. Individuals can change too. For instance, light-skinned people tend to tan when exposed to sunlight, and individuals living at high elevations tend to increase their lung capacity. For these changes (which are not passed on to offspring, unlike what Lamarck proposed), we use the term *acclimation*.

So, adaptations are heritable traits produced by natural selection that increase individual fitness in an environment for which the adaptation is well suited, relative to individuals not having such a trait or having it to a lesser degree.

FIGURE 4.6 **FITNESS IS A MEASURE OF REPRODUCTIVE SUCCESS**

The commonplace use of the term *fitness* refers to physical health—aerobic capacity, physical stamina, muscle tone, etc. (a). When biologists refer to fitness, it has a very different meaning: *biological fitness* is measured by reproductive success, so that an individual's fitness is a measure of his or her contributions of offspring to the gene pool of the next generation (b). It's true that physically fit people are more likely to survive and reproduce and therefore more likely to be biologically fit, but the terms are independent with different meanings.

4.3 DNA AS AN EVOLUTIONARY LEGACY

Although nineteenth-century intellectuals were familiar with the idea of inheritance, how it worked remained a mystery throughout the century. Mendel's great work in the 1860s could have shed considerable light on the subject, but his published work languished in obscurity, without impact until its rediscovery in 1900. And it wasn't until the middle of the twentieth century that it was demonstrated that DNA was the molecule of inheritance.

As we go about our daily lives, we do not normally think of each other or ourselves as phenotypes that are mostly built based on information stored in the DNA sequences of our genes. Nor do we readily admit that we are merely the mortal home for our immortal genes. However, we may from time to time be reminded that the phenotypic differences among us are largely due to allelic differences in our shared genes. And when we look in the eyes of a chimpanzee or gorilla, we may reflect that our differences and similarities are largely attributable to differences and similarities in DNA sequences. These are things that a modern understanding of evolution and genetics entails.

If we restrict our focus to just one generation back—to one's parents—it is easy to accept that the DNA in our genes is a legacy inherited from our parents. But of course our parents received their DNA from their parents, and so our DNA is also a legacy from our grandparents. Of course, we could keep going in this fashion since everyone who is born is the result of an unbroken line of reproductive success. In short, a person's DNA is a legacy from his or her ancestors in the long-ago past. In fact, the evolutionary tree of humans (both modern and extinct species known from the fossil record) goes back about seven million years, and of course those ancestors had their own "non-human" ancestors before that. All along the way, beneficial versions of alleles were favored by natural selection, producing adaptations and contributing to fitness.

4.3.1 Mutations, and how alleles differ from one another

Different alleles for a particular gene differ simply by not having the exact same sequence of DNA bases. It might be as little as one different base at one position, as is true for the two alleles found on chromosome 11 for the human *hemoglobin* protein. Hemoglobin carries oxygen to body tissues. The normal allele produces a version of the protein that results in typical rounded red blood cells that effectively transport oxygen through the circulatory system. The *sickle-cell* allele is a mutant allele that produces a version of hemoglobin that results in sickle-shaped cells that are poor at transporting oxygen. If that were the end of the story in this case, we could say the normal allele is good and the sickle-cell allele is bad. However, sickle cells provide considerable resistance to malarial infections.

In most locations, it is preferable to have two copies of the normal allele, but in areas where there is risk that malaria will attack the blood system, it is preferable to have one normal and one sickle-cell allele. Such individuals have higher fitness because they have reasonable oxygen transport on account of the normal allele *and* resistance to malaria on account of the mutant allele. In the normal allele, the resultant mRNA triplet in the sixth position is GAG, which specifies the amino acid *glutamic acid*, whereas in the sickle-cell allele the triplet in that position is GUG, which calls for the amino acid *valine* (Figure 4.7). Otherwise, the string of 146 amino acids (or $3 \times 146 = 438$ DNA bases) are identical, yet this is enough to change the conformation of the protein, which influences cell shape, thereby producing life-threatening oxygen-transport problems when in homozygous form but malarial resistance when heterozygous with a normal allele.

Any change in DNA that results in the substitution of one base for another, such as the U for an A in the sixth mRNA triplet in the hemoglobin gene, is called a *point mutation*. Most such mutations are likely to produce proteins that are inferior to the normal version. Less often, however, mutations create superior or new proteins, and natural selection can reward these novelties with the result that they spread in the population.

4.3.2 Chromosomal mutations can also contribute to evolution

Point mutations are the most common type of mutation, but there are others. Chromosomal mutations, for instance, are quite different because a whole section of a chromosome, often amounting to thousands of DNA bases, is involved. One type of chromosomal mutation known to be important in evolution is known as a *duplication*. During crossing-over in meiosis, an error occurs by which a

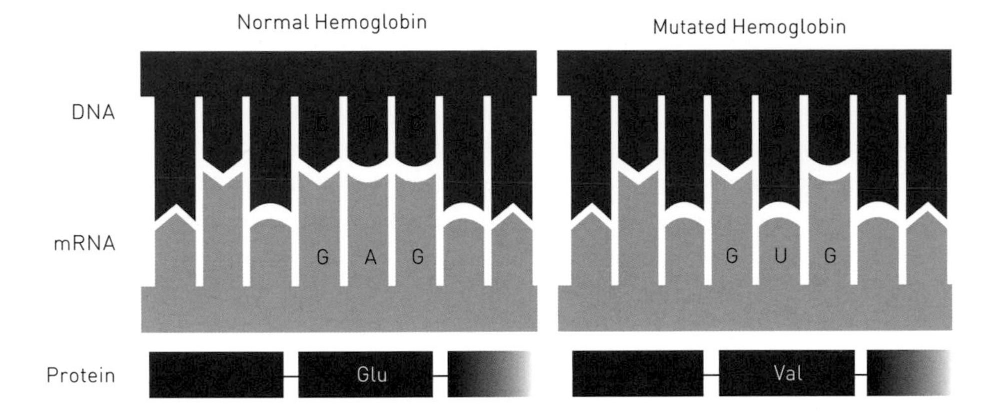

FIGURE 4.7 **SMALL DNA SEQUENCE DIFFERENCES CAN HAVE FAR-REACHING CONSEQUENCES**
Although the difference in DNA sequences between the code that produces normal hemoglobin and the code that produces sickle-cell hemoglobin is only one nitrogenous base, changing that base from a thymine (T) to an adenine (A) codes for the amino acid valine (Val) at that position instead of glutamic acid (Glu). Having a valine in that position instead of a glutamic acid changes the shape and therefore the function of the resulting molecule.

section of the chromosome becomes duplicated. This means that the chromosome with the duplicated section of DNA now includes repeated copies of the genes contained in that section.

Consider what appears to have happened when a section of the X-chromosome containing an *opsin* gene experienced such a duplication. X-chromosomes are found in both male and female mammals, although men have only one copy while women have two. Opsins are pigment proteins found in the cells of the retina that are necessary for color-sensitivity. If you have "normal" color vision, it's because you have a functional allele on chromosome 7 for the gene that encodes short-wave opsins (S opsin, which is blue-violet sensitive), a functional allele on the X-chromosome for the gene that encodes medium-wave opsins (M opsin, which is yellow-green sensitive), and a functional allele at an adjacent position on the X-chromosome for the gene that encodes long-wave opsins (L opsin, which is red sensitive). In this way, you have *trichromatic* (three colors) color vision.

Most mammals have dichromatic color vision, having the S opsin and the M opsin. This means their experience of color is considerably less than yours. Aside from the New World (i.e., American) monkeys (Superfamily Ceboidea), we primates are exceptional mammals in being trichromatic, and it appears to be a duplication that gets the credit. If one compares X-chromosome gene sequences between a dichromatic mammal and a human, it is apparent that there has been a gene duplication in that section of the X-chromosome in humans. When there is a duplication, any subsequent mutation that affects one copy will not necessarily affect the other copy. In this case, the duplication within the X-chromosome created two versions of the M opsin gene. Later, one copy experienced several point mutations that resulted ultimately in converting the extra copy of the M opsin gene into an L opsin gene.

4.3.3 Mutation rates in sperm and eggs

Point mutations that create new alleles can occur in any chromosome during meiosis. Most such mutations are likely to have negative consequences, and accordingly, since they are not adaptive and therefore do not enhance fitness, they are likely to be removed from the gene pool through the process of natural selection. A minority will have positive consequences (as in the point mutations that have converted the duplicated M opsin gene into an L opsin gene), and their prevalence may therefore spread in the gene pool.

Long ago it was recognized that the high rate of sperm production in male meiosis compared to the low rate of egg production in female meiosis would likely have a bearing on the relative mutation rates in the DNA of sperm and eggs. This is because chromosomal copying occurs in males at such a prodigious rate that there is a greater likelihood of copying mistakes that produce mutations. This is true even in the sperm of adolescents who aren't, of course, very old. The amount by which the mutation rate in sperm exceeds that in eggs depends a lot on mating systems (Chapter 6); in those species where there is a high rate of sperm production to meet the demands of polygyny (where one male has multiple female partners), the mutation rate in sperm is even greater. In primates like ourselves, the average mutation rate in sperm production is six or seven times that found during egg production. In addition, there is growing evidence that the mutation rate in sperm increases with the age of the father.

4.4 THINKING OF BREEDING GROUPS AS GENE POOLS

Many significant advances in biology occurred in the first decade of the twentieth century. Mendel's work was rediscovered and its implications were finally appreciated. The ABO blood system was discovered, which had great benefits for medical procedures involving blood. The concept of the "gene" and the discipline of "genetics" became established. And Godfrey Hardy and Wilhelm Weinberg, the former an English mathematician and the latter a German geneticist, independently developed the idea that has come to be known as the Hardy–Weinberg Equilibrium (HWE). The Hardy–Weinberg Equilibrium links genes and evolution. It deals with the frequencies of different alleles for particular genes in populations of sexually reproducing organisms, including humans. Of central importance to the subject is the concept of the gene pool, which we've seen as the tally of all the alleles for all the genes in a sexually breeding population. This term didn't appear until the mid-twentieth century, but it is a concept going back to Hardy and Weinberg. The *equilibrium* is the notion that unless there are evolutionary forces at work, the frequencies of alleles in a gene pool do not change over time, nor do the frequencies of the diploid combinations of those alleles.

4.4.1 Human breeding groups can be thought of as gene pools

Probably everyone knows the term *gene pool*, although it is less certain that people have an accurate sense of what it is. The word "pool" suggests a collection, and indeed the gene pool is a collection of genes. Because there are many genes in sexually reproducing species, and because many of those genes have multiple alleles, a more satisfactory definition is "all the alleles of all the genes in all the individuals in a sexually reproducing population." The fuzziest element of this definition is "population." Certainly, a population consists of members of the same species, but it is commonly the case that there are no sharp geographic boundaries around populations. For instance, if we were to argue that the people of Toronto, Canada, constitute a "population," we would have to acknowledge some indefiniteness to this. Toronto males seeking female mates are certainly most likely to find partners in Toronto, but a few may find a partner instead in New York City, and a rarer few will find one in Hong Kong or West Monroe (which is a relatively small city in the southern United States). One fundamental observation emerges as we consider the population from generation to generation through the production of offspring: we see the individuals changing (i.e., dying and being born) but the genes continuing. In this sense, people are mortal but their genes can be immortal.

There is one additional significant point to be made about gene pools. What is really meant is "allele pool." It is the diversity of alleles that is the measure of diversity in a breeding population. Most people have the same genes but different versions of those genes, which are alleles. It would be conceptually helpful to change the term to *allele pool,* but gene pool is a thoroughly entrenched term.

4.4.2 The ABO blood group gene pool

Although a true gene pool includes all the alleles for all the genes, for something more manageable that still conveys the idea, we sometimes apply the term to a smaller number of genes. We can even speak of the gene pool for one gene. In this sense, it is all the alleles for one gene found in all the individuals of the population. What is of interest to biologists, including genetic counselors, is the relationship between the gene pool (i.e., the frequencies of different alleles in the population) and the genotypes of the individuals in that population. We'll illustrate this relationship between allele frequencies and genotype frequencies by returning to a now-familiar example, the ABO blood group system introduced in Section 3.5.

Estonia is a small nation of about 1.3 million people. As with all regions, the A, B, and O alleles are not equally common in the Estonian population. By sampling, it is known the O allele is the most common one in the gene pool, at 58 per cent, followed by the A allele at 25 per cent, and the B allele at 17 per cent. Can we use this information to determine the genotype and phenotype frequencies in the population? Yes.

We could do this by simply applying mathematical formulae derived by Hardy and Weinberg. But we can do it in a more visual manner by using a modified Punnett square. So far, we only used Punnett squares for matings between individuals—for instance a purple flower with two purple alleles mating with a white flower with two white alleles or a man with two O alleles mating with a woman with one O and one B allele. There are three differences when we use a Punnett square in a gene pool application as opposed to a mating of two particular individuals. First, for a particular mating, there can never be more than two alleles per person (and often there are only two copies of one allele) because people are diploid creatures. But in a gene pool there can be more than two alleles. Second, for a particular mating, the alleles are found in one person's eggs and one person's sperm, but in a gene pool Punnett square, the gametes too are pooled for the analysis. Finally, for a particular mating, the two alleles are always found in a 50:50 ratio, but in a gene pool this is not the case, because the various alleles for a gene are not likely to be equally common in the gene pool.

The Punnett square for the Estonian ABO gene pool is shown in Figure 4.8. As before, the row titles describe the alleles in haploid sperm, and the column titles describe the alleles in haploid eggs. But instead of two equal-sized rows and two equal-sized columns, here we have three rows and three columns (since there are three alleles in the system), and our rows and columns are different sizes because the alleles are not equally common in the gene pool. What does this signify? Instead of being eggs and sperm from two individuals, here we show the proportions of eggs and sperm as they exist in the population. That is, 58 per cent of all eggs (and all sperm) in the population (millions of eggs, trillions of sperm) carry the O allele, 25 per cent carry the A allele, and 17 per cent carry the B allele. The nine interior boxes are the possible results for the offspring. If you look at those nine boxes, you will see six different genotypes (AA, AO, BB, BO, AB, and OO), as expected.

FIGURE 4.8 PUNNETT SQUARES CAN BE MODIFIED FOR POPULATIONS

This is a population Punnett square for the ABO blood group for Estonia. The row titles are the proportions of sperm in the population representing the relative proportions of the A, B, and O alleles in sperm. The column titles are the same, for eggs. There are three sperm choices for this gene and three egg choices, meaning there are nine choices for the resulting offspring (the interior boxes). Even though we are dealing with three alleles in this gene pool, note that the gametes still must be haploid and the interior boxes still must be diploid genotypes. The math, shown in the offspring genotype boxes, is tallied in Table 4.1.

	A (0.25)	B (0.17)	O (0.58)
A (0.25)	AA (0.25 x 0.25)	AB (0.25 x 0.17)	AO (0.25 x 0.58)
B (0.17)	AB (0.17 x 0.25)	BB (0.17 x 0.17)	BO (0.17 x 0.25)
O (0.58)	AO (0.58 x 0.25)	BO (0.58 x 0.17)	OO (0.58 x 0.58)

To determine the proportions of the genotypes in the population, we "do the math." That is, in each box, you will see a simple multiplication formula. These do not have to be memorized! Instead, notice that in each case, we multiply the frequency of the sperm that carry that allele by the frequency of the eggs that carry that allele. These formulae are repeated in Table 4.1, where they are calculated. In this way, having been given the allele frequencies in the gene pool, we are able to determine the genotype frequencies of the people in the population. At this point in history, Estonians are 34 per cent type O, 35 per cent type A, 23 per cent type B and 8 per cent type AB.

In a population that is not evolving, these allele frequencies (58 per cent O, 25 per cent A, 17 per cent B), these genotype frequencies (6 per cent AA, 29 per cent AO, 3 per cent BB, 20 per cent BO, 8 per cent AB, and 34 per cent OO), and these phenotype frequencies (35 per cent type A, 23 per cent type B, 8 per cent type AB, and 34 per cent type O) will not change. However, we find that allele frequencies in some gene pools do change over time, and this demonstrates that such populations are evolving.

Genotype	Genotype frequency in the population	Phenotype	Phenotype frequency in the population
AA (1 box)	0.25 × 0.25 = 0.06 (6%)	A (3 boxes)	6 + 29 = 35%
AO (2 boxes)	0.25 × 0.58 = 0.145 0.58 × 0.25 = 0.145 (29%)		
BB (1 box)	0.17 × 0.17 = 0.03 (3%)	B (3 boxes)	3 + 20 = 23%
BO (2 boxes)	0.17 × 0.58 = 0.10 0.58 × 0.17 = 0.10 (20%)		
AB (2 boxes)	0.25 × 0.17 = 0.04 0.17 × 0.25 = 0.04 (8%)	AB (2 boxes)	8%
OO (1 box)	0.58 × 0.58 = 0.34 (34%)	O (1 box)	34%

TABLE 4.1 Based on the allele frequencies of O = 0.58, A = 0.25, and B = 0.17 for the Estonian population, the resultant genotype and phenotype frequencies in the population can be calculated. "Boxes" refers to the nine boxes in the Punnett square.

4.5 THE EVOLUTIONARY COSTS OF SEX

Why is there sex? An automatic answer is to produce offspring. Although this is an incomplete answer, it is not an incorrect one. But it is important to remember that reproduction can and does occur without sex. In fact, asexual reproduction existed for more than a billion years before sexually reproducing organisms evolved, and there is still more asexual reproduction in the natural world than sexual, just not in multicellular organisms. Interestingly, in those organisms that can reproduce either way (Section 2.2.4), the method that appears to be favored is asexual, at least most of the time.

So, the more profound question is, what are the costs and benefits of sexual reproduction, in comparison to asexual reproduction, and why do so many species use sex if it is costly? We'll consider the costs of sex first, and then in the next section we'll consider the benefits.

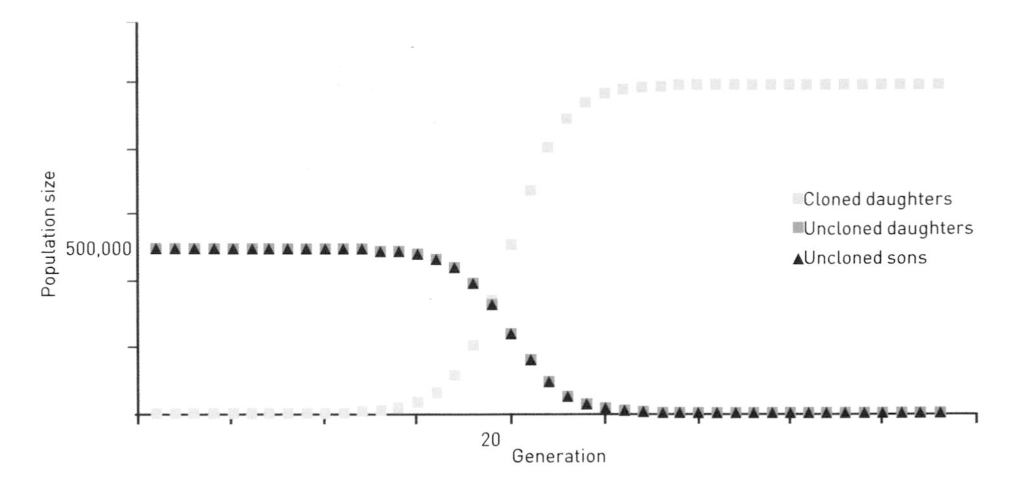

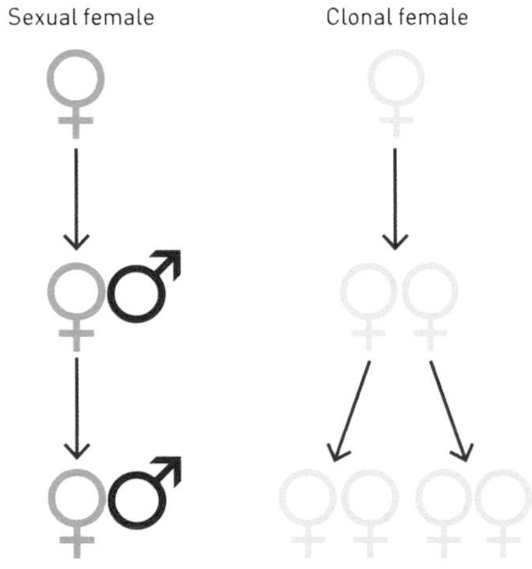

FIGURE 4.9 ASEXUAL REPRODUCTION IS NUMERICALLY MORE EFFICIENT THAN SEXUAL REPRODUCTION
Consider the numerical argument for asexual reproduction. At generation 0, there are equal numbers (500,000 of each) of sexually reproducing males (triangles) and sexually reproducing females (dark squares, overlapped by the males), and there is a mutation that arises in one female that allows her to clone herself asexually (light squares). After 19 generations, the cloning females pass the sexual females in population size, and in the next generation the cloned females exceed the numbers of sexual individuals, both male and female. By generation 40, the sexual population is extinct.

4.5.1 The numerical argument that sex is costly

Imagine a sexual population of one million individuals. Imagine also that each female produces two offspring and that resource limitations never let the population exceed one million. Now imagine that a mutation arises in one female such that she clones herself. Every time she produces an offspring, 100 per cent of her genes, including the one that enables asexual cloning, get passed on. For the sexual females, it is different. First, every time each of them produces an offspring, only 50 per cent of her genes get passed on to that individual. Second, the sexual females continue to produce males, which tend to invest minimally in offspring. This is known as the twofold *cost of sex*. Now, assuming the environment can only support one million individuals and assuming that the cloned individuals and the sexual individuals survive equally well, what happens to this population over time, all else being equal (Figure 4.9)?

In the nineteenth generation, the number of cloned (and therefore cloning) females exceeds the number of sexual females. In the twentieth generation, the number of cloned females exceeds the total number of sexual individuals, both male and female. By the fortieth generation, the entire population is made of cloned females, making the sexual population (including males) extinct. This twofold cost of sex is also referred to as the *numerical argument*.

4.5.2 Sexual reproduction also involves search costs

Costs related to reproducing sexually are not limited to the twofold cost represented by the numerical argument. *Search costs* are costs associated with finding a mate. In abundant or highly social species where densities are likely to be high, this is not difficult. Nor is it difficult in those species that occupy special habitats where many individuals of the same species are likely to encounter one another. Other organisms exist in low densities, however, such as where food availability dictates a thinly dispersed population to ensure each individual can find enough to eat or where populations have declined to the point where there are just not that many individuals left. In these species, searching is likely to be more costly. These are also species where we would expect organisms to invest in long-distance communication, such as the production of airborne signals that act like out-of-body hormones, known as pheromones, or the use of sound signals. We might also expect the females of such species to have the capacity for long-term sperm storage.

4.5.3 Sexual reproduction can entail health, injury, and mortality costs

There can be significant *survival costs* associated with sex. As we will see in Chapter 5, in many species there is real male–male combat in which injury or even death can be the result. Australia's brown marsupial mouse (*Antechinus stuartii*), which superficially looks like a rodent

FIGURE 4.10 **THE PURSUIT OF MATING OPPORTUNITIES CAN BE LIFE-SHORTENING**

Male brown marsupial mice (*Antechinus stuartii*) devote everything to mating when the breeding season arrives, allowing their internal systems to break down and resulting in an early death as the mating season draws to a close.

but is really a marsupial, is well known for the costs males suffer in connection with mating. When the time for breeding arrives, males single-mindedly pursue sex, mating with as many females as they can, copulating for long periods, and fighting for the rights to do so. They thoroughly exhaust themselves, and they invest nothing in body maintenance. The fur begins to fall off, they bleed internally, their immune system shuts down, and they suffer gangrene, dying a young death (Figure 4.10). Within a few weeks, every single male is dead. Females live more than one year.

Testosterone (Section 9.1.1) is a hormone known as an androgen, meaning it is associated with male physiology. Females have testosterone as well, but in lower concentrations and commonly in systems where it is neither produced nor consumed so rapidly. Testosterone itself is known to be life-shortening. This is not only because it leads to risky behaviors that often have unhappy endings but also because of costly influences on the body itself. In humans, for instance, the circulating testosterone levels commonly found in males reduce the efficacy of the immune system. (One result of this phenomenon is that women respond to vaccines better than men do.) A large study comparing longevities of normal males with castrated males showed that the latter group has considerably greater life expectancy.

In fact, as we will see in more detail in Chapter 5, in many species males allocate great amounts of energy and time in the pursuit of mating opportunities, including investing in all sorts of otherwise unnecessary things: antlers that are grown and shed seasonally, as in moose; loud songs that increase the exposure to predators, as in frogs; and bright plumages, the pigments of which are costly to assemble, as in birds, to name a few examples.

Although females do not suffer the high testosterone costs experienced by males who pursue stereotyped male mating behaviors, sex can still amount to a survival or health risk for females as well. As we'll see in Chapter 7, the optimal behaviors with respect to frequency of mating and number of partners can be different for males and females. In the context of male-perpetrated harassment, the most direct impact on females is physical injury. This can be injury incidental to the harassment, such as damage to the female reproductive tract, but there are many organisms where female death is the result of aggressive and coercive male behavior.

Even if physical injury is not the consequence of male-perpetrated harassment, there can be substantial stress and energy depletion costs suffered by females. These can result in weight loss, body condition declines, and decreased immune responsiveness. Frequent mating, or efforts expended to avoid unwanted mating, can expose females to a greater risk of predation. And, there are lost opportunity costs associated with unwanted mating: the time and energy involved could be applied to other pursuits, including rest and foraging for food.

For both males and females there is also the disease cost associated with sexual reproduction. Pathogens living in or on either sex can be transferred to the partner, especially in species that employ internal fertilization. Asexual breeders, obviously, do not suffer from sexually transmitted infections.

4.6 THE EVOLUTIONARY BENEFITS OF SEX

Sexual reproduction must have substantial benefits that counteract the costs of sex enumerated in Section 4.5. Otherwise, all organisms would be asexual, and males would disappear. Yet most multicellular organisms employ sexual reproduction, and many of those species are about 50 per cent male. So why sex? It would be incorrect to call upon the pleasures of sex as an answer, because as we argued in Section 2.1, the pleasure rewards are only a proximate explanation, and here we are seeking the ultimate explanation.

As a starting point for consideration, we might contemplate those species that are facultatively sexual, like the water fleas (Figure 2.7) that reproduce asexually most of the time during benign seasons but that switch to sexual reproduction when seasonal change makes conditions challenging. This is a pattern that is exhibited in a variety of species in a variety of groups—most reproduction is asexual, with sexual reproduction being an infrequent mode of reproduction that occurs in response to an environmental stimulus, commonly a deterioration of conditions or food supply. In contrast, in the organisms that are most familiar to most people—insects, crustaceans, vertebrate animals, and others—sex is commonly the only method of reproduction (obligate sex).

4.6.1 Advantages associated with a diverse gene pool

Reflect for a moment on the hypothetical cloning female we considered in the numerical argument (Figure 4.9). Although the cloning female wins after 40 generations, how would we describe the resultant population? It is one million individuals that are genetically exactly the same. In contrast, the sexual population it replaced had been a population of one million unique individuals. These are dramatically different populations!

Let's imagine also that this hypothetical species is diploid and that it has 10,000 genes at 10,000 loci. In the asexual population, the genes in the gene pool will be represented by only one or two alleles. In the sexual population, at least some of the genes in the gene pool may very well have several alleles; it will be a richer gene pool. In addition, in the sexual population, each gene will find itself in a unique genomic environment (because the alleles for the other genes at the other 9,999 loci will be highly diverse). In the asexual environment, in contrast, there is only one genomic environment, because all the individuals are clones.

This diverse gene pool in the sexual species has a tremendous influence on the diversity of offspring. Because of meiosis, a parent's sperm or eggs contain only 50 per cent of its genes (which is a cost), but each sperm or egg contains a unique sample of those genes (which is a benefit); no two sperm or two eggs are going to be genetically identical. As a result, no two offspring will be identical. Because environments are complicated and because they are subject to change, producing genetically diverse offspring increases the likelihood that at least a portion of them will be well suited to that environment and therefore able to cope and reproduce. In this way, parents that reproduce sexually are more likely to succeed in passing on at least some of their own genes to the next generation.

4.6.2 Inbreeding reveals the value of gene mixing through sex

Inbreeding is sexual reproduction that occurs between relatives, and where it occurs, it tends to eliminate the genetic benefits of sex. Because relatives—especially siblings—are likely to share many of the same alleles, the genetic content of their eggs and sperm are likely to be more similar than non-relatives. When relatives mate, it tends to result in increased homozygosity and decreased heterozygosity. That is, every time there is reproduction between relatives, it increases the likelihood that their offspring will have two identical alleles for a gene, as opposed to two different ones. Such individuals have less genetic diversity overall, and if they happen to have two alleles that are dysfunctional, problems result. Accordingly, most species have behaviors and other attributes that maximize outbreeding. Incest, which is mating between close relatives, is widely avoided in nature.

Royal families have always acted to concentrate power and wealth. This posed a problem when the time came for heirs to marry, because outbreeding either shared some of that wealth with another family or failed to bring into the family as much wealth as marriage to a rich relative might. Because of this, many royal families practiced inbreeding. One extreme example was the Habsburg royal family that reigned in Spain when it was the most expansionist nation on Earth (1516–1700).

The *family tree* for the last Spanish Habsburg, Charles II, is shown in Figure 4.11. A healthy outbred family tree brings in spouses "from the outside" whenever it is time for a marriage. A close look at Charles' family, which spans about 200 years, shows a high degree of inbreeding. There are at least three uncle–niece marriages and two first-cousin marriages here, as well as additional marriages between relatives. Charles II suffered many disabilities, including substantial cognitive shortcomings, and he died without children at age 39. It is almost

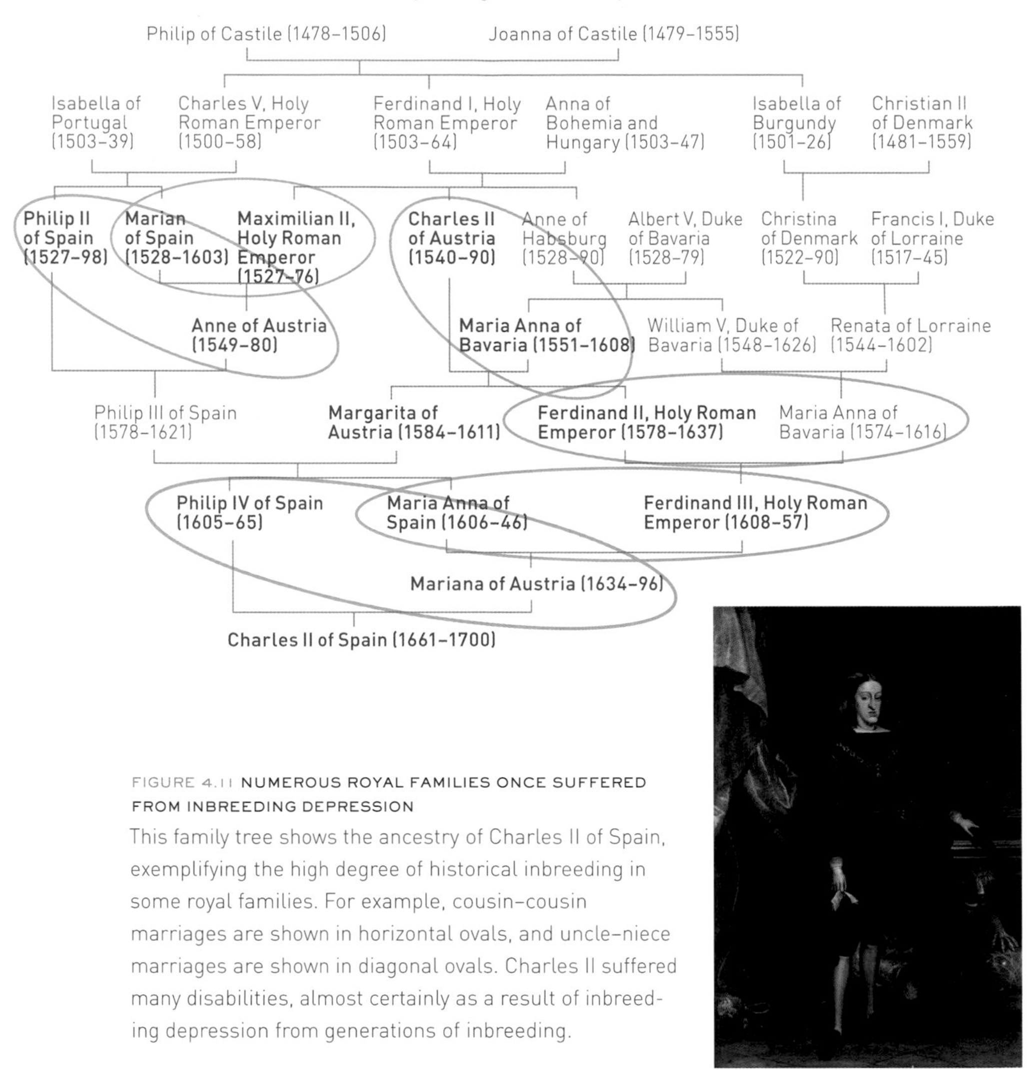

FIGURE 4.11 NUMEROUS ROYAL FAMILIES ONCE SUFFERED FROM INBREEDING DEPRESSION

This family tree shows the ancestry of Charles II of Spain, exemplifying the high degree of historical inbreeding in some royal families. For example, cousin-cousin marriages are shown in horizontal ovals, and uncle-niece marriages are shown in diagonal ovals. Charles II suffered many disabilities, almost certainly as a result of inbreeding depression from generations of inbreeding.

certain that he suffered from inbreeding depression, which is a reduction of fitness associated with breeding between relatives. Remember that although inbreeding increases the amount of homozygosity, homozygosity is not bad per se. But, increased homozygosity increases the likelihood an individual will get two copies of the same deleterious alleles. When this happens, it means that such alleles will have negative consequences, because the deleterious effect will be expressed in the phenotype. In contrast, when deleterious recessive alleles that are paired in a heterozygous state with a dominant allele, the recessive allele is not expressed.

Small populations of animals commonly show evidence of inbreeding depression, especially when humans reduce natural habitats to small fragments. A well-documented example of this is the cougar (*Puma concolor*) population in Florida, whose numbers were reduced to fewer than 40 individuals in the 1970s. This population became very inbred, leading to abnormalities and low reproductive success. In 1995, wildlife managers captured eight Texas females and transplanted them to the Florida population. This assisted the species' recovery in Florida, not so much by the addition of eight individuals, but more because those eight individuals entered the breeding population and introduced new alleles and therefore greater genetic diversity than there had been. The population has increased several-fold since then.

4.6.3 The Red Queen hypothesis and Muller's ratchet

Theoretical biologists have devoted much thought to ideas related to the evolution of, and benefits of, sex as a mode of reproduction. One popular idea is the Red Queen hypothesis. This borrows the Red Queen from Lewis Carroll's *Through the Looking Glass,* a classic children's story. In it, the protagonist Alice meets up with the Red Queen. The queen implores Alice to hurry, and they run for a long time. When they stop, they are still in the same place, and Alice observes that where she is from, you get somewhere if you run. To this, the queen replies: "Now, here, you see, it takes all the running you can do, to keep in the same place."

Running to stay in the same place is used here as a metaphor for evolution. The high rate of extinction in the fossil record suggests that if organisms don't keep evolving, they will lose the place they've achieved among life's biodiversity and will become extinct. This idea therefore embraces the idea that organisms must continually adapt simply to survive, especially when faced with predators, parasites, and competitors that are also becoming better adapted through evolution. Since rapid genetic response to ever-changing conditions and ever-evolving threats is likely to be beneficial, sexual reproduction has advantages, since it produces unique genotypes, and hence phenotypes, with every generation—unlike asexual reproduction.

Muller's ratchet is the idea that when a deleterious mutation occurs in an asexual population, the population is stuck with that mutation (unless it mutates back to the original). Over time, this is likely to happen for multiple genes, and the resulting collection of sub-optimal alleles is called *genetic load*. An analogy for this process suggested by evolutionary biologist Matt Ridley is the continual making of photocopies from photocopies, with a loss of information each time, eventually leading to a very poor image. Sex, including those relatively infrequent instances among species that are usually asexual, has a capacity to eliminate these sub-optimal alleles. This is because of the impact of meiosis. During the making of gametes, one allele from each gene is packaged into the egg or sperm; about half the time it will be the "good" allele, meaning that gamete can go on to produce an individual free from the sub-optimal allele.

4.7 THE ROLE OF SEX IN THE CREATION OF SPECIES

The creation of new species is termed speciation. Asexual species evolve by changing over time through mutation or conjugation (Box 2.1) and can eventually produce new species without benefit of sex. However, for sexually reproducing species, sex is often a key part of the speciation process. This is not simply because sex is the mode of reproduction. Instead, it is because reproductive isolation (Box 4.2) between individuals of different populations facilitates the establishment of two species where there had previously been one.

4.7.1 Two different patterns of species evolution

As a first step, a distinction must be drawn between two different patterns of species evolution, known as anagenesis and cladogenesis (Figure 4.12). In the case of anagenesis (meaning "creation upward"), an ancestral species accumulates changes gradually over time, primarily through mutations and natural selection, such that at a later point in

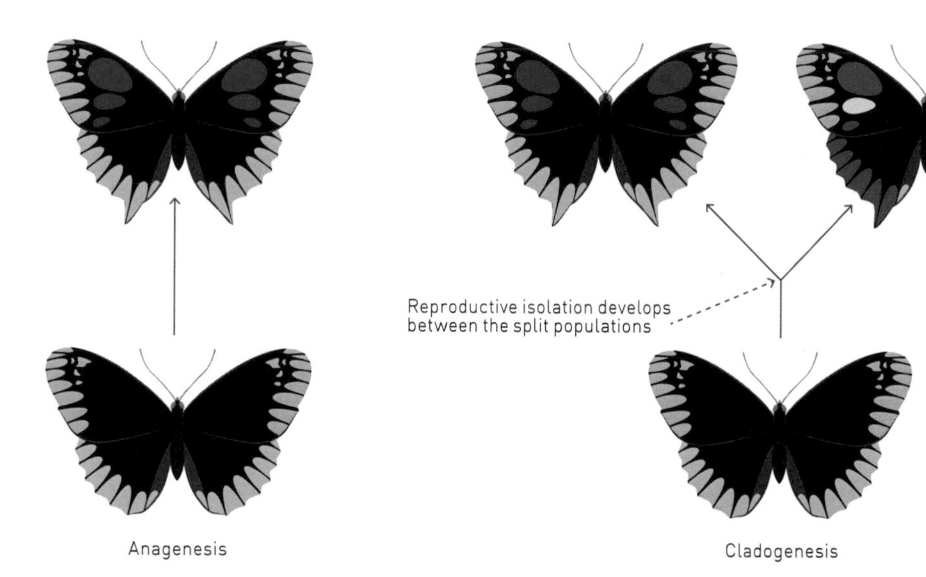

FIGURE 4.12 **ANAGENESIS AND CLADOGENESIS ARE CONTRASTING PATTERNS OF CREATING NEW SPECIES**
Although both anagenesis (left) and cladogenesis (right) produce new species, many authors only refer to the latter as speciation because of its multiplying effect. In cladogenesis, two species are derived from a single ancestor, known as their *common ancestor*.

time it is sufficiently different to warrant being called a new species. The ancestral form is extinct in the sense that it no longer exists, but the gene pool did not become extinct; instead, it merely changed over time to produce a transformed species. In anagenesis, the process begins with one species and ends with one species. Sex is important because reproduction occurs along the way, but it is not otherwise a key factor in the formation of new species, in contrast to cladogenesis.

In cladogenesis (meaning "creation of branches"), the gene pool of an ancestral species—referred to as the common ancestor—becomes split in two, and thereafter changes accumulate such that at a later point the two are sufficiently different from one another to warrant being called different species. One branch may be more or less the same as the original ancestral species, although given the passage of time, it is equally or more likely that both parts have transformed. In cladogenesis, the process begins with one species but ends with two. In sexually reproducing species, over time the two derivative parts of the original gene pool become reproductively isolated; even if the

BOX 4.2 SEX AND THE DEFINITION OF "SPECIES"

It is not a conceptual challenge to accept that a garter snake and a humpback whale are different species. But what about the garter snake and the ribbon snake shown in Figure 4.13?

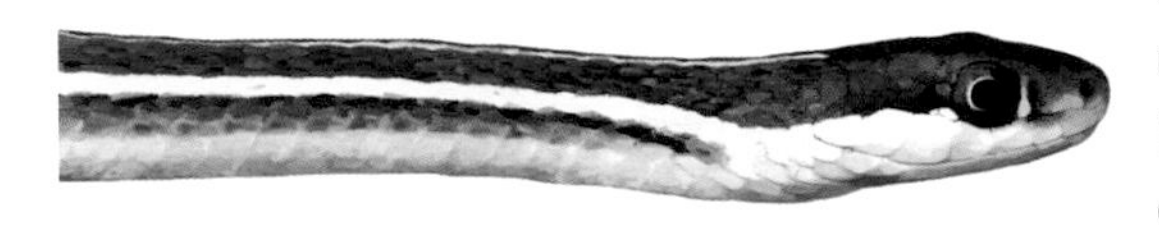

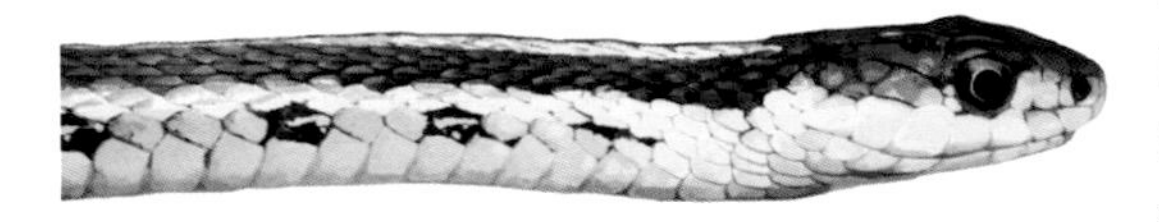

FIGURE 4.13 **THE BIOLOGICAL SPECIES CONCEPT RELIES ON A BREEDING TEST**

Eastern ribbon snake (*Thamnophis sirtalis*) (top) and eastern garter snake (*Thamnophis sauritus*) (bottom) of eastern North America are similar in appearance and have similar genes. However, even though they often share the same habitat, they do not interbreed, and therefore we consider them different species according to the biological species concept.

Drawings by Peter Mills

Developing a satisfactory definition for what constitutes a species has proven to be difficult for biologists. Originally, species were distinguished on the basis of their morphology (known now as the *morphological species concept* or, in the case of fossils of extinct species, the *paleontological species concept*). Yet, this has proven to be inadequate in many cases due to variability within populations. On the one hand, two populations that are quite different genetically can look very similar and can therefore suggest to the taxonomist that they are in fact one species. On the other hand, two populations can look quite different in one or more distinct ways, yet they are really quite similar genetically, and the mistake made here is that they are construed to be two species.

The most widely used definition, the biological species concept, relies on sex for its operational application: species are groups of interbreeding natural populations that are reproductively isolated from other groups. Note the meaning of *interbreeding* here. It is not the same as inbreeding, which is mating between relatives. Instead, interbreeding simply means breeding between. For species to be considered separate according to the biological species concept, there must be reproductive isolation. This means that if two individuals cannot produce offspring in nature, for whatever reason, they are reproductively isolated and therefore not of the same species. The importance and mechanisms of reproductive isolation are dealt with in more detail in Section 4.7.3.

So, the preferred species definition relies on a "breeding test." As such, it is clear the biological species definition is of no help whatsoever for asexual species. A second problem for this definition is that geographically distant populations that might be genetically the same (and therefore truly the same species) never meet, and therefore never put their status as one species or two to the reproductive test. For these distant populations, for extinct species known only as fossils, and for asexual species, we need to resort to other definitions, most notably the morphological species concept.

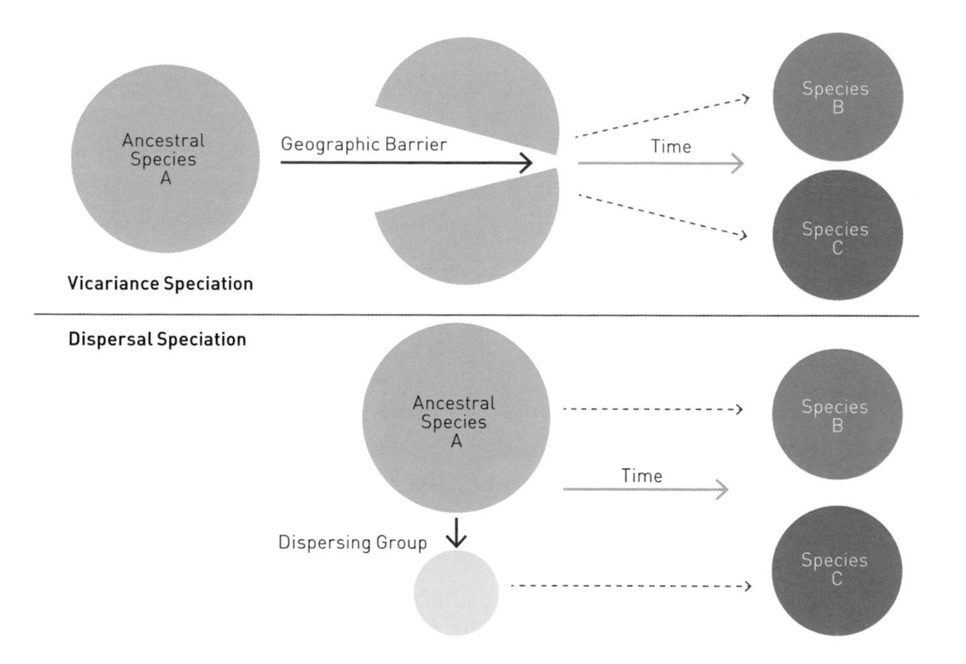

FIGURE 4.14 ALLOPATRIC SPECIATION OCCURS WHEN AN ORIGINAL POPULATION IS SPLIT IN TWO
This can occur either by a geographical event (vicariance speciation) or by a biological event (dispersal speciation), followed by a period in which the two derivative populations diverge genetically to become reproductively isolated.

two gene pools re-contact each other, they will remain two separate species because of reproductive barriers that have arisen in the intervening period.

4.7.2 How do gene pools become split at the start of speciation?

The usual means by which the gene pool of a sexually reproducing species becomes split in two is through geography. This can occur either because of a biological event affecting the organism or because of a geographical event. In both cases, an event occurs that separates the original gene pool into two separate, derivative parts that are no longer in contact (Figure 4.14).

The most likely biological event that results in such a split is known as *dispersal*. Most animals disperse, especially in the case of young individuals seeking a place to occupy, but such dispersal commonly involves only relatively short distances, within the normal range of the organism. There are exceptional cases. For instance, in 1877, a few African herons known as cattle egrets (*Bubulcus ibis*) managed to cross the Atlantic and successfully colonize South America. Biologists call such a new group in a new environment a *founder* population. It has not been documented whether the cattle egrets that are now abundant in South America have genetically diverged from the ancestral population in Africa in the intervening 140 years. However, in other cases where humans have purposefully introduced such a founder, divergence has been documented. For instance, house sparrows (*Passer domesticus*) were introduced to North America in the nineteenth century, and these populations have diverged genetically from the ancestral European population.

When a gene pool split occurs because of a geographical event, it is referred to as *vicariance*. For instance, a major river may change direction as it erodes an ever-changing path, and by doing so, it may divide a population in two. The Colorado River is a sufficiently large river that has helped carved out the even more significant Grand Canyon. There are several lizards and small mammals there that were once single species but are now pairs of similar but different species, one on each side of the canyon. Geographic barriers can also split marine populations. The Pacific Ocean and Caribbean Sea were connected as recently as three million years ago. Then, a period of mountain building separated the Pacific from the Caribbean, and many marine species that had occupied the area were also split in two, a Caribbean derivative and a Pacific one. Those gene pools have now been separate for three million years, and there are numerous cases of cladogenesis in this region as a result.

Whether cladogenesis follows a dispersal event or a vicariance event, both are referred to as allopatric speciation (*allo* referring to "other," and *patric* referring to "homeland"). Biologists have demonstrated that physical separation is not absolutely essential to split gene pools, and that therefore some speciation is *sympatric*, but allopatry is almost certainly the more common process. In any event, the barrier between the two derivative populations, whether produced by dispersal or vicariance, terminates gene flow between them.

This simply means their gene pools are independent of each other because they no longer breed together.

4.7.3 Pre-zygotic and post-zygotic reproductive barriers

When the gene pool of a sexually reproducing species splits in two following a dispersal or vicariance event, gene flow ceases to occur between the two derivative gene pools. However, the two are not immediately different species. After all, immediately following the event the two derived gene pools are more or less genetically the same. Genetic change over many generations within one or both populations following the event must occur for speciation (i.e., cladogenesis) to result. Sometimes the two derivative populations re-contact each other. Whether they merge into one gene pool again depends upon whether enough time passed for them to have accumulated enough genetic change to remain reproductively isolated, despite being no longer physically isolated.

When two derived species that share a common ancestor experience re-contact, what prevents them from interbreeding again to form a single, merged gene pool? That is, why doesn't gene flow resume? The short answer is because time and genetic change have established reproductive barriers. There are two main categories of barriers, based on whether a zygote is ever formed between individuals of the two gene pools.

Pre-zygotic barriers ("before the zygote") are those that prevent the creation of a zygote. This occurs when the barrier impedes mating or prevents mating from producing a zygote. Post-zygotic barriers ("after the zygote") are those where mating and fertilization occurs and a zygote is therefore formed, but the ultimate result is reproductive failure or much-reduced reproductive success.

There are five categories of pre-zygotic barriers. In *habitat isolation* (also known as ecological isolation), the two populations have evolved to occupy different habitats and therefore never encounter one another. In *temporal isolation* (where *temporal* refers to timing), the two populations have evolved to have different breeding seasons, and so when one population is engaged in mating, the other is not, and vice versa. In

FIGURE 4.15 **HYBRIDS ARE OFTEN STERILE DUE TO MEIOSIS CHALLENGES**

Horses (*Equus ferus*, left image) have 64 chromosomes, meaning their eggs and sperm each have 32. Donkeys (*Equus africanus*, right image) have 62 chromosomes, meaning their eggs and sperm each have 31. When a donkey (usually the male) mates with a horse (usually the female), they do produce a viable offspring, a mule (bottom image). Mules have 63 chromosomes. There are cases where mules are fertile, although it is very rare, because meiosis with an odd number of chromosomes usually results in eggs and sperm that do not contain a correct complement of chromosomes.

behavioral isolation, the two species have behavior patterns (especially mating rituals) that have evolved to be different, eliminating the possibility of mating. In *mechanical isolation,* mating has become a physical impossibility; for instance, the intromittent organ of the male does not match the reproductive tract of the female. Finally, in *gametic isolation,* mating occurs but the sperm cannot fertilize the egg.

There are three categories of post-zygotic barriers. *Hybrid inviability* occurs when the zygote is unable to develop into a healthy adult due to too many genetic differences between the sperm and egg. *Hybrid sterility* means that the offspring develops but is itself sterile. When a

donkey (*Equus africanus*) mates with a horse (*Equus ferus*), for instance, the mule that is produced is sterile (Figure 4.15); this is because horses have 64 chromosomes (32 pairs) and donkeys have 62 (31 pairs), meaning the mule has 63. Finally, *hybrid breakdown* occurs when the first generation hybrid is both viable and fertile, but in subsequent generations there is widespread inviability or sterility. As we noted in Section 2.2.3, sometimes hybridization produces viable, all-female, asexual populations.

CHAPTER 4 SUMMARY

- The Age of Enlightenment, coupled with the development of the scientific method and an improving knowledge of the wider natural world, led to the nineteenth-century recognition of evolution as the over-arching theory of the life sciences.
- Charles Darwin's *On the Origin of Species* (1859) is likely the most influential academic book ever published.
- There are multiple lines of evidence supporting the Theory of Evolution, from paleontology to comparative genetics.
- The primary evolutionary mechanism is *natural selection,* by which individuals better suited for survival and reproductive success contribute genetic information disproportionately to subsequent generations.
- Biological *fitness* is measured by reproductive success, so that an individual's fitness is a measure of his or her contributions of offspring to the gene pool of the next generation.
- DNA is the molecule that not only encodes information about making genes but also represents the legacy inherited from a long line of ancestors.
- Mutations in DNA change the coding, which in turn can result in changes in the resulting proteins; evolution acts on these changes, which can result in enduring changes to a species' gene pool if they are beneficial to survival or reproductive success.
- The Hardy–Weinberg Equilibrium is an expression of stable allele and genotype frequencies in a non-evolving population, meaning that changes in those frequencies in a population indicate evolution is at work.

- Sex can be costly in several ways, including various risks and even shortened lives, but it is also beneficial by generating unique phenotypes and by minimizing the impact of deleterious mutations.
- Where the gene pool of a sexual species is split and there is intervening time during which the two resulting gene pools diverge, barriers to successful sexual reproduction between the two also develop that result in the establishment of two species.

FURTHER READING

Coyne, J.A., & Orr, H.A. (2004). *Speciation.* Sunderland, MA: Sinauer Associates.

Darwin, C. (1859). *On the origin of species by means of natural selection, or the preservation of favored races in the struggle for life.* London: John Murray.

Dawkins, R. (2009). *The greatest show on earth: The evidence for evolution.* New York: Free Press.

Heitman, J. (2015). Evolution of sexual reproduction: A view from the fungal kingdom supports an evolutionary epoch with sex before the sexes. *Fungal Biology Reviews, 29*(3–4), 108–117. https://doi.org/10.1016/j.fbr.2015.08.002

Miller, K.R. (1999). *Finding Darwin's God: A scientist's search for common ground between God and evolution.* New York: Cliff Street Books.

Ridley, M. (1993). *The Red Queen: Sex and the evolution of human nature.* London: Viking Books.

Tibayrenc, M., Avise, J.C., & Ayala, F.J. (2015). In the light of evolution IX: Clonal reproduction: Alternatives to sex. *Proceedings of the National Academy of Sciences of the United States of America, 112*(29), 8824–8826. https://doi.org/10.1073/pnas.1508087112

Zimmer, C. (2011). *Evolution: The triumph of an idea.* New York: Random House.

5 Sexual Selection

KEY THEMES

Sexual selection is a subset of natural selection where the sole context is mating. There are two branches. *Inter-sexual* selection involves mate choice, usually by females, and *intra-sexual* selection involves competition between members of the same sex, usually males. Not all competition is direct, however. In most species, females invest considerably more in reproduction than males. As a result, females tend to succeed by being more discriminating than males in selecting mates. In some species, however, there is sex role reversal where males invest more in reproduction than females.

Following the publication of Darwin's *On the Origin of Species*, nineteenth-century naturalists and other thinkers had many observations and comments to express, and they ranged from extreme antipathy to excited approval. Over time, the ranks of scientists accepting Darwin's propositions increased. The idea that evolution had happened (and continues to happen) was the quickest to win converts. Widespread acceptance of natural selection as the primary means of evolution took longer, but this idea too eventually prevailed in most scientific circles.

One of the immediate objections to Darwin's Theory of Evolution was an apparent paradox related to sex. It was easy to call upon many examples from nature where individuals, especially males, had features that could not possibly increase survival. For instance, the tail of the male Indian peafowl (*Pavo cristatus*), known as a peacock, is an extravagance that is costly to produce and is clearly a significant burden when it comes to hiding from predators or flying (Figure 5.1). Surely, the abundant examples of such excessive, ornate, and evidently costly features in males were enough to demonstrate that Darwin must be wrong. After all, natural selection would be expected to eliminate the alleles that produce such life-threatening phenotypic traits. Darwin himself worried about this; in an 1860 letter to America's leading botanist, Asa Gray, he said, "the sight of a feather in a peacock's tail, whenever I gaze at it, makes me sick!"

But that was in 1860, and Darwin was a deep thinker. He applied himself to the issue in the decade following publication of *On the Origin of Species*. In 1871 he published another influential book, which had two themes, as revealed in its title: *The Descent of Man, and Selection in Relation to Sex*. Both parts were controversial. For the first part, it came as no surprise that there was much resistance to the idea of humans having evolved from non-human ancestors. For the second part, Darwin presented the idea of sexual selection, a profoundly important sub-category of natural selection. This, too, met with resistance, since Victorian society was rather repressed when discussing matters of sex, especially when Darwin indicated that the sexual behavior of female organisms played a role of equal evolutionary importance to that of males.

FIGURE 5.1 ELABORATE TRAITS LIKE THE PEACOCK TAIL CHALLENGED EARLY EVOLUTIONISTS

The tail of the male Indian peafowl is costly in multiple ways: costly to produce, costly by making the bird conspicuous to predators, and costly by making flight difficult. Sexual selection theory, however, explains why this trait is important for the male to achieve matings.

5.1 SEXUAL SELECTION IS A SUB-CATEGORY OF NATURAL SELECTION

In Chapter 4, the argument for natural selection was set out by reference to five observations and three inferences. In natural environments, parents produce an excess of offspring (observation 1). Those offspring need resources, but resources are not infinite (observation 2). Populations generally remain stable in size (observation 3). These observations indicate there is an ongoing struggle for existence (inference A). Within populations, individuals vary in their phenotypic characteristics (observation 4) and much of that variation is inherited (observation 5). Survival and reproductive success depends in part on these variable inherited traits, leading to differences in fitness among individuals (inference B). The proportions of all the alleles for all the genes in the breeding population, known as the gene pool, will change accordingly (inference C).

Because sexual selection is a sub-category of natural selection, all the points in this summary apply to sexual selection as well. But, sexual selection is narrower than natural selection per se, being exclusively restricted to matters related to mating and reproductive success.

5.1.1 Comparing examples of natural selection and sexual selection

It is helpful to use hypothetical examples to contrast natural selection that deals with non-sexual matters against sexual selection. We'll summarize two examples of non-sexual natural selection and then contrast them with two examples of sexual selection.

(a) *Example 1: Natural selection where the selective agent is another organism.* A non-native predatory fish is introduced to a lake where small, native, bottom-dwelling fish reside. The small bottom-dwelling fish have not been exposed to such predatory fish before. There is some variation in the eye position of the bottom-dwelling fish, ranging from eyes located on the side of the head (lateral position) to eyes located higher up (dorsal position). Those with more dorsally oriented eyes detect the new predators more easily, so they react more quickly and are therefore less likely to get eaten. Since they survive better, they produce more offspring (higher fitness). Their eye position is genetically determined, and so over time these alleles become more concentrated in the population and lateral eyes therefore become increasingly rare. In fact, not only do dorsal eyes become more common, they become ever more dorsal (an adaptation to the new predator).

(b) *Example 2: Natural selection where the selective agent is non-living.* A small predatory mammal has brown fur during the summer, but when it produces new fur in the autumn, it is white, camouflaging the predator in the snowy landscapes of winter (an adaptation). Among individuals, there is variation in the timing of these changes of fur coat and in the intensity of the white, with some individuals retaining smudges of brown. Climate change is dramatic in this environment, however, and snow becomes rare and never lasts long on the ground. Those individuals whose fur changes later in the autumn and whose fur retains more brown smudging have better camouflage. They are better able to sneak up on their prey, making them better fed and healthier parents. They therefore produce more offspring (higher fitness), their alleles spread in the gene pool, and the population therefore becomes better suited to its new (non-snowy) environment because a greater number of individuals within the population have less white fur (an adaptation).

(c) *Example 3: Sexual selection where the selective agents are members of the same species and same sex.* Male deer do not provide any parental care; all parenting is done by the females. Male deer grow antlers every summer, but there is variation in the size of such antlers. In the autumn, when the sexes are prepared to mate, males achieve the most reproductive success by mating with as many females as possible. So, the males compete for access to the females by fighting with each other. Those males with the bigger antlers (a mating adaptation) fight more successfully and therefore mate with more females (achieving higher fitness). This means their alleles that generate larger antlers get passed down to the next generation, changing the gene pool to one where antlers are, on average, bigger.

(d) *Example 4: Sexual selection where the selective agents are members of the same species but the opposite sex.* During the breeding season, male songbirds earn the attention of females through birdsong. Song repertoires and song rates vary in the population. Birdsong conveys information about the genetic composition of the male, where healthier birds generally have more complex song repertoires and higher singing rates. Females accordingly choose males that sing more songs and sing more frequently than other males (a mating adaptation). Accordingly, the good singers breed more successfully (achieving higher fitness), and therefore the alleles of the good singers increase in the gene pool, increasing the song repertoire capacity and singing frequency of the population.

These four hypothetical cases exemplify the four major categories of natural selection as defined by the selection agent. In natural selection per se, the agent can be another species (a *biotic agent,* case [a]), or it can be a non-living component of the environment (an *abiotic agent,* case [b]). In sexual selection, the agent must be other individuals of the same species, known as *conspecifics,* and it can either be members of the same sex where the focus is competition for mating opportunities (case [c]) or members of the opposite sex, where the focus is mate choice (case [d]).

5.1.2 Re-formulating the natural selection argument for sexual selection

Because sexual selection is a form of natural selection, it has to follow the same five-observation, three-inference argument. Having now considered examples of both, here we modify the generalized natural selection argument for sexual selection.

The argument follows (see also Table 5.1). Note that only two of the five observations are altered (in *italics*). Parents produce an excess of offspring (observation 1) in *mating systems where females are usually the limiting sex* (observation 2), yet populations generally remain stable (observation 3). These observations indicate there is an ongoing struggle *among males for access to mates* (inference A). Within populations, individuals vary in their phenotypic characteristics *related to mating, such as female mating preferences* or *competitive ability among males* (observation 4) and much of that variation is genetic (observation 5). Reproductive success depends in part on these variable inherited mating traits, leading to differences in fitness among individuals (inference B). The population changes over time because the alleles of those individuals that mate more successfully become enriched in the gene pool (inference C).

In short, throughout all time in the evolutionary history of sexually reproducing species, individuals with alleles that contribute relatively more to successful mating must disproportionately populate the next generation with their own alleles. This ensures that the gene pool is one where alleles that favor sexual activity leading to successful reproduction are prevalent. Still, even today, it is much easier for most

	Natural Selection	Sexual Selection
Observation 1	Organisms overproduce offspring	Organisms overproduce offspring
Observation 2	The availability of resources is limited	***The availability of females is limited***
Observation 3	Population sizes are generally stable	Population sizes are generally stable
Inference A	There is a struggle for existence	***There is a struggle (among males) for mating access to the limited resource (females)***
Observation 4	Individuals vary in their traits	Individuals vary in their ***mating*** traits
Observation 5	Much variation is heritable	Much variation is heritable
Inference B	Survival depends upon these inherited traits, leading to differences in success that are related to that variation	*Reproductive* success depends upon these inherited ***mating*** traits, leading to differences in ***mating*** success related to that variation
Inference C	More successful individuals within the population contribute more to the next generation, changing the gene pool in favor of their own alleles	More successful ***reproducers*** within the population contribute more to the next generation, changing the gene pool in favor of their own alleles

TABLE 5.1 Comparison and contrast of the five observations and three inferences for natural selection per se and sexual selection, with changes noted in bold italics. This analysis stereotypes male and female roles, although the stereotypes do prevail in most cases of obvious sexual selection.

people to imagine natural selection at work in producing adaptations in organisms for increased survival than it is to imagine selective forces producing phenotypes that value reproductive success so highly. But remember this: in the end, reproductive success is much more important for genes than survival of the individual that houses those genes. Consider Box 5.1 for an extreme example.

BOX 5.1 SUICIDE SEX: BIG COST FOR BIG PAYOFF

If the mating habits of the European praying mantis (*Mantis religiosa*) cannot convince you that the triumph of mating is the ultimate success, perhaps nothing will. It is not true that copulation is always a male's final act in life, but it is certainly frequently true, since female mantises exhibit sexual cannibalism. The male typically mounts the female's back, clasping her, and then deposits sperm in a structure at the tip of her abdomen. Sometimes, especially if he is a small male, the female will twist around and consume his head during the act. Regardless, he continues mating, although in this case it will indeed be his last act. She is likely to finish consuming him after mating is over (Figure 5.2).

Sexual cannibalism is not restricted to mantises; black widow spiders (*Latrodectus variolus*) are another example. For both the mantis and the spider species that engage in this type of mating, the male has indeed lost his life. But, by having successfully mated, his genes will make it to the next generation through the eggs he has fertilized. Experiments have shown that poorly nourished females have a greater likelihood of cannibalizing their mates. In any event, his offspring may do better given their mother has been well nourished by eating his body. In one species at least, the orb-web spider (*Argiope bruennichi*), the offspring of mothers who ate their male partners survived longer.

FIGURE 5.2 **SEXUAL CANNIBALISM STILL REWARDS THE MALE VICTIM**

A female praying mantis (*Mantis religiosa*) cannibalizes her male partner following mating. His head is already consumed, although mating continues. His life is over, but his genes will make it to the next generation due to this opportunity to fertilize her eggs.

Simon Kovacic / Alamy Stock Photo

FIGURE 5.3 **ARMAMENTS AND FIGHTING BEHAVIORS ARE USUALLY MALE ATTRIBUTES**

The red deer (*Cervus elaphus*) (a) has been the subject of long-term studies that analyze male–female differences in reproduction and their relationship to sexual selection. Only the stag (male), shown here, grows antlers. Most direct fighting between members of the same species is done by males, usually over females or resources that are of value to females, as in these Scottish mountain hares (*Lepus timidus*) (b).

[5.3b] © Andy Rouse / naturepl.com

5.1.3 Manifestations of sexual selection

Sexual selection that operates through competition within one sex (usually males) is known as intra-sexual selection. Weaponry is abundantly common in the natural world. Much of it is for dealing with other species, either in an offensive or a defensive manner, and both sexes benefit from having weaponry for non-sexual conflicts. In a sexual capacity, however, armaments overwhelmingly are exhibited in male bodies rather than female ones. These are for battles over mates, and they include things like larger male bodies as in the northern elephant seal (*Mirounga angustirostris*), antlers and large horns as in deer and sheep, leg spurs as in many chicken-like fowl, and many other types of weaponry. With these armaments, males are far more likely than females to directly compete with members of their own sex by engaging in fighting (Figure 5.3). We will see that in rare cases, it is the females that sport weapons for battling in the sexual realm (Section 5.4).

Sexual selection that operates through mate choice (usually where females are choosing males) is known as inter-sexual selection, not to be confused with the term intersexual as is used for ambiguous gender (Section 9.5). Inter-sexual adaptations include behaviors, but they also include associated physical apparatus, commonly referred to as ornaments. Ornaments are highly diverse: examples are bright colors, as in the male guppy (*Poecilia reticulata*) and elaborate feathers, as in many male birds (see Figure 5.1). Other inter-sexual traits include songs, used by many male birds, frogs, insects, and even bats, and they also include dances and other performances, as observed in male jumping spiders. It is not difficult to appreciate the value of armaments, but ornaments are more plainly useless—except in the context of mating.

5.2 WHY ARE FEMALES USUALLY THE "LIMITING SEX"?

If you think about the argument for sexual selection, you can see it rests upon a disparity in the second observation: that individuals of one sex (usually females) are less available for mating than the other sex (usually males); we commonly say that females are usually the limiting sex, in that females are the sex that limits the amount of sexual activity. Why is this?

5.2.1 Females usually invest more in reproduction

In more than 90 per cent of mammal species, males provide no parental care. In fact, in most animal groups, females are far more likely to be the major investor in offspring. In most male mammals, the only investment in reproduction is in furthering their own mating success. In other words, all they contribute is sperm—no resources, no defense of mother and young, no parenting. Members of the deer family fit this pattern, and the European red deer (*Cervus elaphus*) has been the subject of a long-term study that has investigated such patterns of parental investment (Figure 5.3). Males invest a lot in mating effort, competing with each other for access to females, but make no investment in parenting. When a male "services" a female, he provides her with semen carrying millions of sperm. But because this doesn't take long, he can service more females later that day and will do so if he can, so his investment per female is low.

In contrast to males, consider what the female red deer contributes to the enterprise: (a) the egg, (b) her womb for a gestation period of about eight months, (c) nourishment of the fetus so it can reach a birth weight of perhaps 15 kg, (d) nourishment of the calf (lactation) for about two months after birth, and (e) protection and guidance for almost a year. These investments are very costly for the female in three ways: nourishment of the young for about a year, endangerment to herself from predation given the burden she assumes as a mother, and what biologists refer to as *lost opportunities*.

Lost opportunities constitute a stark difference between males and females. A male takes a few minutes to mate with a particular female, and he may briefly deplete his sperm supply but he makes thousands of sperm per minute. Obviously, he cannot mate with two females at once, and there is a brief period where he either cannot service a second female or if he does, it will be characterized by a depleted supply of sperm to fertilize her. Yet this is a negligible lost opportunity compared to the female. In contrast, once she is pregnant, there is no benefit to further mating. She is committed for the next year to that pregnancy, which will likely yield one young. If times are tough, she might also skip the following season and not breed again for two years instead. That constitutes a very significant lost opportunity—at least a year's worth of additional or alternative mating prospects.

Repeatedly, biologists have found that the greater the disparity between investments in reproduction, the more likely it is that sexual selection will be significant. The relationship between parental investment and sexual selection is shown in Figure 5.4. This stereotypes organisms by showing that females invest more in reproduction than do males; there are exceptions to this (Section 5.4), but it is generally true. Where females invest a lot in reproduction and males do not, females tend to have a lower potential reproductive rate and a corresponding lower amount of sexual activity, opposite to males. As a result, at any one time there tends to be a surplus of males available for mating and a relative shortage of available females. Even though the sex ratio in the population may be 50:50, biologists describe this operational sex ratio (the number of males ready to mate to the number of females ready to mate) as male-biased.

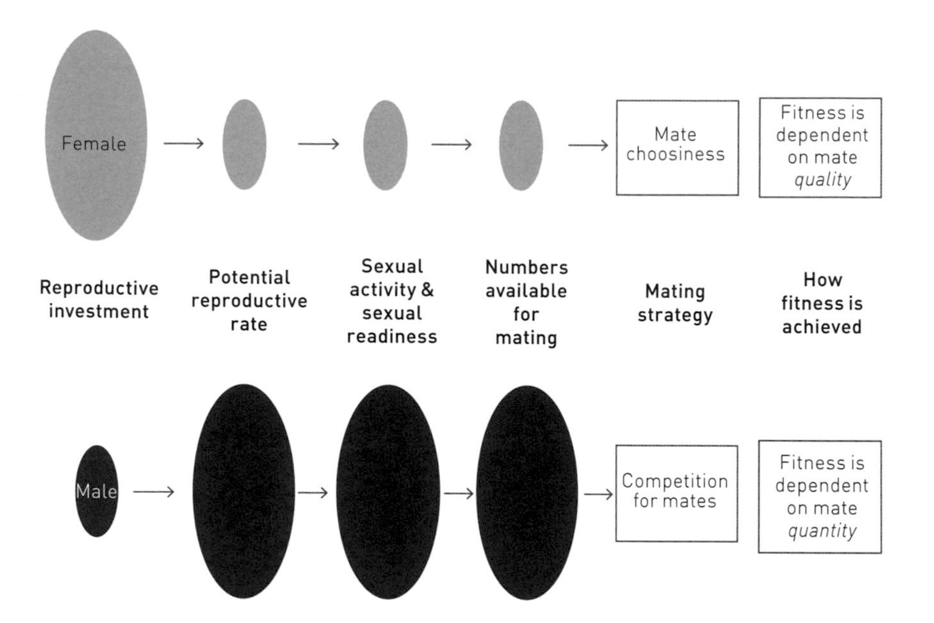

FIGURE 5.4 **MALE-FEMALE DIFFERENCES IN REPRODUCTIVE INVESTMENT HAVE MANY CONSEQUENCES**

This flowchart connects parental investment (left edge of panel) with sexual and mating attributes (middle of panel) and with mating strategy and fitness consequences (right edge of panel). The relative size of the ovals contrasts the differences between the sex that invests a lot in reproduction (usually females, as shown here) and the sex that invests less (usually males, as shown here).

It should never be underestimated how important the fact is that all organisms alive today have an astonishingly auspicious ancestry. Every organism living today (including you!) has had an unbroken string of successful forebears during the past 3.5 billion years or so. And, since sex evolved around 1.2 billion years ago, this is true for both the maternal line and the paternal line of sexually reproducing organisms. This means that *all* ancestors, both male and female, mated and produced at least one offspring, which itself survived to reproduce. This is why for modern organisms that rely upon sex for reproduction, the genes inherited from a long line of ancestors that enjoyed a perfect history of reproductive success ensure sex remains profoundly important in the present.

What the analysis of the male–female disparity in parental investment shown in Figure 5.4 tells us is that mate choice is normally of particular importance for females and competing for access to mating is typically important for males. Accordingly, in the classical view, fitness is best achieved for females by choosing high *quality* males, and fitness is best achieved for males by competing for a high *quantity* of mating. This is known as the *quality–quantity dichotomy*. This dichotomy represents an extreme, of course, and we will shortly consider ways that the disparity in parental investment can be reduced or even reversed. And in Chapters 6 and 7, we will see that mate choice can involve multiple partners.

5.2.2 Does inter-sexual selection produce payoffs?

When Darwin proposed the theory of sexual selection in 1871, Wallace, with whom he had co-presented the Theory of Evolution by natural selection in 1858, was skeptical. Wallace could accept that male embellishments that facilitated male–male competition like antlers (armaments) were a product of selection, but he could not accept that male embellishments like songs and mating dances (ornaments) existed because females assess males on their basis and choose mates accordingly. This remained the major theoretical difference between the two men's views for the rest of their lives.

The concept of sexual selection was mostly ignored for a century following Darwin's 1871 publication, but in the 1970s it experienced a resurgence of attention that has led to a rich body of research on the subject. Numerous corollaries have been developed: mating systems (Chapter 6), sexual conflict (Chapter 7), and mating strategies, which are various sets of behaviors with matching physical traits that work together to achieve mating success. Many research programs in the past 40 years have demonstrated that both male–male competition and female choice pay off reproductively for both sexes and are therefore profoundly important selective forces in nature.

The part of sexual selection theory that Wallace and others objected to was the half now called the female choice hypothesis, also called inter-sexual selection. Following the resurgence of interest in sexual selection, several research programs were designed to test this idea, and since then, the importance of female choice has been supported in countless studies of both invertebrates and vertebrates.

The long-tailed widowbird (*Euplectes progne*) was the subject of one of the first sets of experiments. This songbird nests in African grasslands and savannahs. During most of the year, males and females look similar and rather ordinary, being brownish with average proportions, although males are moderately larger. During breeding, the males change dramatically, trading their dull brown plumage for one that is jet black with patches of red and cream. In addition, they grow a very long tail, which, at as much as 50 cm, is dramatically longer than the female tail, at about 7 cm. The males display using a tail-encumbered flight over that part of the grassland they claim as their own, trying to attract females to the nests they have started at multiple places in their territories.

The researchers experimentally investigated the relationship between a male's tail length and the number of active nests (that is, the number of nests occupied by female mates) in his territory. They studied 36 males and divided them into 4 groups: (1) ones whose tails were shortened, (2) ones whose tails were cut and re-glued (to control for cutting effects), (3) ones who were merely captured and had a small identification band put on the leg (to control for handling), and (4) ones who had the tail pieces cut from group 1 birds inserted into their own tails, lengthening them. The results are shown in Figure 5.5. There were small differences in the number of nests among the first three groups, which were not statistically significant, but the males whose already-long tails had been lengthened had significantly more nests—more than twice as many. Follow-up empirical work showed that in populations with unmanipulated tail lengths, males were likely to have an extra mate for every three centimeters of extra tail length, so that the male with the longest tail had the most nests.

Both the experimental and empirical studies showed that females were choosing males with the longest tails. We must defer for the moment why that might be advantageous for these discriminating females. For males, tail length—which encumbers their flight and doubtless makes them vulnerable—has a reproductive payoff. Their alleles are passed on to the next generation with greater frequency than their shorter-tailed brethren. What stops tail length from becoming ever greater? Likely, it is the countering action of natural selection: beyond a certain point, the cost to the males in terms of

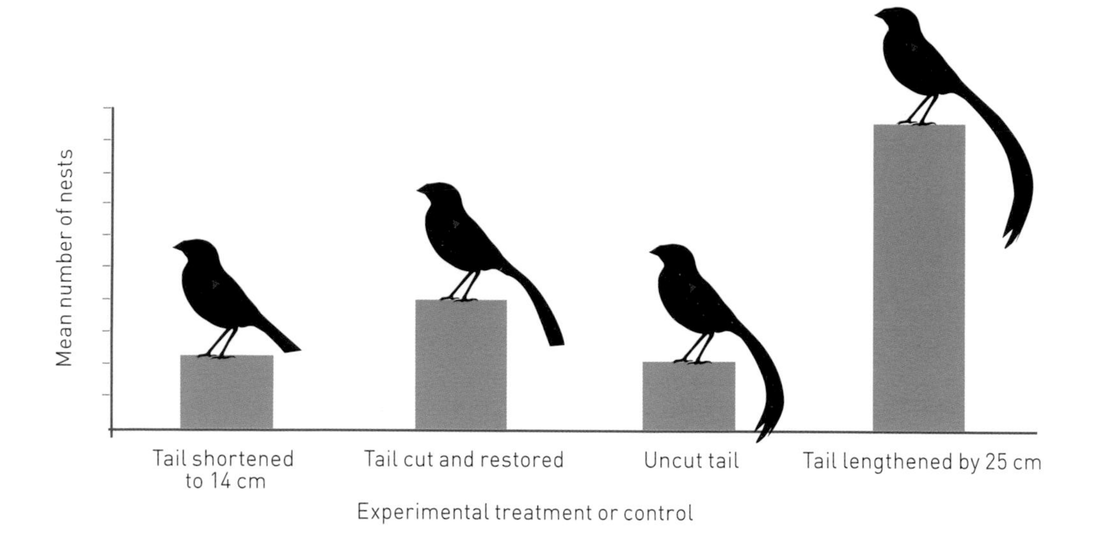

FIGURE 5.5 **ELONGATING A MALE WIDOWBIRD'S TAIL IMPROVES HIS MATING SUCCESS**

The reproductive success of male African long-tailed widowbirds (*Euplectes progne*) has been tested on samples whose tails were experimentally shortened (left bar), left the same length (middle two bars), or lengthened (right bar). Even though males in the middle two groups had "normal" length tails for the species, and even though the artificially lengthened tails made the males in the lengthened group don't fly as well, the males with the unnaturally lengthened tails attracted the most females to their territories, and therefore had the most nests and produced the most young.

[5.5a] Adapted from Andersson (1982)

[5.5b] Eleanor H. Hattingh / Shutterstock.com

survival is so great that it outweighs the mating benefits of longer tails. But the females are clearly responding to tail length, even preferring males that have tails that are unnaturally long, known as a *supernormal stimulus*.

This study also showed a pattern that has been found in many species where sexual selection is operating on sexual traits such as tail length. Males as a group experience much higher variance in reproductive success than females. That is, most females in these species are able to breed, but often, a significant proportion of the males don't get to breed at all and a small proportion of males get to do a disproportionately large amount of the breeding.

Testing what is in it for the females has been more difficult to measure, but a long-term study of pronghorn (*Antilocapra americana*) (Figure 5.6), a fleet-footed antelope-like animal of North American grasslands, demonstrates that mate choice is indeed important. Pronghorn can run very fast—capable of reaching 90 km/h for short distances. During the breeding season, females employ several different strategies as they sample which male(s) to mate with. Behavioral studies show that females sample the capacity of different males to enforce a "zone of tranquility" in his territory or around his harem of females. Some females are known as *inciters*, inciting conflicts between rival males and then mating with the winner. Both methods test males and reward them with mating opportunities.

Female pronghorns have a relatively brief period of sexual receptivity and fertility known as estrus. During the mate-selecting period prior to estrus, female pronghorn expend considerable energy in their search for the right mate. If all males' sperm offered the same benefits, this would not be the case. By monitoring the success of different males and females, this research has shown that the young sired by the chosen males survive better and achieve independence sooner, so it does indeed pay off for the female to be so discriminating.

What about humans? Although males provide parental care, unlike most mammals, it is patently true that in most societies for most of history, females are more burdened with reproductive investments. Indeed, there is evidence that human females are the choosier of the two sexes when it comes to mating. See Box 5.2 for one example.

FIGURE 5.6 **FEMALE PRONGHORN PREFER A MALE THAT PROTECTS THEM FROM HARASSMENT**

Pronghorn (*Antilocapra americana*) females sample the capacity of different males (the larger individual, second from left, with horns) to enforce a "zone of tranquility" around his harem. In some cases, females incite conflicts between males with an apparent goal of seeing which male will prevail. After breeding, the harems disband and the rest of parenting is up to the female.

Green Mountain Exposure / Shutterstock.com

5.2.3 Does intra-sexual selection produce payoffs?

The widowbird and pronghorn examples show that inter-sexual selection—where female choice is the main determinant—has payoffs for the males who seek to be chosen and for those females who make the choices. What about payoffs in cases of intra-sexual selection? Here, we'll return briefly to the red deer we introduced in Section 5.2.1.

The most successful male red deer have a lifetime reproductive success of as many as 25 young. Meanwhile, their less successful brethren produce few young—sometimes none. These results are dramatically different, providing a classic case of high variance in male reproductive success. Reproductive success in males is dependent upon their ability to establish and maintain harems of females, and this is related to their fighting ability (i.e., intra-sexual selection).

BOX 5.2 CHOOSINESS IN HUMANS: FEMALES CAN SMELL GENETICALLY SUITABLE PARTNERS

Some fascinating research about human female mate choice has revolved around a set of genes known as the major histocompatibility complex (MHC) genes. They are a group of genes that represent the genetic code for the creation of cell surface proteins that fight pathogens. In the 1970s it was discovered that in house mice (*Mus musculus*) MHC genes play a role in mate choice exercised by females. That study found that mice mated assortatively. That is, the MHC alleles of the chosen male partners tended to be more different from the MHC alleles of the female choosers than would be expected from random mating. This is an example of negative assortative mating, where individuals avoid mating with individuals who are like themselves. The adaptive advantage of this pattern of mating is that the offspring will have increased MHC heterozygosity, and this translates into an immune system that will be better able to combat pathogens. This relies on the notion of a *genetic complementarity* advantage, where a suitable male for a particular female is one whose alleles are most different from hers.

In the 1990s, evidence was presented that humans also have negative assortative mating for MHC alleles. In a now classic study, male and female Swiss volunteers were categorized for the MHC alleles they had using genetic techniques. Male volunteers wore T-shirts for two consecutive nights, sleeping alone and avoiding perfumed products. The following day, female volunteers were asked to rate for attractiveness the odors of six shirts (Figure 5.7). As predicted, the females rated higher the shirts of those males who were most dissimilar to them in terms of their MHC alleles (Figure 1.8). It appears, therefore, that the negative assortative mating in humans for MHC alleles is also mediated by smell. Women also indicated that they were reminded by the preferred odors of actual or former mates, suggesting their tendencies had been in operation in previous mate decisions.

There was one confounding factor that disabled a women's capacity for accurately assessing men in this manner. Women taking oral contraceptives—which generally prevent them from ovulating and which alter the cycling of several hormones—did not prefer the odors of males with different MHC alleles (see Figure 1.8).

Because of this, their reproductive lives last for the relatively short period of their physical prime, during the period from about age 6 to 11, although they are unlikely to be dominant for all those years. Males therefore exhibit high variability in their sexual, and hence reproductive, success.

FIGURE 5.7 **HUMAN FEMALES USE ODOR TO ASSESS ATTRACTIVENESS IN A MATE**

This woman is sniffing a T-shirt worn by a man for 48 hours, rating it for its olfactory attractiveness based on her unconscious assessment of the MHC alleles in the genotype of the wearer. (See also Figure 1.8 for the results.)

Source: © Alex Mills

The results of this study surprised the public, both for the unexpected role of odor in human mate choice and for the unanticipated interference of the pill in the functioning of that system. Humans do have many scent-producing glands, although we don't usually think of ourselves as being olfactory creatures. But this pattern should not be surprising because having MHC heterozygosity is important. Research has shown links between higher-than-ideal MHC similarity in partners and several measures of reproductive failure: (1) recurrent spontaneous abortions, (2) poorer-than-average outcomes at fertility clinics, and (3) longer intervals between successive births. While oral contraceptives give females sexual freedom and give couples considerable insurance against pregnancy, it appears a downside is interference associated with appropriate mate choice for females seeking a partner with whom to have children.

Males, it seems, are not making choices on the same basis. In another study that presented human males with sweat odors from a variety of females, males did not rate the attractiveness of those odors in terms of MHC allele similarity. Why this is, is not known, but it is consistent with the principle that overall, males are considerably less choosy than females.

In contrast, many female red deer begin breeding at about age three. They tend to live a bit longer than males, and they breed throughout their lives. Their breeding success is not primarily determined by fighting ability; instead, it is dependent upon the health of mother and her offspring, as well as by the constraint that she can produce at most two young per year. (One is the norm, and they occasionally skip a year too.) By having approximately 1 offspring per year, an exceptional female can produce as many as 10 in a lifetime. In addition, and unlike the case with males, very few of the female population produce no young at all. In short, in a population of red deer, there are more mothers than fathers. Genetic and ethnohistoric analysis indicate the same is true for human populations, although not to the same extent.

5.3 HOW MATE CHOICE BASED ON ORNAMENTS INCREASES FEMALE FITNESS

It's not difficult to see how armaments and other tools of competition can be favored by selection if they lead to competitive success that translates into mating opportunities. It has always been more difficult to accept the value of ornaments—ridiculously long tails, bright colors, and behaviors associated with such ornaments such as songs, dances, courtship flights, etc.—especially when they clearly impair the ability to survive. Accordingly, biologists have spent considerable time thinking about, and testing, ideas related to the establishing of ornaments and their importance in the mating game.

There are three main theories. Two of these are based on benefits that are indirect benefits because they are genetic. That is, provided the choosing sex attends to the sexually selected ornaments and makes mate choices based on those ornaments, her offspring will benefit genetically. These are known as the runaway hypothesis and the good genes hypothesis. The third idea is based on direct benefits. That is, the benefits are not genetic; commonly they are good resources (food, foraging territory, nest sites) or other non-genetic benefits like protection. This is the good resources hypothesis.

5.3.1 Genetic benefits in the runaway hypothesis: "Sexy sons"

The *sexy son* term was introduced in the 1970s, but it was a new phrase for a part of the older runaway hypothesis, initially proposed in 1930 by R.A. Fisher. Fisher's was one of the few publications dealing with sexual selection in the otherwise empty century following Darwin's book on the subject. Unlike the good genes hypothesis (Section 5.3.2) where the genetic benefits appear in both sons and daughters, in this case sons only benefit, and a mother choosing "sexy partners" benefits through the sons she produces by those matings.

In cases where a couple produces both sons and daughters, it is important to remember that genes from each parent continue in all offspring, both sons and daughters. In a population where males seek to be chosen by females on the basis of genetically based male ornamentation, those females who choose males with the best ornaments benefit by having sons with approximately that same level of ornamentation. (The inherited ornamentation will not show up in daughters because the genes are only expressed in males.) In terms of fitness, it pays off for the mother to make such choices because of the large numbers of grandchildren she can have through her ornamented (i.e., sexy) sons.

How is this a "runaway" system? Recall the long-tailed widowbird (Section 5.2.2) as we re-construct a hypothetical history, keeping in mind that choice exercised by females will be instinctive, not consciously analytical. At an earlier stage of evolution, females choose males that are healthy, and that includes ensuring a partner has a tail in good condition since it is important for flying. If the male's tail quality has a genetic basis, the female's acumen in assessing and preferring a superior tail also has a genetic basis (i.e., both the male trait and the female preference have a genetic basis), and the groundwork is laid for a runaway selection. Even though the ornamentation shows up in sons and the preference for such ornamentation shows up in daughters, the genes for both ornamentation and preference become non-randomly associated together in the genome, and they exist in both sons and daughters. Females assess males on the basis of their tails, thereby (a) concentrating in offspring more alleles related to longer tail production in males and (b) concentrating in offspring more alleles related to female inclination to choose males on the basis of their tail. Over generations, the tail gets longer and longer, and the preference for long tails gets stronger and stronger. This is the runaway process. However, a point will be reached where the benefits of sexy sons (in this case, long-tailed males) are outweighed by the costs of such an impediment (tails that are too long for essential behaviors like flying) and an equilibrium is reached.

In this way, the genes under selection in runaway selection are for sexiness in the chosen sex and preference of sexiness in the choosing sex. This hypothesis does not lead to selection for genes otherwise related to health, survival, parenting skills, etc.

5.3.2 Genetic benefits in the good genes hypothesis: Better survival

The good genes hypothesis is a term thoroughly entrenched in the scientific literature. However, it would be better to refer to it as the "good alleles hypothesis" for the same reason that it would be preferable to speak of the "allele pool" instead of the gene pool (Section 4.4). This is because this hypothesis is based on comparing the fitness values of one allele versus other alleles for a given gene.

The good genes hypothesis links overall genetic superiority with the ornament under selection. This means that the ornamental traits chosen by females are honest indicators of the male's genetic quality, whose genetic contributions will tend to produce offspring of superior genetic quality. Whereas the genes selected in the runaway hypothesis are beneficial by producing sexy sons, the genes in the good genes hypothesis are beneficial by producing offspring (sons and daughters) that are genetically superior in a variety of ways. This idea was proposed in the 1980s, shortly after the runaway hypothesis was re-introduced in the scientific literature.

Studies that support the good genes hypothesis are those where a relationship between the sexually selected trait (long tails, large song repertoire, intense colors) and an increase in offspring survival is demonstrated. For instance, in some songbirds it has been shown that females prefer males with more complicated songs and larger repertoires (i.e., the sexy trait, since it has reproductive value, not survival value) and the offspring from such matings have stronger *immunocompetence*, which means the ability to resist disease. In the odd-looking stalk-eyed fly (*Teleopsis breviscopium*, Figure 5.8), females choose males with longer eyestalks, and the resultant offspring have a lower incidence of genetic disorders.

FIGURE 5.8 **SEXUALLY SELECTED TRAITS CAN SIGNAL GENETIC QUALITY**
The stalk-eyed fly (*Teleopsis breviscopium*) has eyes at the ends of long stalks. The stalks are particularly elongated in males, and this is a sexually selected trait. Males that have longer eyestalks mate more successfully and also are superior genetically: females that rely on this trait in choosing partners have offspring with fewer genetic disorders.
iStock.com / Parkpoom

The handicap principle is a version of the good genes hypothesis. This is the view that the sexually selected ornament must be costly to the owner for it to be able to convey honest information about the genetic quality of that male. The message the handicap delivers is something like "I am so genetically superior that I can survive with this burdensome tail." It must be a real handicap because if a male could cheat by sporting the ornamental handicap and yet have inferior alleles, then it would not be of use to the female. It is easy to see how the long tail of the widowbird is costly, both to produce and as an impediment. So too are the effort and exposure related to singing, as in frogs, crickets, birds, and others. It also turns out that sporting bright colors can be difficult in some cases, such as where the male must find enough of the right food to assemble those pigments. For instance, the orange beak of the well-known zebra finch (*Taeniopygia guttata*) is made using *carotenoids*, substances that are in limited supply in the environment.

5.3.3 Benefits in the good resources hypothesis: Honest promises

Some sexually selected traits honestly signal the genetic quality of the holder, which we've called indirect benefits. Yet in other cases, the sexually selected traits honestly signal a promise of valuable resources for the female, known as direct benefits. In other words, in species where there are both sexually selected traits and paternal investment, links have been found between the two. This leads to the good resources hypothesis. Songbirds have been good study organisms for this hypothesis because males provide considerable parental care.

The male nightingale (*Luscinia megarhynchos*) is considered one of the world's best songsters, although song quality varies among individuals. Males also assist at the nest by bringing food to the young, known as *provisioning*. Recent work has shown that males that sing more elaborate songs in more complex and orderly sequences provision better at the nest. Knowing the level of song complexity allowed the researchers to predict the degree of provisioning, and presumably female nightingales can do the same.

5.4 SEX ROLE REVERSAL

So far, we've observed that the greater the disparity in parental investment, the more likely it is that sexual selection will be operating, either through inter-sexual mate choice or intra-sexual competition, or both. And, because producing sperm is biologically less expensive than tasks like nourishing a developing fetus, if one sex is investing little, it is likely to be males. Accordingly, in most cases it will be males that are competing and females that are choosing, with the result that males are the more likely sex to have armaments or ornaments and to have high variability in reproductive success.

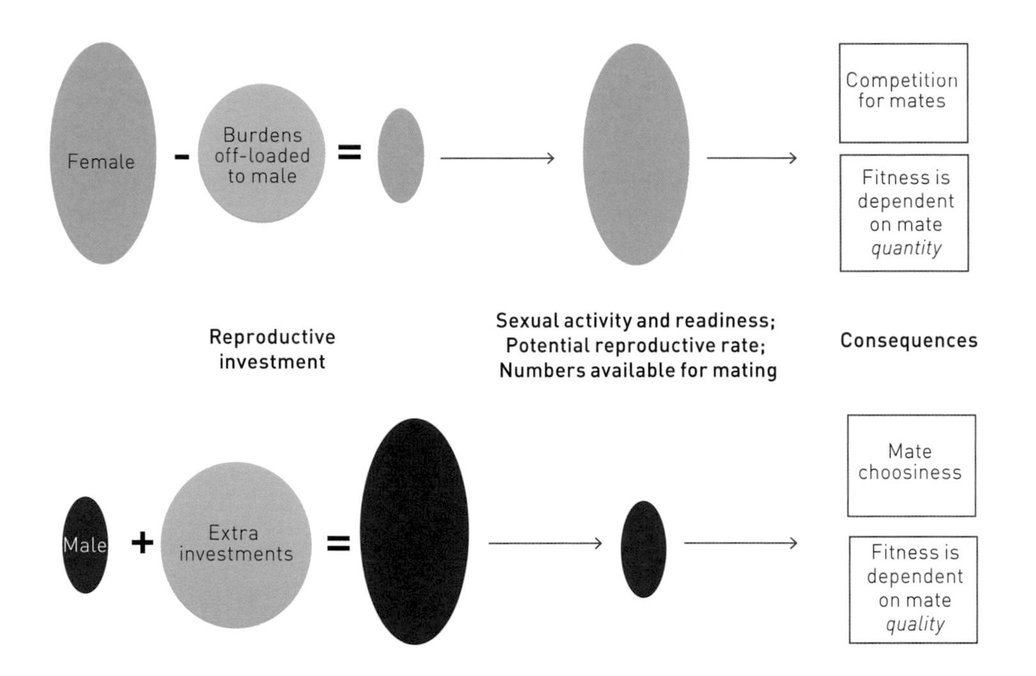

FIGURE 5.9 **SHIFTING THE BURDENS OF REPRODUCTIVE INVESTMENT CAN REVERSE SEX ROLES**

It is easier and relatively common for males to invest less in reproduction (Figure 5.4). However, in cases where males provide extra investments—commonly by accepting some of the burdens of reproduction normally assumed by females—the disparities can even out or, as here, males can invest more than females. In such cases, it is the female that is more sexually active, with a higher potential reproductive rate. The operational sex ratio becomes female-biased, and it is the females that exhibit intra-sexual competition for mates and the males that exercise mate choice.

But not all species follow this pattern. There are many cases where species have breeding systems in which males invest heavily in reproduction, although such systems are relatively rare in many groups, including mammals. If some of the burdens that females normally assume in reproduction can be off-loaded to males, or if males can provide other investments such as care or resources for the female or young, the investments can be more nearly equalized. Less often, the relative investments can become reversed, with males investing more heavily (Figure 5.9). In these cases, we have sex role reversal: females make relatively low parental investment and consequently exhibit high potential reproductive rates and high levels of sexual activity and readiness. The operational sex ratio is therefore female-biased.

Recall what we know happens in situations with low male parental investment and a male-biased operational sex ratio: sexual dimorphism that includes armaments and ornaments in males, as well as male–male competitive behaviors and male performances, songs, and displays directed to choosy females. Using this, we can make predictions about male and female differences when males invest more in reproduction. That is, we will expect higher levels of aggression among females, more female–female combat, females with larger body size, and females with armaments for battle and ornaments to impress choosy males. Furthermore, we will expect females to experience higher variance in reproductive success than males, with some having rather dramatic success and others having virtually no success.

5.4.1 Sex role reversal: Adjusting investments made by males and females

Most birds share parenting duties. If there is one sex that goes it alone, not surprisingly it is usually the female. However, there are super-dad cases, such as the group of marsh birds known as the jacanas (Figure 5.10). Females lay the eggs, but thereafter the male incubates the eggs and raises the young once they hatch. Meanwhile, the female, who is about 60 per cent larger than the male and is more brightly colored, defends a large nesting territory, aggressively driving away other females. She does her best to attract a harem of males to this territory, each of whom will be given his own clutch of eggs. As such, this species exhibits considerable paternal care, in contrast to those species where maternal care is typically much greater. (Don't confuse paternal care with parental care; the former refers to care by fathers, and the latter to care by parents.)

In the wetland habitats occupied by jacanas, there is a high rate of egg loss due to predation and fluctuating water levels. Theorists suggest that this role-reversed system, where females are bigger, more aggressive, more territorial, and less parental, has evolved to maximize egg production, and that of course has to be the responsibility of the female. This is a reminder that female birds are unlikely to ever be quite so free of parental investment as males can be when they only donate sperm. Eggs, with their investment of yolk, their relatively large

FIGURE 5.10 **SEX ROLE REVERSAL IS DEPENDENT UPON REPRODUCTIVE INVESTMENT LEVELS**
This African jacana (*Actophilornis africanus*) chick has hatched from an egg incubated by his father, and it will continue to be tended to by the father, which is the adult bird it is following. If the chick is a female, it will become the larger, more aggressive, more territorial sex, and if successful, she will have as many as five males in her harem.
© Lou Coetzer / naturepl.com

size, and the time it takes to produce them, are always going to be expensive to produce.

There is no strong evidence that females in sex-role reversed birds have unusual testosterone levels, but the males do have exceptionally high prolactin levels. As we'll see in Chapter 9, in humans, prolactin is associated with milk production and delivery; in birds, it is associated with incubation.

The prevailing pattern in species with reversed sex roles is this major adjustment between the sexes in terms of reproductive investment. Consider another well-known case. As with all species, in the sea horse (*Hippocampus* spp.), males produce sperm and females produce eggs. But a casual observer watching them mate could easily mistake males for females and vice versa. Breeding commonly begins with a "dance," in which they both participate. Mating culminates with the female inserting her *ovipositor* (literally, "egg-depositor"), located near the base of her tail, into the male's pouch, located on his belly. She releases her unfertilized eggs into his pouch. The male then releases sperm into his own pouch, and the sperm fertilize the eggs. (Many female insects have ovipositors but they are used to deposit *fertilized* eggs into a substrate where the eggs will hatch.)

The fertilized sea horse eggs embed in the wall of the pouch, where the developing offspring receive many benefits from the male, including nourishing secretions, immune protection, and oxygen. Although this paternal care in the absence of maternal care seems to completely reverse the roles of males and females, studies show it is more of an equalizer than a reverser; the yolk in the eggs is very costly for the female to produce, so parental investments are similar.

Although far less common, there are many cases where the father provides the primary care to the young. Among animals, the most common examples are in fish and frogs. In South America's Darwin's frogs (*Rhinoderma darwinii*), the male collects the fertilized eggs and keeps them in his mouth (actually, the vocal sac in his throat) where they develop as tadpoles, getting protection from the father and gaining nourishment from both paternal secretions and the yolk they received from their mother.

5.5 NOT ALL INTRA-SEXUAL COMPETITION INVOLVES FIGHTING

The most direct way of competing is by fighting, and the most unambiguous way to win is to kill one's opponent. In red deer, for instance, fighting is what males do to secure mating opportunities, although death is a very rare outcome. In fact, a battle is often preceded by a kind of postured standoff where the potential combatants assess each other. Sometimes at this stage one will back away, in which case the competition is won without any real contact.

In fact, studies of intra-sexual selection reveal that there are many ways to compete for mating success. In this section, we review some of those diverse ways. Although the choosing sex sometimes plays a role in these competitions—by inciting battles, as in the case of pronghorn, or in making mate choices dependent upon the outcomes of competition—in this section we restrict our examination to true intra-sexual competition.

5.5.1 Scrambles are a form of indirect competition

Where competition involves no physical contact, we say it is indirect competition. Most children who celebrate Easter are familiar with the traditional egg hunt. Here, adults (or Easter bunnies) hide painted or chocolate eggs. Children involved in the hunt are then assembled, and at a particular time the hunt begins, during which the children engage in a high-energy scramble to try to find as many eggs as possible. Hitting each other and tripping (direct competition) are not part of the system, at least if the rules are followed, but speed and searching efficiency are (indirect competition).

FIGURE 5.11 COMPETITION FOR MATES IN MANY SPECIES IS A SCRAMBLE
Some damselflies are scramble competitors, and there is strong selection among males for flight efficiency in finding and "hooking up" with females. In these azure damselflies (*Coenagrion puella*), the male (upper) has secured the female at the neck by using the claspers at the end of his abdomen. The female then brings the tip of her abdomen forward to where she picks up sperm from the male near the base of his abdomen.
iStock.com / Andyworks

In mating where there is no direct competition and where the resource (as in available females) cannot be monopolized (unlike in a harem), the contest is known as a scramble competition. Courtship in such species can be a rather an undignified affair. In damselfly mating, a male literally "hooks up" with a female (Figure 5.11) by attaching claspers at the tip of his abdomen to a spot located at the female's neck, often by coercion (see Chapter 7). In damselfly species that mate using scramble competition, males first rush to find females prior to hooking up in this manner. Whereas larger size is usually an advantage to males that truly fight, as in direct competition, it appears that smaller size can be advantageous in some scramble competitions; smaller male damselflies are more nimble fliers and are better able to succeed in mating.

Just as direct competitors have armaments that result in sexual dimorphism, as in deer antlers, scramble competitors can also have sexually dimorphic hardware that assists them in scrambling. Male mosquitoes (Family Culicidae) are scramble competitors, and finding females is their main goal during their short adult lives. Male mosquitoes find mates by using their highly sensitive *Johnston's organ*, located in the antenna. This is a sensory organ that is highly sensitive to wing beat frequencies and is in fact tuned to the wing beat frequency of conspecific females so that males can find them from some considerable distance away. Females also have Johnston's organs, but they are more highly developed in males because males rely on them for scramble competition, resulting in sexual dimorphism.

5.5.2 Endurance and subterfuge are also forms of indirect competition

When we examined the costs of sex in Section 4.5, we reviewed the miserable ending to the life of the male marsupial mouse. This is a classic case of endurance competition, another form of indirect competition. How does a male win the competition to mate in these cases? By sheer endurance—and the costs are always great and usually life-shortening.

Rhesus macaques (*Macaca mulatta*) are primates that live in multi-male, multi-female groups where promiscuity is the norm. Much copulation occurs between partners that establish consortships. These are not monogamous pair-bonds, but they are male–female associations with a high degree of interactive behavior. Males succeed best when they achieve top body condition as the mating season approaches and then by maintaining as many consortships as possible, which can include high rates of copulation. By the end of the mating season, the top males are now in the worst body condition, having lost considerable weight, but they will have fared well in the endurance rivalry by siring numerous offspring.

Subterfuge is another indirect form of competition. As we will see in Chapter 7, in some species, some males mimic females to be able to get close to potential mates. Other males known as sneakers employ the strategy of being tiny and sneaky, mating quickly when opportunity arises. We'll discuss this as well in Chapter 7.

5.5.3 Sperm competition

Telling someone you study sperm competition is likely to produce odd looks, but it has become a huge area of research over the past few decades. It has provided profound insights into male–male competition, as well as into female complicity in such contests.

Competition for access to mates commonly does not stop once a female has made her choice, because females will often mate with more than one male. Sperm competition is the continuation of the intra-sexual competition at the level of fertilization; it is a contest between the sperm of two or more males to fertilize the eggs of a single female. The driving factor in sperm competition is nothing new: male fitness is dependent upon fertilizing eggs, and if two or more males have access to those eggs, it is a significant matter to all males what the outcome is. It can be a significant matter to the females too, as we will see in Chapters 6 and 7. For both sexes, it can be the difference between getting one's alleles into the next generation, or not.

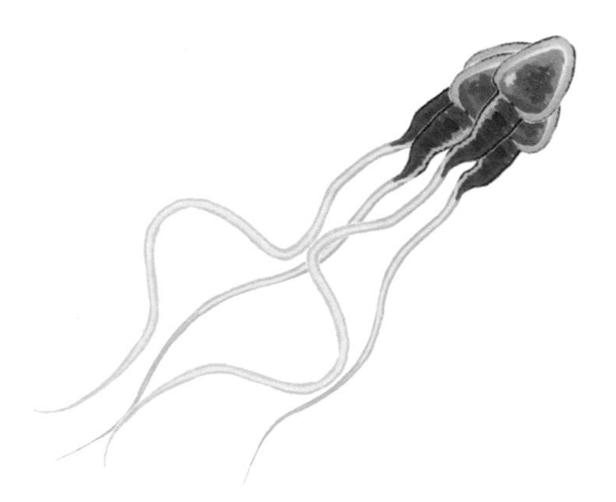

FIGURE 5.12 **SPERM MOVE MORE EFFICIENTLY BY AGGREGATING**
Sperm in aggregations like the mouse sperm shown here swim more quickly and reach the eggs sooner. In promiscuous species, individual sperm aggregate with others from the same male, but when semen is experimentally mixed in monogamous species, the individual sperm aggregate indiscriminately.
Drawing by Peter Mills

Sperm competition can end up amounting to a race to the egg or eggs. It's been known for some time that sperm within the ejaculates of many species aggregate together, and in some cases the aggregates swim faster than do sperm alone. In rats and mice, the head of the sperm are even equipped with hooks that are used to stabilize these clusters (Figure 5.12). This apparent cooperation among these sperm sounds like the antithesis of competition, but a closer look proves otherwise. Deer mice (*Peromyscus maniculatus*) are a promiscuous species where females commonly mate with more than one male and where genetic analysis

BOX 5.3 THE HUMAN MIND AS A SEXUALLY SELECTED TRAIT

The conventional explanation for the evolution of the complex human brain is that it increasingly allowed us to solve problems in challenging and changing landscapes, leading to increased survival. With this large problem-solving brain came other spinoff benefits: language for storytelling, appreciation of beauty, athletic prowess as a form of display, music, dance, a sense of humor, and even moral behavior.

In 2000, psychologist Geoffrey Miller published the provocative book *The Mating Mind: How Sexual Choice Shaped the Evolution of Human Nature*. He challenged the view that these non-survival attributes of humans were mere by-products of a brain that was shaped by evolutionary forces for its survival value. Instead, Miller argued that these human attributes were not spinoffs but were features that had been selected, not through "natural" selection but through sexual selection (Figure 5.13). In other words, these non-survival attributes of the brain have given their owners a mating advantage over millennia, with the result that they have become common attributes in the population.

FIGURE 5.13 **SOME ATTRIBUTES OF THE HUMAN MIND ARE LIKELY SEXUALLY SELECTED**
Some argue that biologically "useless" behaviors like athletics, music creation, and other art forms are not useless under the rules of sexual selection. Instead, they are interpreted as adaptations that influence mate choice, honed by selection over countless generations in the human past.

Ornaments are more often exhibited in males than in females, but this is due to the common scenario where females invest more in parenting and are therefore choosier (Figure 5.4). In humans, where there is substantial parental investment on both sides and where brain attributes are apt to show up in the brains of both sexes, Miller did not argue that creativity, athletic prowess, humor, moral refinement, and other such traits were features of men rather than women. On the contrary, these attributes are found in both sexes.

This idea has strong appeal. After all, we've spent this chapter demonstrating that the sexual values of some traits are as important or more important than the survival values of other traits. "Useless" features like splashes of color and unwieldy tails are of profound sexual importance. Similarly, "useless" human attributes, like the ability to capture nuanced ideas in a poem, may similarly have profound sexual importance. It is well recognized that beautiful human bodies are sexually attractive, but it is also easy to demonstrate from survey data that humans acknowledge that athletic ability, a sense of humor, moral goodness, and many other features unrelated to issues of survival are also reported to be attractive or even erotic features in a potential mate.

shows that litters can be sired by multiple fathers. Experiments show that when the semen of two males is mixed, the sperm preferentially cluster with sperm from the same male. In contrast, in monogamous species like beach mice (*Peromyscus polionotus*), where sperm competition is not expected, when the semen of two males is mixed experimentally, the sperm cluster indiscriminately.

Most theorists would consider human courtship, in its almost infinite variety, to be a manifestation of sexual selection. Human courtship primarily operates inter-sexually, but perhaps also intra-sexually, at least indirectly. And, in cases of long-term monogamy, which is common in human societies, we might expect sexually selected traits to show up in both sexes. For a novel idea about the evolution of the human cognitive capacity, see Box 5.3.

CHAPTER 5 SUMMARY

- Sexual selection is a subset of natural selection where the forces of selection are members of the same species and where the context is mating.
- Intra-sexual selection influences reproductive success through competition between individuals (usually males) for access to mates, favoring competitive adaptations.
- Inter-sexual selection influences reproductive success through the exercise of mate choice, usually by females, favoring attractive adaptations.
- Females commonly commit greater investments in producing young, meaning that females are more likely to maximize reproductive success by exercising mate choice.
- Males commonly make lower investments in producing young, meaning that males are more likely to maximize reproductive success by competing for access to as many mates as possible.
- There are many examples from a diversity of animals that demonstrate greater reproductive success for individuals succeeding in intra-sexual competition or exercising mate choice.
- In many species, females choose males on the basis of ornaments such as bright colors, physical embellishments like long tails, and showy displays.

- Ornaments upon which mate choice is exercised often are honest indicators of male quality, such as superior genes, more suitable genes, or strong immunity.
- Benefits flowing from sexual selection mechanisms include superior genes in offspring, as well as resources that improve offspring success.
- Sex role reversal, in which females are the more competitive sex and males are more likely to exercise mate choice, occurs in cases where reproductive investments of males are greater than those of females.
- Not all inter-sexual competition is direct; competition can be indirect through endurance, through a scramble competition, or by sperm competition within the female reproductive tract.

FURTHER READING

Andersson, M. (1982). Female choice selects for extreme tail length in a widowbird. *Nature* 299, 818–820.

Chaix, R., Cao, C., & Donnelly, P. (2008). Is mate choice in humans MHC-dependent? *PLOS Genetics, 4*(9), e100184. https://doi.org/10.1371/journal.pgen.1000184

Darwin, C. (1871). *The descent of man and selection in relation to sex.* London: John Murray.

Gibson, R.M., & Langen, T.A. (1996). How do animals choose their mates? *Trends in Ecology & Evolution, 11*(11), 468–470. https://doi.org/10.1016/0169-5347(96)10050-1

Hamilton, W.D., & Zuk, M. (1982). Heritable true fitness and bright birds: A role for parasites? *Science, 218*(4570), 384–387. https://doi.org/10.1126/science.7123238

Marlowe, F.W., & Berbesque, J.C. (2012). The human operational sex ratio: Effects of marriage, concealed ovulation, and menopause on mate competition. *Journal of Human Evolution, 63*(6), 834–842. https://doi.org/10.1016/j.jhevol.2012.09.004

Miller, G.F. (2001). *The mating mind: How sexual choice shaped the evolution of human nature.* New York: Anchor Books.

Szykman, M., Van Horn, R.C., Engh, A.L., Boydston, E.E., & Holekamp, K.E. (2007). Courtship and mating in free-living spotted hyenas. *Behavior, 144*(7), 815–846. https://doi.org/10.1163/156853907781476418

Trivers, R.L. (1972). Parental investment and sexual selection. In B. Campbell (Ed.), *Sexual Selection and the Descent of Man, 1871–1971* (pp. 136–179). Chicago: Aldine-Atherton.

Zuk, M. (2002). *Sexual selections: What we can and can't learn about sex from animals.* Berkeley: University of California Press.

6 Mating Systems

KEY THEMES

Mating systems are ways groups are organized with respect to sexual and reproductive behaviors. Some systems are promiscuous, involving multiple partners and no bonds. Monogamous systems involve pairings between one male and one female, although there is commonly extra-pair mating. Polygamy is distinguished by multiple partners for at least one sex, and sometimes both, with social bonds. Males reap benefits from multiple partners mostly by producing more offspring, but females reap benefits from multiple partners in various ways, including additional resources and genetic diversity in offspring.

As a reflection of Western culture, the early years of television endorsed a norm—that the natural family (perhaps even the natural state of being) was a unit containing a dad, a mom, and children, the so-called *nuclear* family. Despite a rapidly changing view of what constitutes a family, this view with a monogamous male–female pairing remains the prevailing model, entrenched in socially traditional institutions, including mainstream religions. Yet anthropologists and social historians remind us that the 1950s norm of the human family in Western societies, including the mating arrangements within it, has been and continues to be far from universal.

A *mating system* is the way a sexually reproducing population is structured with respect to who mates with whom, and it ranges from complete monogamy to unrestrained promiscuity, with a diversity of non-monogamous systems in between (Figure 6.1). Studies of mating systems in nature show that the system that prevails in a particular species is highly influenced by sexual selection (Chapter 5). Analysis of mating systems cannot be adequately conducted without reference to parental investment, variance in reproductive success, and inter-sexual and intra-sexual selection. It has also been noted that the prevailing mating system for any species, including humans, can change from place to place, as differences in resource availability and population density vary among those places.

Over past centuries, the mating system formalized in Western societies through marriage has been one of committed, monogamous, heterosexual pairings. The reality has changed dramatically in recent decades, first with the liberalization of commitment and more recently with the inclusion of same-sex couples, yet two-person arrangements remain the socially endorsed system. It is easy to see that when human mating systems are formalized differently in different cultures, that contrast can be the source of cross-cultural tension. Likewise, when norms around mating systems shift within a society, it can cause controversy. This is especially so in dynamic societies like ours that seek to accord equal rights to males and females, including in matters related to family structure and sexual freedom. In this chapter, we consider the biology of mating systems. As we consider humans along the way, we do our best to distinguish the biology from the culture,

FIGURE 6.1 **MATING SYSTEMS REFLECT THE STRUCTURE OF SEXUAL RELATIONSHIPS**
Polygyny, characterized by one male and multiple females, is a very common mating system in the animal kingdom.

keeping in mind that biology is not necessarily a prescription for the way things have to be.

6.1 THERE ARE FIVE MAJOR TYPES OF MATING SYSTEMS

There are five major mating system types (Figure 6.2): monogamy, polygyny, polyandry, polygynandry, and promiscuity. The first distinction among them involves the concept of the *social bond*. Mating systems that include social bonds are those where there is some degree of relationship among the members of the mating group, whether it is two individuals or several. This means that the relationship between two mating individuals lasts longer than the act of copulation (or the equivalent of copulation in organisms that use other systems, such as external fertilization) and the brief lead-up to that act. Promiscuity is distinguished from other mating systems by involving sexual reproduction that occurs without social bonds. It usually involves multiple sexual encounters with multiple partners, although to say such pairings are random would be incorrect; choosiness can still play a role in promiscuous systems. As you might expect, promiscuous mating systems do not involve partners that share parenting duties. There is either no parental care or the care provided is *uniparental care*, which is usually though not always maternal.

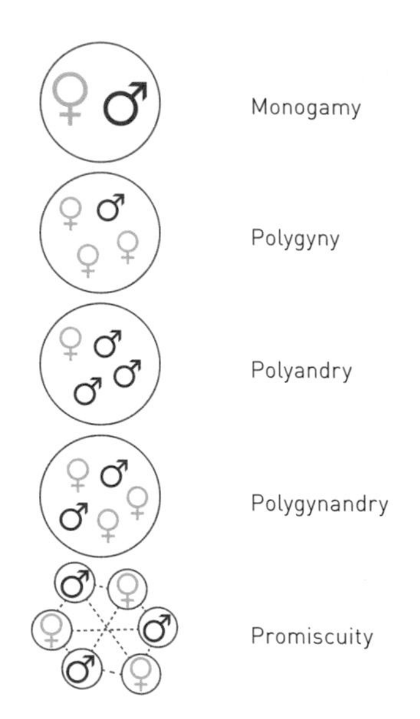

FIGURE 6.2 **THERE ARE FIVE MAJOR MATING SYSTEM TYPES**
Polygyny, polyandry, and polygynandry are all forms of polygamy. Promiscuity is characterized by having no bonds formed outside the mating event.

In contrast is monogamy ("one partner"), the pairing of one female with one male. Obviously, all fertilizations involve a single male and a single female, but for mating to constitute monogamy there must also be a pair-bond that exists for more than just the duration of copulation. When we consider monogamy in detail we will distinguish among different types: genetic versus social monogamy, and serial versus perennial monogamy.

Polygamy ("multiple partners") describes the remaining three types of mating system—polygyny, polyandry, and polygynandry. These are mating groups of one or more members of one sex with multiple members of the opposite sex. As with monogamy, this category requires social bonds that last longer than the duration of mating events. Sometimes, polygamy is incorrectly treated as synonymous with polygyny ("multiple females"), but this is only the most common form of polygamy in nature. Polygyny is a very common mating system among mammals, as we will see, but it is not the only polygamous system. Polyandry ("multiple males") occurs where the group consists of one female and multiple males; African jacanas (Figure 5.10) employ such a polyandrous mating system. Polygynandry ("multiple females and males") is characterized by social bonds among two or more males and two or more females. The numbers of each sex do not have to be equal, and in fact usually there are more females in polygynandrous breeding groups than males.

Cross-culturally now and in the past, the majority of human societies have permitted polygynous relationships, where one male has more than one female partner with whom he regularly has sexual intercourse. It is also true, however, that most mating among humans, even in polygyny-permissive societies, occurs within monogamous couplings. Polyandrous mating systems are much rarer than polygynous ones, both in nature and in human societies; however, this does not necessarily mean that it is rare for females—humans or otherwise—to have more than one partner outside the structure of mating systems, as we will see. This is because although it is almost universal in nature to have structured mating systems, individuals will often seek to optimize their reproductive opportunities by mating outside the social partnerships they have established.

6.2 MONOGAMY

Monogamy is the norm for most individuals in developed human societies, although many of those societies endorse polygyny in one way or another for those who can afford it, and mate fidelity is far from perfect. In some societies, monogamy is also a religious imperative. Yet it is certainly not the prevalent mating system in a survey of sexually reproducing organisms. Estimates of the numbers of mammal species that are truly monogamous range around 2 to 5 per cent, and even among our own primate group the figure is only around 14 per cent. Humans are most often considered polygynous when all societies that have been studied are taken into account, although in those societies, it is only the relatively powerful males that have polygynous breeding groups, and the majority of "ordinary" males have monogamous relationships. Many of these males would prefer to have more wives if they could afford it, and the monogamous relationships are often only socially monogamous—not sexually monogamous in the sense of fidelity.

6.2.1 Distinguishing among types of monogamy

Whereas monogamy is relatively rare among mammals, it is quite common among birds. The stock image of a father and mother bird attending to young in a nest reflects the reality of mating systems among many types of birds. However, among most songbird species that have been researched, mate fidelity is not strongly developed. Mating that occurs outside the pair-bond is known as extra-pair copulation (EPC) or extra-pair mating (EPM) (analogous to extramarital sex in humans). EPC sometimes results in extra-pair fertilization (EPF), leading to extra-pair paternity (EPP), where the offspring are labeled extra-pair young (EPY)—five acronyms, but one concept.

We consider how this pattern among birds came to be known in Box 6.1, but for the moment just consider the pervasiveness of the pattern. One study reviewed 29 species of apparently monogamous European and North American songbirds and reported that among all nests, 39 per cent contained EPY, and that among all offspring, 23 per cent were sired by a male that was not the "father" attending the nest. The list includes many familiar monogamous species, such as the northern cardinal (*Cardinalis cardinalis*). This system of social monogamy involves a pair-bond between one male and one female, but mating is not exclusively restricted to those two individuals. It is different from *sexual monogamy*, which is characterized by the pair-bond and the absence of mating with other individuals outside that pair. More often, instead of sexual monogamy, biologists use the term genetic monogamy, which is sexual monogamy where there is genetic proof that a female's offspring are sired by her male partner. Social monogamy is so widespread that any monogamous songbird found in the temperate zone that does not have extra-pair mating in its behavioral repertoire is a conundrum worth investigating.

A minority of birds do appear to be genetically monogamous; puffins (*Fratercula arctica*, Figure 6.3), for instance, appear to be faithful mates, notwithstanding that they often breed in dense colonies. Among primates, Azara's owl monkeys (*Aotus azarae*), found in South American rainforests, appear to be genetically monogamous, as do North American coyotes (*Canis latrans*)—at least the urban coyotes that the researchers focused on. Notably, genetically monogamous mammals exhibit biparental care, where males are also attentive parents.

FIGURE 6.3 A MINORITY OF BIRDS ARE GENETICALLY MONOGAMOUS
Most birds with biparental care are monogamous, but only a minority, like the Atlantic puffins (*Fratercula arctica*) shown here, are evidently truly genetically monogamous. Most North American songbirds, for instance, exhibit high rates of extra-pair copulation.
iStock.com / jacquesvandinteren

One other sub-category of monogamy is serial monogamy, a term that is applied to humans as well as other species. Serial monogamy can co-exist with either social or genetic monogamy. This involves a monogamous pair-bond that is not permanent, in which case a member of the pair forms a new monogamous pair-bond after the first pair-bond dissolves. Many migratory birds are serial monogamists, having one partner during one breeding season and a different partner the next year. The other sub-category of monogamy is perennial monogamy, or mating for life, where pairs form "permanent" monogamous bonds. Note that within perennially monogamous species, there can still be high rates of extra-pair paternity. At one time, Australia's black swan (*Cygnus atratus*) was presented as the ideal in monogamy, because pairs stay together for years. While this is true, it is now known that about one in six black swan cygnets (offspring) has extra-pair paternity.

BOX 6.1 TESTING FOR EXTRA-PAIR PATERNITY IN BIRDS

Theory predicted that there could be benefits for both males and females if they sometimes mated with individuals outside the monogamous pair-bond, but before modern day genetic techniques became available, this was difficult to demonstrate. When these techniques were first applied to the issue in the 1980s, however, even theorists were surprised by the rates of extra-pair paternity they were finding, especially in monogamous birds.

Working in conjunction with ecologists, geneticists compared the DNA of mothers, fathers, and offspring, applying the same techniques that came to be used as evidence in paternity suits in family courts. Using bacterial "DNA scissors" known as *restriction enzymes*, researchers cut up DNA samples from different individuals. These enzymes only cut the DNA at specific places characterized by particular sequences, so that each individual's DNA ends up in a unique number of pieces of various sizes. Then, the DNA pieces are put in a gel through which an electric current is passed, and since the DNA has an electric charge, the DNA moves through the gel. Larger pieces move more slowly than smaller pieces, meaning that larger pieces will have moved a shorter distance in the gel when the current is turned off. The resultant pattern is called a *Southern blot*, and the process is known as *gel electrophoresis*.

For a given stretch of DNA, each individual will have a unique pattern on the Southern blot (Figure 6.4) based on the number of pieces and the sizes of those pieces. Offspring will have unique patterns too, of course, but the pieces that make up their gel signature should match either the pieces from the mother or the father—approximately 50 per cent from each.

6.2.2 Hypotheses for monogamy

It is worth mentioning that in some cases males can only mate once, as in western honeybees (*Apis mellifera*). Mating occurs in flight, and the males, known as drones, compete for the attention of the queen. A drone lucky enough to get an opportunity to mate dies right afterward, since the act of sperm delivery includes the ripping of his intromittent apparatus from his body. Honeybees are not a monogamous species, however; the polyandrous queen mates with as many as a dozen drones during mating flights (widowed after each encounter!), storing sperm from multiple individuals for future use.

Polygyny is sometimes thought of as the default system among sexually reproducing organisms, at least from the point of view of the male whose fitness, all else being equal, rises with each partner he can

If a piece in an offspring does not have a match in the mother's Southern blot, then it should be found in the Southern blot of the putative father. If not, then this offspring has extra-pair paternity, meaning it was sired by a male other than the putative father. More recently, with further improvements in technology and declining costs, DNA among putative parents and offspring can be exactly sequenced (Section 3.6.4), and then compared.

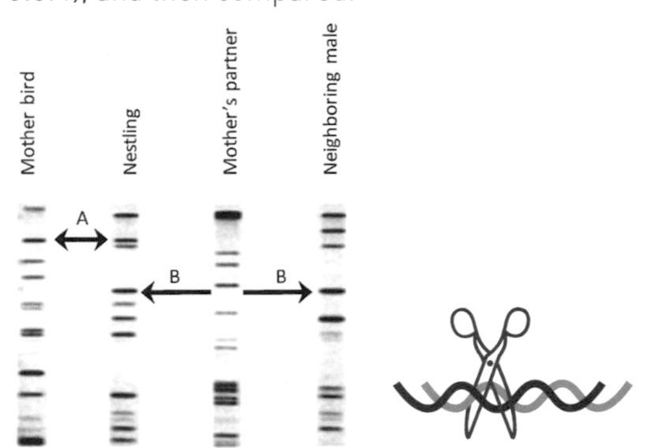

FIGURE 6.4 **DNA ANALYSIS CAN DETECT CASES OF EXTRA-PAIR MATING**

This is a *Southern blot* resulting from *electrophoresis* of DNA from four individuals of the same bird species. Each band represents a piece of DNA that has been cut by a kind of molecular scissors known as a *restriction enzyme*. Because of the uniqueness of individuals' DNA sequences, these scissors cut the DNA of an individual into pieces of unique size that end up in specific locations on the blot. The mother bird is the female in attendance at the nest. The nestling is one of the young birds in the nest. The mother's partner is the male in attendance at the nest, and since these birds are socially monogamous, he is the putative father. The neighboring male is a male of a monogamous pair with its own nest nearby in the ecosystem. By comparing the patterns, it is clear that the nestling is indeed the offspring of the mother (see band at arrow A for example). But it is also clear that the nestling was sired by the neighboring male (see band at arrow B for example) and not by the putative father.

mate with. If a male can control more than one female—either by directly controlling females who for one reason or another are spatially clumped or by defending rich resources that females need for reproducing—polygyny is the likely outcome. Yet there are many species that have monogamous mating systems, and consequently the hypotheses advanced to explain monogamy always include some component that argues that monogamous mating systems actually yield higher fitness than polygynous ones.

The most persuasive underlying cause for monogamy is that biparental care is, if not necessary for successful reproduction, highly advantageous over single parenthood. This is known as the mate assistance hypothesis for monogamy. Although a male may always achieve more fertilizations by mating with multiple partners, if those

FIGURE 6.5 TWO PARENTS ARE OFTEN BETTER THAN ONE
In starlings (*Sturnus vulgaris*), nests where both parents contribute to egg incubation (light-shaded columns on the right) have warmer incubation temperatures, shorter incubation periods, higher hatch rates, larger hatchlings, and higher rates of fledging (i.e., successfully leaving the nest) than nests where the mother alone is the incubator (darker columns on the left).
[6.5a] Adapted from Reid, Monaghan, & Ruxton (2002)

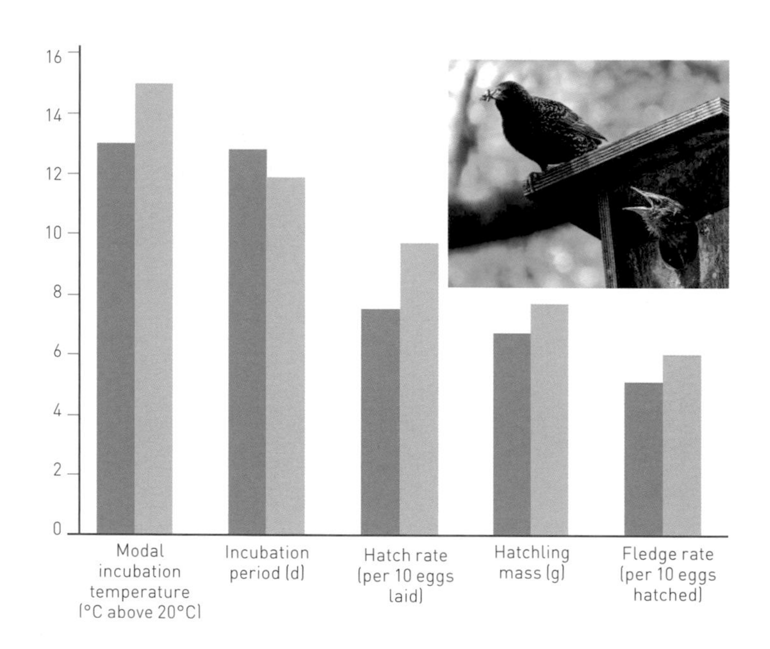

fertilizations do not produce as many successful offspring as when he commits to a monogamous partnership, then his fitness will be lower. So, monogamy prevails.

What types of care do males provide that increase the reproductive success of monogamous enterprises? Food is an obvious one for carnivorous species like red foxes (*Vulpes vulpes*); it is much less common in herbivorous species. The warm-bloodedness of birds and mammals means that male care can include warmth, as in California mice (*Peromyscus californicus*); nests where the male has been removed produce significantly fewer young. Experiments reveal that when egg incubation is a shared responsibility of both the male and female bird, the eggs are rarely cooled and hatching success is therefore much higher than when incubation is the sole duty of the mother bird (Figure 6.5). Other species of birds are able to overlap two nests, with the male assuming responsibility for the young from the first nest while the female proceeds to look after the eggs from the second.

A common contribution by males to paternal care is *protection*, either from predators or from competing males that attempt to increase the likelihood of mating with the female by killing her off-

FIGURE 6.6 **MALES WILL OFTEN GUARD FEMALES THEY HAVE MATED WITH**
While the female green darner (*Anax junius*, right) lays her eggs in the water, the male (left) guards her by retaining the mating hold on her using the claspers at the tip of his abdomen, which connect to a spot behind her eyes. This is a form of mate guarding known as "contact guarding."
Tim Zurowski / Shutterstock.com

spring. (We will return to the subject of infanticide in Chapter 7.) Among prosimian primates (the group of primates with the oldest fossil record), females of those species that travel with their young are usually accompanied by the father, in contrast to those females that leave their young in a protected nest, who travel alone.

There are other monogamy hypotheses proposed as alternative or contributing forces. In the mate guarding hypothesis, males must remain with the female to guard against the real possibility that she will otherwise seek or engage in extra-pair copulations for her own advantage (Figure 6.6). In the female-enforced hypothesis, females enforce monogamy in one of two ways. If males don't remain, females in some species abandon the reproductive enterprise (by abandoning either the eggs or the young). In other species, females harass their partners when the male pursues other females. There is widespread evidence for mate guarding among animals, and female enforcement also has some evidentiary support.

Finally, there is the celebrity couple hypothesis. This rests upon the foundation that there are superior individuals, both genetically and in health. If the male is going to be a particularly good provider and the female is going to be particularly fecund, once they pair up they are unlikely to do better by seeking matings with other, presumably inferior, individuals.

6.3 POLYGYNY IS THE MOST COMMON FORM OF POLYGAMY

Polygyny, polyandry, and polygynandry differ from monogamy by involving more than two individuals in the mating group, and they differ from promiscuity by having bonds among the mating individuals that last longer than the mere duration of the mating event itself. In this section, we will not include social monogamy with significant extra-pair mating. However, it is good to be mindful of social monogamy when reading this section, since many of the advantages that are demonstrable in overt polygamy are similar to the reproductive advantages that ensue through the covert "cheating" that exists in socially monogamous species.

Under what circumstances do we find that polygamy prevails as a mating system? There are two answers: when it is advantageous for both sexes, or when it is advantageous for one sex to have multiple partners *and* circumstances are such that that sex can commandeer either resources or individuals of the opposite sex.

6.3.1 Two models for polygyny

Where females are clustered in space, the likelihood of polygyny is high. Females may cluster because they are related, they may cluster to pool their resources against predators, or they may cluster because they are cooperative hunters, as in lionesses. The likelihood of polygyny is high when females cluster because they represent a rich opportunity for males: access to mating with a group of females has such high potential reproductive rewards that males are likely to compete strongly for that access, resulting in polygyny. A male that can exclude other males from this profitable mating is said to be employing

FIGURE 6.7 **SOME FORMS OF POLYGYNY RELY ON DEFENDING FEMALES**

A gorilla (*Gorilla gorilla*) troop is typically a female-defense polygynous system, with a silverback male (left) and several breeding females, as well as young.

Rolf Nussbaumer Photography / Alamy Stock Photo

female-defense polygyny. The mating system of gorillas (*Gorilla gorilla*) is best described this way, with one silverback male as the focal individual, three or more breeding females, and various young (Figure 6.7).

Even where females are not normally clustered in space, or where it is difficult for males to control female movements, polygyny can be very common if a male can control a limited resource needed by females. This is known as resource-defense polygyny. Again, if this is the case, males are highly likely to compete for control of that resource. Typical examples are male control of a birthing beach in elephant seals (*Mirounga angustirostris*), male control of safe roost sites in Jamaican fruit bats (*Artibeus jamaicensis*), and male control of nectar-producing flowers in the case of purple-throated carib hummingbirds (*Eulampis jugularis*).

The patterns can look similar, since male control of the limited resource required by females tends to cluster those females. Consequently, researchers acknowledge that sometimes it is difficult to distinguish whether it is primarily defense of females or defense of resources that is driving the pattern. Also, it can shift from one to the other, depending upon seasonal changes in the patterns of key resources, depending upon the age of the males, and depending upon the numbers of females that are available for mating. In any event, polygyny is not just driven by male coercion. Sometimes at least, females are complicit, as demonstrated theoretically in the polygyny threshold model (Box 6.2).

BOX 6.2 THE POLYGYNY THRESHOLD MODEL

There are many species, including humans, where some individuals mate in monogamous relationships and other individuals mate in polygynous ones. If a male provides resources or parenting, why would a female ever opt for polygyny over monogamy? In the context of resource defense, the answer is provided by something known as the polygyny threshold model. That is, there is a threshold, on one side of which a female will choose monogamy and on the other side of which she will choose polygyny.

The polygyny threshold model is shown in Figure 6.8. It may look more complicated than it is. It merely represents choices available to a female as she sets out to breed. The y-axis is her expected reproductive success and the x-axis is the quality of the territories controlled by individual males among which she can choose. The two curving lines represent a monogamous option (choice #1) and a polygynous option (choice #2). Not surprisingly the two lines on the graph go upward to the right, meaning that female reproductive success (y-axis) increases as territory quality (x-axis) improves. Also not surprisingly, the monogamous option is the higher line, meaning, all else being equal, her reproductive success is more likely to be higher if she does not have to share the male. But, not all territories are equal, and sometimes her better choice is polygyny.

If there are two territories, an inferior one and a superior one shown by the two vertical lines, the polygyny threshold is represented by the horizontal line at (b). Here, her choices are the same (the threshold), because at line (b) her expected reproductive success is the same for the inferior territory in monogamy and the superior territory in polygyny. At that point, whatever benefits she gets from being in the superior territory is offset by having to share that male's contributions with another female. If she could find an unmated male with a superior territory, that would be her best choice, because then she could have the reproductive success shown by horizontal line (a). Certainly, she will not want polygyny in an inferior territory because that will yield lower reproductive success, i.e., at line (c). In this way, one can see that in some cases, a female will elect polygyny over monogamy if the resources that come with that polygyny more than offset the costs the female will suffer by sharing the male with another female.

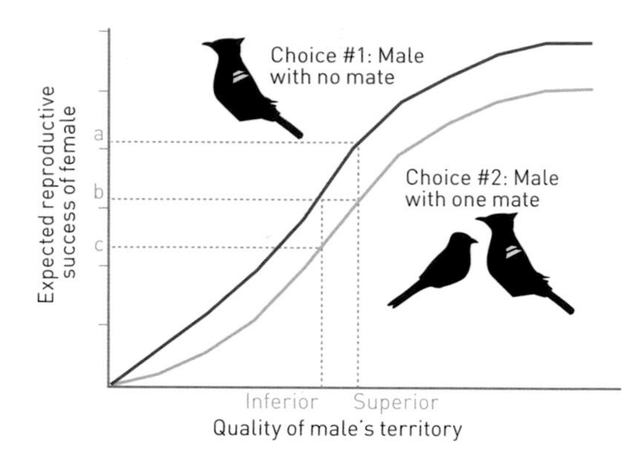

FIGURE 6.8 **POLYGYNY CAN BE A BETTER CHOICE FOR FEMALES**

The polygyny threshold model demonstrates how an already-mated male may be a better choice for a female than an unmated male. The threshold is shown at point (b) on the y-axis, where the combination of inferior territory and monogamy (choice #1) produce the same reproductive success for the female as the superior territory and polygyny (choice #2).

FIGURE 6.9 **HUMAN POLYGYNY IS WIDESPREAD BUT MOSTLY RESERVED FOR POWERFUL MALES**
A polygynous Egyptian family ca. 1890, showing a man with three wives and two servants.

6.3.2 Human polygyny

Although polygamy as a form of marriage is not legal in most nations, polygyny is widely practiced in many countries where it is either legal or not criminalized, especially in Africa, the Middle East, and parts of Southeast Asia. In addition, historical analyses of human mating systems reveal that polygyny has been a widespread system throughout history (Figure 6.9) and that its current illegal status in most countries is largely grounded in intolerance of the practice by major religions. An ambitious and relatively complete review of human societies, past and present, revealed that 588 were polygynous and only four were polyandrous. The number of monogamous societies was 639, but in 71 per cent of those, there was occasional polygyny.

Indeed, many men through history have had large harems of reproductive-age women, regardless of whether most men in those societies were in monogamous marriages. In some instances, membership in these harems involved marriage; in others, it did not. Without exception, however, the harem-holding men were relatively powerful and wealthy—a pattern that continues today. It is therefore tempting to interpret human polygyny in the context of resource defense. Even in cultures where polygyny is illegal, as in the primary cultures of Europe and North America, wealthy men in formal monogamous relationships have long had a history of having multiple sexual partners—mistresses, concubines, and courtesans—who were social subordinates whom they could nonetheless support financially. And of course, prostitution in all places is dominated by male consumers who engage female prostitutes.

Human polygyny has been subjected to considerable academic analysis. One large, recent study considered international diversity in degrees of polygyny and then considered the "ecological" traits of those cultures where polygyny was more common. These researchers derived a *human polygyny index* (HPI), which is determined by calculating the ratio of the variability in the number of sexual intercourse partners males have annually to the variability in the number of sexual intercourse partners that females have annually: i.e., (male heterosexual partners) ÷ (female heterosexual partners). To make the numbers comparable, the researchers used a common measure of variability known as the *standard deviation*. So, the HPI ratio compared the standard deviations of males and females in terms of annual numbers of sexual intercourse partners. An HPI of greater than 1.0 indicates polygyny: males in the population experience more variability than females, meaning that males are more likely to have either fewer, or more, sexual intercourse partners than females.

For the 48 countries studied, the mean HPI was 1.90, revealing that overall, modern human societies continue to have a polygynous sexual structure, notwithstanding the widespread social endorsement of monogamy. Generally, a higher HPI is associated with greater income inequality between males and females, greater acceptance of social inequality, younger age at first marriage for females, and greater fertility rates (births per woman). In addition, higher HPI values are associated with lower life expectancy for both sexes overall, with scarcity of resources, and with higher levels of pathogen stress. This last correlate, pathogen stress, may be linked to either good genes or good resources, because both can counter susceptibility to disease.

Another line of evidence that supports the claim that human societies are more likely to be polygynous than polyandrous, especially historically, is analysis of worldwide allele patterns on Y-chromosomes, which only males have. When genes are sampled from all continents and the diversity of DNA sequences are examined on the Y-chromosome and compared to DNA sequences on other chromosomes, the most plausible explanation for the pattern is that humans have been polygynous for much of their history, with a more recent shift toward monogamy.

6.4 THE BENEFITS FOR FEMALES OF MATING WITH MULTIPLE MALES

In monogamy and polygyny, females tend to mate with one male. In the three systems we have yet to consider more fully—polyandry, polygynandry, and promiscuity—females typically mate with more than one male. However, we've also seen that in many monogamous systems (i.e., socially monogamous systems), there is considerable extra-pair mating that occurs. Researchers have also found that a similar pattern can occur in polygyny—mating by both males and females outside the polygynous group, known as *extra-group copulation* (EGC). In short, mating with multiple partners in nature, even of a promiscuous nature (i.e., without any pair-bonds), is far more common for both males and females than strict adherence to the "rules" of a mating system.

The fact that females are inclined to mate with multiple males was hard for biologists of earlier generations to accept, because the common view was that males but not females were the party that pursued sexual relations and because there seemed to be no biological advantage for females to mate with multiple partners. With a liberalization of perspective and advances in research methods, human females have been added to this group, since non-monogamous behavior has proved to be more common among human females than previously realized. This is true even before the great social changes beginning in the 1960s that were bolstered by women's liberation and by improved and more available methods of contraception. This has been a profound recognition, given how deeply embedded the social rules of female chastity had been—now at least partly eclipsed by more sexually liberal values.

With the development of genetic techniques like those featured in Box 6.1, biologists who study mating systems experienced a paradigm shift as it became clear that true monogamy was in fact so much less common than had been believed. With that realization, biologists asked,

FIGURE 6.10 **SOCIAL STATUS CAN BE A COMPONENT OF EXTRA-PAIR MATING**
Black-capped chickadees exist in socially complex flocks during the non-breeding season. During the breeding season, females of monogamous pairs will sometimes leave the pair's territory to solicit matings with other males; in one study, such males always had higher status in non-breeding flocks than her own mate. Studies of songbirds show that behaviors that seek such extra-pair mating are usually secretive.
Steve Byland / Shutterstock.com

how complicit are females in these patterns? We've already examined the widespread prevalence of extra-pair mating among temperate zone songbirds. We also now know that either sex can initiate these trysts; in one study investigating the familiar North American black-capped chickadee (*Poecile atricapillus*, Figure 6.10), females were the more frequent initiators, leaving their own territory to mate nearby with another male—who was *always* of higher status than her own partner, based on winter flock structure. Studies of other groups, including insects, fish, reptiles, and mammals, also show that female non-monogamy is widespread and is commonly initiated by females. So, before resuming our review of mating systems, we first must address what the benefits of non-monogamy are for females.

Choosing the best term for these female behaviors is a challenge. Some researchers apply the label *female promiscuity* to describe these behaviors, but we have already employed the term promiscuity in its more customary and limited sense for cases where organisms do not establish pair-bonds. Others describe these female behaviors as *polyandrous,* but again we have used polyandry in its more particular and customary mating system sense. Consequently, as we consider female strategies that involve mating with multiple males regardless of the particular mating system, we will refer to them as *non-monogamous,* and this includes cases of social monogamy.

6.4.1 Non-monogamous females hedge their bets against infertility

The fertility insurance hypothesis is an explanation of female non-monogamy that depends upon the real risk to females of relying upon a partner who is either *azoospermic* (sterile on account of having no sperm) or who, at least temporarily, suffers from sperm depletion, presumably by having mated recently with one or more other females. There is considerable evidence that this is a real motivator of non-monogamy among many diverse species. Fallow deer (*Dama dama*) employ a polygynous mating system, like the red deer featured in Chapter 5, where many males do not breed but several males in their prime breed with most of the females. Females usually produce just one young per year, and most females mate with just one male. However, some mate more than once, especially later in the breeding rut or when having mated an old male, in both of which cases sperm depletion is a real risk.

Sperm depletion has also been documented in several reptile species. Red-sided garter snakes (*Thamnophis sirtalis*) are an interesting study species for this question because the males produce sperm in the fall, store them over winter, and then use this finite supply as the snakes leave the hibernacula in the spring. One study that vasectomized some of the male sample found that females that received spermless semen re-mated significantly more often than the control group. The mechanism allowing female detection of the absence of sperm is unknown in reptiles, but there are many female invertebrates that can also sense if they have received sperm, influencing their decisions about re-mating. Experiments using wild songbirds that prevented the social partners of females from delivering sperm when they mated (by experimentally restricting the flow of sperm delivered) showed that females had fertile eggs nonetheless—entirely fertilized by neighboring males.

Having multiple male partners can also be worth it for species that reproduce using external fertilization. For instance, in the African foam-nesting frog (*Chiromantis xerampelina*), simultaneous matings involving from 1 to 12 males reveal that fertilization success from the female's perspective is positively correlated with the number of males. Female salamanders are known to mate with a second male for the purpose of filling their sperm storage devices, a system known as *topping up.*

6.4.2 Non-monogamous females reap genetic benefits

The value to females suggested by the fertility insurance hypothesis can be assessed quantitatively: how many more offspring does she produce when she mates non-monogamously? In contrast, hypotheses related to genetic benefits are assessed more qualitatively. Admittedly, genetically superior individuals may survive and breed better, which ends up being a quantitative success, but that is more challenging to determine. Regardless, recall that genetic benefits are known as indirect benefits, as opposed to direct or material benefits, like food or protection.

In Chapter 5 we examined the concept of mate choice and its central role in sexual selection, usually in the context of a female making choices about which male with whom she will mate. That inquiry was essentially focused on the selection of a single mate. We expand this discussion and consider female mate choice with respect to multiple mates. Both involve choice. Not surprisingly, similar ideas are invoked in both cases, and consequently there is some overlap in terminology and analysis. Here, we will use the term "model" rather than "hypothesis"—simply to distinguish the application of these ideas to the related concepts in sexual selection in Chapter 5 and non-monogamous mating in this chapter.

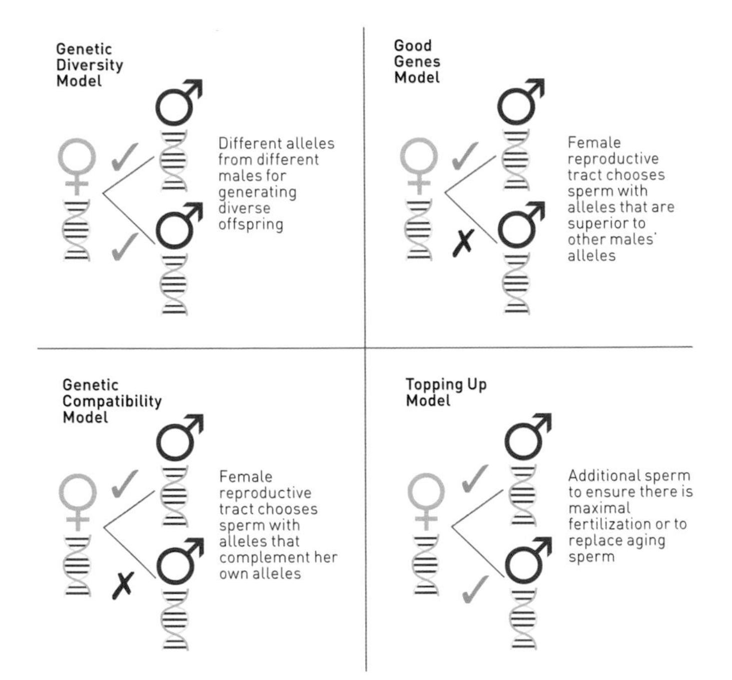

FIGURE 6.11 NUMEROUS MODELS EXPLAIN FEMALE NON-MONOGAMY
Three models propose explanations for the genetic benefits of non-monogamous mating for females, and the topping up model is an additional non-genetic explanation. For the genetic diversity, good genes, and genetic complementarity models, the benefits are directly genetic. For the topping up model, the benefit is simply to have enough viable sperm.

There are three sub-categories of ideas related to genetic benefits for the female as a consequence of mating with multiple males: (1) obtaining more genetically diverse offspring (the genetic diversity model), at least where mothers produce multiple offspring, (2) obtaining superior alleles for her offspring (the good genes model), and (3) obtaining the most suitable sets of alleles—given the alleles she herself has—for her offspring (the genetic complementarity model) (Figure 6.11). In (2), the alleles she accesses are superior against an average standard, whereas in (3) the alleles she accesses are superior in the context of her own set of alleles. A fourth genetic benefit of female non-monogamy is more pragmatic and has nothing to do with the alleles per se and everything to do with sperm viability, where females obtain fresh sperm from a partner in cases where sperm have a relatively short "shelf life." This is a version of the topping up model. Some female butterflies, for instance, are inclined to continue or resume mating if the sperm they are storing are reaching the limits of continued viability. Note that these various genetic models explaining female non-monogamous mating strategies need not be mutually exclusive.

The genetic diversity model is encapsulated in the idiom "Don't put all your eggs in one basket." If females use sperm from different males for fertilizing their eggs, females will have a more genetically varied progeny and hence offspring of more varied phenotypes. This does not mean that the offspring of such a brood or litter will be overall superior. In fact, we'd expect them to have the same average fitness as a brood of young sired by one male. Full siblings from monogamous matings share about half their alleles, since each parent provides a "random" sample of half of his or her genes in the gametes. In contrast, half-siblings (who share a mother, for instance, but two different fathers) share only about a quarter of their alleles. Where a female produces a diversity of half-siblings from two or more fathers, by being collectively more varied, at least some of her offspring are more likely to be able to cope with the diversity of challenges presented by the environment, an argument that connects to the Red Queen hypothesis (Section 4.6.3). For example, western honeybee (*Apis mellifera*) queens mate with multiple males. Research has demonstrated that different male lines have different levels of resistance to the hive-threatening foulbrood bacterium, meaning that by mating with at least several males, at least some of the queen's worker daughters will express foulbrood resistance when the colony is threatened.

The good genes model aligns with the good genes hypothesis that was introduced in Chapter 5, where discriminating females select genetically superior males for the genetic benefits their offspring will realize. Applied here, it is similar. If the female can make the appropriate mate selection, her offspring will overall be genetically superior than they otherwise would be. The challenge for the female may be that a male with good genes may be difficult for her to target or to assess based on his phenotype. Or she will not know whether she might have an opportunity to mate with an even better genetically superior male in the near future. In this scenario, multiple mating will increase her odds of making the best genetic choices in the interests of her offspring, provided she can select certain sperm preferentially from the choices within her reproductive tract. Females that do this are said to be exercising cryptic female choice. Cryptic female choice is more likely to exist in organisms where sperm can be stored—a group that includes many insects, reptiles, and birds (and a few types of mammals).

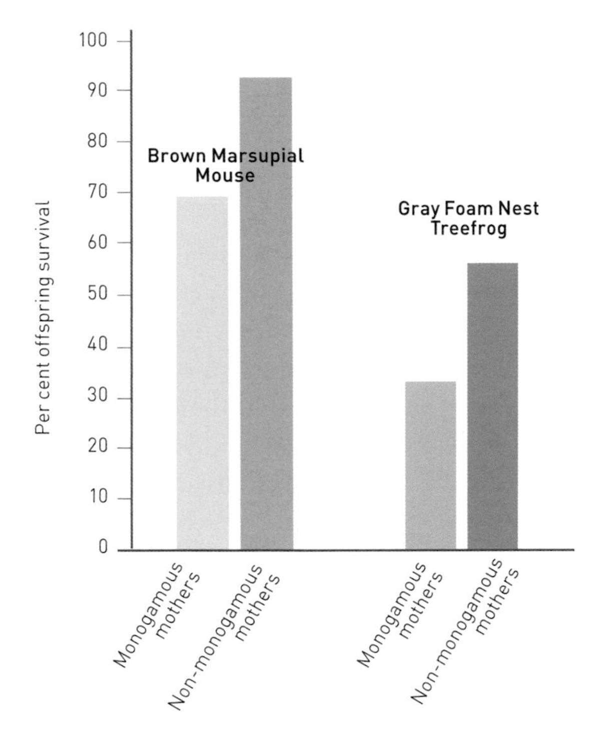

FIGURE 6.12 **FEMALE FITNESS COMMONLY IMPROVES WHEN MATING IS NON-MONOGAMOUS**

In multiple species, the rate of offspring survival increases when the mother mates with multiple partners, even when there is no paternal care, indicating the benefits are genetic. Two examples are Australia's brown marsupial mouse (*Antechinus stuartii*) and Africa's gray foam nest tree frog (*Chiromantis xerampelina*).

Adapted from Byrne & Whiting (2011) and Fisher, Double, Blomberg, Jennions, & Cockburn (2006)

The Australian brown marsupial mouse (*Antechinus stuartii*) is a species where males "mate to death" and accordingly provide no parental care (Figure 4.10). Females mate with either one or more than one male. Research that considered how well the offspring survive to the point of weaning has shown that significantly greater proportions of young from the non-monogamous mothers survive (Figure 6.12). There are numerous other organisms where similar studies show superior survivorship of the offspring where mothers have mated non-monogamously. What is not clear from these studies is the mechanism. Is it because the offspring are more diverse, as in the genetic diversity model? Is it because the female is somehow choosing the sperm of superior males, as in the good genes model (an example of cryptic female choice)? Or is it because she is choosing the most suitable sperm for her own genotype, as in the genetic complementarity model (another example of cryptic female choice)?

This third idea, the genetic complementarity model, relies upon the idea that based on an individual female's set of alleles, certain males will be a better choice as mates than other males. This does not argue that some males are inherently genetically superior but rather that for a given female there are better and worse choices based on the male's complement of alleles. This hypothesis relies on the idea that offspring viability and fitness are related to genetic heterozygosity. Organisms that have greater heterozygosity at multiple loci, a condition known as heterosis, are usually at a fitness advantage, since they have a greater number of alleles in total and therefore a greater repertoire of genetic recipes for making proteins. It follows that an individual that chooses a mate with different alleles will be more likely to have heterozygous offspring, and so we say such mates are genetically complementary. (Some authors use the term *genetically compatible*, but this can have a second, different meaning and should be avoided.) Inbreeding depression (Section 4.6.2) describes the opposite situation, because inbred individuals have increased levels of homozygosity. When a female chooses a male whose alleles are complementary to her own set, this is also known as a *sire-dam interaction* (where "dam" is the female counterpart of "sire").

One of the favorite systems for considering genetic complementarity in both humans and non-humans are the MHC genes introduced in Box 5.2. There are several MHC loci, meaning an individual can have more than two copies of MHC genes. The greater the number of these loci that are heterozygous, the more MHC alleles an individual has, and the greater the protein arsenal that individual has in battling pathogens.

Inbred individuals are likely to have more homozygosity at MHC loci, in contrast to outbred (i.e., heterotic) individuals, and reproductive females should of course aim for the latter. More than one study of European songbirds has found that extra-pair young are more heterozygous than within-pair young, suggesting that a female's social partner is more closely related than her extra-pair partner. One study examining European blue tits (*Cyanistes caeruleus*) found that the offspring sired by extra-pair males tended to have higher levels of overall heterozygosity, and the daughters, at least, survived better. Numerous studies using rodents, fish, lizards, and birds indicate that females make mate choices based on individual odors that indicate complementary MHC alleles.

The argument that it is beneficial to have substantial heterozygosity at MHC loci is compelling, but choosing males to generate heterozygous offspring in general, not just at MHC loci, is also logical. One experiment using Pacific crickets (*Teleogryllus oceanicus*) produced interesting results that support the idea of both cryptic female choice and the genetic complementarity model. Females were mated to one of four options: (1) two of her brothers, (2) two non-brothers, (3) a brother and then a non-brother, or (4) a non-brother and then a brother. Female hatching success was then measured. One might expect that the female mating with the two non-brothers would have the highest hatching success, but the females mating with a brother and a non-brother (in either order) were equally successful; it was only the female mated to two of her brothers that had low success. This suggests that within the reproductive tracts of the females that mated with a brother and a non-brother, the female preferentially used the sperm of the non-brother to fertilize her eggs.

6.4.3 Non-monogamous females can reap direct benefits

When biologists speak of direct benefits, they refer to non-genetic things that benefit the female, such as food, protection, and even tranquility. This sounds rather mercenary by human standards: females consenting to having sexual relations with males in return for benefits. Yet, even among humans these patterns are demonstrable. Without doubt, there will be exceptional individuals, but studies of personal ads seeking partners reflect the predicted patterns. Men tend to seek physical attractiveness and youth in partners more than women do, while women seek things that signify resources more than men do. As men's resources increase with age, so too do men's expectations increase. In contrast, as women's physical attractiveness wanes, so too do their expectations wane. We've already examined some of these economics in previous chapters, especially Chapter 5, when females exercise mate choice by preferring males that can re-balance the reproductive investment equation. And, when we examined polygyny

FIGURE 6.13 **NUPTIAL GIFTS ARE USED IN MATING TRANSACTIONS** This male scorpion fly (*Panorpa cognata*) has captured a prey item it will use to negotiate a transaction—the transfer of this nuptial gift to the female in return for mating privileges.

in context of the polygyny threshold model (Box 6.2), we noted that females sometimes choose polygyny if the resources that come with it more than offset the costs of sharing a male with one or more other females.

Many male invertebrates "purchase" sex by providing nuptial gifts, which are usually nutritious (Figure 6.13). A female benefits by taking in these extra reserves, as do her eggs. In experiments using the nursery web spider (*Pisaura mirabilis*), food-limited females mated with more males than did well-fed females. All else being equal, females experienced improved fecundity and hatching success up to the third mating, but with additional matings beyond that, where the benefits became minimal, females exhibited costs of reduced fecundity, reduced hatching success, and even smaller offspring size. Why this is the case is not known, but evidently repeated mating also has physiologic costs.

Nuptial gifts in return for mating opportunities are not the only way a female can benefit from mating with multiple males. The unobtrusive European bird the dunnock (*Prunella modularis*) will copulate as much as 100 times per day during the peak of egg laying. It commonly mates in a two-male form of polyandry known as biandry, although monogamy is also common. Females in these biandrous relationships lay more eggs and successfully raise more offspring than females mated monogamously. This is largely attributable to having the provisioning support of two males, both of whom have a reasonable expectation of some paternity, albeit shared with the other male. There is likely a low limit on manipulating males in this manner: the less likely it is that the male is a sire of some of her offspring, the less likely he will invest in the enterprise, leaving that female for more promising fitness returns for himself elsewhere.

Although human males express numerous traits related to protecting their paternity, such as jealousy and possessiveness, some human societies have traditions and beliefs that confuse the issue of paternity, which works to the advantage of females and their offspring. For instance, some pre-technological societies maintain that it is possible to have more than one biological father, the core belief being that all men who have intercourse with a woman during pregnancy share in the paternity of the resultant child. This is known as partible paternity. Such a biologically incorrect but culturally entrenched belief may represent a female strategy that ensures greater resources for her offspring, and this explanation is borne out when the well-being of children is considered. Among the Bari of Venezuela, a society with culturally endorsed polyandry, children with more than one such "biological father" survive to age 15 at a significantly better rate than children with only one biological father, and this difference is attributable to the secondary provisioning by the extra father.

One other context of non-monogamy in females is succumbing to harassment or coercion by males. Indeed, the term convenience polyandry is applied where females end up mating with multiple males to avoid their harassment. Giving in to coercion is not proactively adaptive for females, but it may sometimes be in her best interests if the costs of resistance are high. Theoretically, it occurs when the costs of resisting male advances exceed the costs of mating with such males. Costs of resisting include injury and energy waste, although the most significant cost of such mating can be that the female is prevented from exercising choice, something that can be of profound importance to her.

So, coercive mating aside, non-monogamy is frequently in the best interests of females. This does not contradict our examination of sexual selection in Chapter 5, where it was argued females should be and often are choosy. Instead, it is because their choosiness can be extended to several males.

6.5 POLYANDRY

In polygyny the *focal individual* is the male and in polyandry the focal individual is the female. Yet these two forms of polygamy are hardly equally common. This is because males and females are not merely opposites, like right hand and left hand or black and white. They are alternative ways of being, but they are asymmetric, as we've already examined. What are the different advantages of these two forms of polygamy for the focal individuals? For males, the answer is the familiar "quantity" answer: to have more female partners results in a greater yield of offspring for him. For females, the answer is more complex, but it could be any combination of the considerations just examined in Section 6.4 related to her own reproductive success.

6.5.1 Polyandrous mating systems are much less common than polygynous ones

The relative rarity of polyandry compared to polygyny is often explained by reference to Bateman gradients. A Bateman gradient for an individual is the number of offspring it produces versus the number of mates it has (Figure 6.14). In female mammals, commonly one male partner will be sufficient to result in a pregnancy that meets the physical limit of the womb. Such females have a zero Bateman gradient because additional matings do not yield additional offspring. But for males, their fitness increases each time they mate with a new partner,

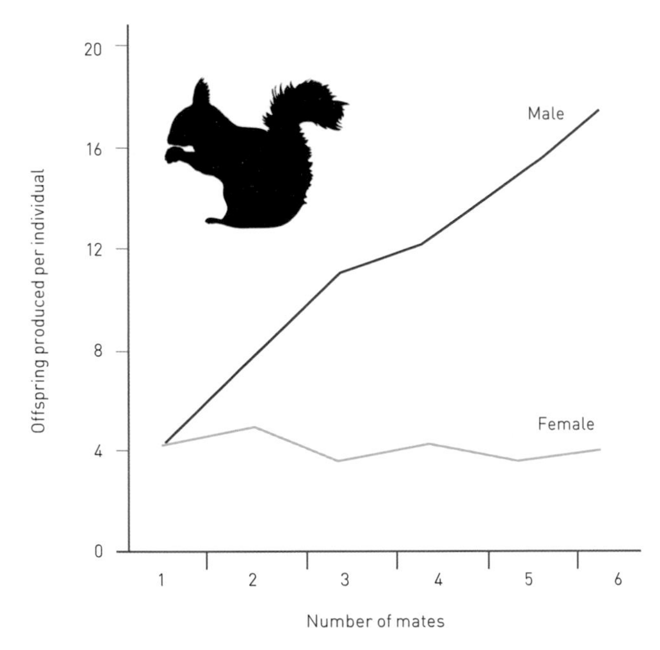

FIGURE 6.14 **BATEMAN GRADIENTS REVEAL FITNESS CONSEQUENCES OF HAVING ADDITIONAL MATES**
The typical male-female differences between the number of mates and the number of offspring produced are illustrated by these classic Bateman gradients for a hypothetical squirrel species. Where females (yellow line) have a finite reproductive capacity based on large investments, the relationship tends to have a zero slope, meaning her productivity does not increase with number of mates. Where males (red line) have a greater reproductive capacity because they succeed using sperm, productivity increases with the number of mates (a nonzero slope). This is partly why polygynous mating systems are so much more common than polyandrous mating systems.

and so they have nonzero Bateman gradients. This is, of course, consistent with the asymmetry of reproductive investments and commitments we've already examined. The zero-slope female line is theoretical; in some species females can have a modestly positive (upward) slope due to genetic benefits (e.g., Figure 6.12) or a negative slope when there are energetic or other costs to mating excessively.

In organisms that provide parental care, polyandry can exist if it is possible for the female to offload much of the parenting onto the male *and* if she is likely to be able to commandeer resources, known as *resource-defense polyandry*. (Researchers have not generally reported instances of commandeering males, which would be considered *male-defense polyandry*.) This combination of offloading to the male and commandeering resources represents a considerably greater challenge than the development of polygyny based on female defense or resource defense. Resource-defense polyandry is mostly reported in fish and birds.

One of the best-studied cases of resource-defense polyandry is the spotted sandpiper (*Actitis macularius*), a common shoreline species in

North America. It exhibits all the characteristics of sex role reversal: the female is larger than the male (by about 25 per cent), and the female is territorial and more aggressive, returning earlier in the spring to compete with other females to control key beaches for nesting. A successful female can have as many as five males in the territory she defends against other females, for each of whom she lays four large eggs. Naturally, the male is the primary parent in this scenario. Why did polyandry develop in some sandpipers? It is difficult to say, but the fact that the young are *precocial*—meaning that they are immediately able to walk and feed themselves by foraging with a parent—likely contributes to the explanation. However, it cannot be a complete answer, as there are species that have precocial young where sex roles have not been reversed.

There is also a second relatively rare type of polyandry, known as *cooperative polyandry*, that is characterized by a female having multiple male partners among whom there is no direct conflict. Most examples are birds, but it has also been identified among some small South American tropical primates known as tamarins and marmosets (Family Callitrichidae), as well as among some fish. Cooperative polyandry presents a conundrum—not only because it is usually males that increase their reproductive success by adding extra females, not the other way around, but also because one or more males could be raising young that are not their own.

Long-term field work studying saddleback tamarins (*Saguinus fuscicollis*) indicates that raising young in these species is particularly demanding, and those rare individuals that mate monogamously are never successful unless they have the assistance of older, grown offspring. Evidently, more than biparental care is usually necessary to successfully raise young. If there are no older offspring around that can help, which is not uncommon, a pair will take on another male. Both males mate with the female at approximately the same frequency and both provide approximately the same amount of paternal care. Presumably each has about a 50 per cent likelihood of siring the young. More generally, cooperative polyandry is expected where (a) the prospects of reproductive success in monogamy are low, (b) resources are limited or young are particularly demanding, making parental care of vital importance, and (c) the production of additional young, as would be provided in polygyny, would be a futile waste of resources.

As indicated earlier, the human polygyny index study indicated that polygyny is far more common in human history and even among current human societies than polyandry. Obviously, there are exceptions such as the Bari people (Section 6.4.3), where a relatively poor resource base exists and where there is a belief in partible paternity. In fact, there are obvious ecological parallels to the conditions faced by tamarins.

6.6 POLYGYNANDRY

Polygynandry is a breeding group of two or more males and two or more females. Well-known examples of polygynandrous systems are those of the African lion (*Panthera leo*) and the bonobo (*Pan paniscus*), a close relative of the common chimpanzee (*Pan troglodytes*). Both these species form breeding groups among which there are multiple sexual relationships. However, there are many less well-known examples, especially among mammals. In fact, genetic techniques applied to ecological systems are demonstrating that numerous systems that were once thought to be polygynous (or polyandrous) based on observations are proving to be polygynandrous. That is, genetic analysis like that shown in Figure 6.4 and more recent DNA sequencing studies demonstrate that breeding groups are sometimes larger than had been thought. Regardless, we will expect that the driving forces are not different from some of the ideas already presented: more offspring for males, more parenting assistance for females, and more genetically diverse offspring for both.

6.6.1 Reproductive skew

Even where polygynandrous breeding can be interpreted as being cooperative, not all is golden in these breeding groups. There is still competition involved, and not all members have equal mating success. In fact, there is a large area of theoretical biology that examines how the total amount of breeding in a group gets partitioned among the members, known as reproductive skew. Two significant factors that influence the breeding success of individuals in groups are the competitiveness of each member and the degree of relatedness among members.

Among both sexes there can be dominance hierarchies. Although it is relatively common to find high-status alpha males using aggression to make sure they have preferential access to the females over subordinate males, dominant females can also limit the reproductive success of other females in cases where it is important for their own welfare. After all, individual females may do better in terms of either available resources or assistance from male mates if other females are not competing for them.

In the alpine marmot (*Marmota marmota*), a European rodent, harassment of subordinate females by dominant females can be devastating for the pregnancies of the subordinates due to stress hormones, and many pregnancies don't come to term. This harassment may even delay the onset of sexual maturation in harassed younger females. In another polygynandrous mammal, the golden lion tamarin (*Leontopithecus rosalia*) (Figure 6.15), it appears to be a different mechanism. Here, the widespread reproductive failure among the subordinates appears to be due to the young being abandoned by those subordinate females to mitigate aggression from dominant females. Where dominant and subordinate females are related, as in some alpine marmot groups, the degree of female–female aggression is much less.

One might well ask why subordinates remain in the group. There could be multiple reasons. Some species can only survive by living in groups, there may not be "room" in other nearby groups, and subordinates may be waiting for an opportunity to move up the pecking order when death removes a more dominant individual. Subordinates are also sometimes related to dominant individuals, and so their participation may be a form of cooperative breeding (Section 6.6.2).

As we will see in Chapter 7, sometimes females suffer the risk of having males kill their offspring if those offspring are sired by other males. One advantage of polygynandry for females is that it is likely to generate confusion or uncertainty among male members as to paternity, this decreases the likelihood of such infanticide. This

FIGURE 6.15 **POLYGYNANDROUS GROUPS MAY STILL HAVE A PRIMARY BREEDING PAIR**

The golden lion tamarin (*Leontopithecus rosalia*) is an endangered South American primate that has a polygynandrous breeding system. Still, a dominant male and a dominant female usually do most of the successful breeding.

DJP3tros / Shutterstock.com

paternity uncertainty can increase male provisioning of offspring if those males "believe" the offspring may be their own, as is the case with the dunnocks (Section 6.4.3).

6.6.2 Cooperative breeding

Cooperative polyandry is a form of cooperation, but it is characterized by a female who actively mates with her male co-parents. In contrast, cooperative breeding is a social system where, in addition to care being provided by parents, offspring are also cared for by other individuals, known as *helpers*. Usually, helpers are related to the parents, often the young adult offspring from previous breeding events. Cooperative breeding has been identified in numerous groups, including insects, fish, birds, and mammals, and it is not restricted to polygynandrous scenarios. Even monogamous pairs can have helpers. But since polygynandrous systems involve numerous individuals of different ages, arranged in dominance hierarchies, and exhibiting variable degrees of relatedness, these circumstances collectively provide rich opportunities for helping. Barbary macaques (*Macaca sylvanus*) are such an example of polygynandrous cooperative breeders among primates. Male or female adults without offspring of their own will commonly "babysit" the offspring of active parents, an act known as alloparenting.

Cooperative breeding that is undertaken by relatives is thought to be evolutionarily stable because although some individuals do not get to breed and therefore do not manage to directly transmit their alleles into the next generation, they can do so indirectly through the success of parents, brothers, sisters, and even other relatives. This reproductive success is known as inclusive fitness, and its persistence as part of a mating system is supported by a special sub-category of natural selection known as *kin selection*, where behavior traits that promote the success of relatives are favored.

6.7 PROMISCUITY

As an adjective, *promiscuous* means to be undiscriminating, and when applied to human sexual behavior, it describes a pattern of transient sexual encounters—so-called one-night stands. As a mating system applied to animals in general, however, it would be wrong to assume that these matings are always undiscriminating, especially among females for whom the quality of DNA they get from their mates' sperm may make a real difference to their own fitness.

In the scientific literature, biologists are not always strict in applying the term *promiscuous* to mating systems that have no social bonds. Because they wish to impart additional information about such bondless mating, they instead refer to them as polygynous, polyandrous, or polygynandrous. One can see the advantage of using these terms because of the additional information imparted. For instance, there may be no social bonds in a particular case, justifying the use of the term *promiscuous*. Yet if that system has a male-biased operational sex ratio, a high variance in male reproductive success, and females that mate with only one male, it has features of a polygynous system. In describing promiscuous mating systems, therefore, polygyny, polyandry, or polygynandry are sometimes used instead, even though there are no social bonds and even though the intention is not to suggest the system in question is other than a promiscuous one. Here, we abide by the "rule" and use *promiscuity* in cases with no pair-bonds, but, if relevant, we indicate whether it is polyandrous or polygynous in nature, as the case may be.

6.7.1 Factors associated with promiscuity

Promiscuity is far more likely in species that provide no parental care. Turtles, for instance, bury their eggs and leave them to be incubated by environmental heat. When the young hatch, they are independent; the adults don't form pair-bonds, and genetic analyses of nests commonly show that females have mated with multiple males, indicating a promiscuous/polygynandrous mating system. All mammals and all birds provide parental care (subject to one bird exception, where the "turtle method" is used). And although there are promiscuous birds and mammals, monogamy, polygamy, or polygynandry is considerably more likely in these two groups. This terminology may be difficult to accept, given that many socially monogamous species engage in extra-pair matings; when pair-bonded humans do this, we may very well label them as promiscuous.

6.7.2 Scramble competitions are usually promiscuous mating systems

In Section 5.5.1, we examined scramble competition. The examples considered there included cases where individuals of the same sex competed for mates by "scrambling" to find partners as efficiently as possible. These systems are usually promiscuous ones. Scramble systems are often characterized by relatively short, intense periods of breeding, as when certain frog species assemble quickly in spring to take advantage of temporary pools formed from melted snow. Species that breed in this manner are known as explosive breeders.

6.7.3 "Lek polygyny" is a promiscuous system resembling hook-up culture

Lek polygyny is a term that is thoroughly entrenched in the ecological literature, but it is really a form of promiscuity since it does not involve social bonds. In addition, it is commonly polygynandrous rather than polygynous, as recent genetic work shows both males and females in some lekking species often mate with more than one partner. "Lek polygynandrous promiscuity" would be a mouthful, but it would be a more accurate description.

A *lek* is sometimes described as a mating arena—a place where males aggregate to engage in competitive display (Figure 6.16). The components of performances vary among species, but they include visual displays (e.g., colors, dances, and other sexually selected adornments) and acoustic displays (e.g., "songs"), as well as male–male aggressive interactions. They can be entertaining places to visit. There are four *Bradbury criteria* for a mating system to be truly a lek: (1) there is an arena where males display, (2) the females that attend the arena exercise mate choice, (3) the mini-territories in the lek defended by the displaying males contain no resources of use to the females, but some clearly are of higher status than others, and (4) there is no male parental care.

Male reproductive success is often highly variable in lekking species. In a study of Ecuador's wire-tailed manakin (*Pipra filicauda*) leks, on average only 3 male birds sired more than 80 per cent of all offspring, even though leks contained from 4 to 12 males. Lekking is the system used by several grouse and sandpipers in northern temperate zones, but there are also lekking insects, amphibians, reptiles, and mammals.

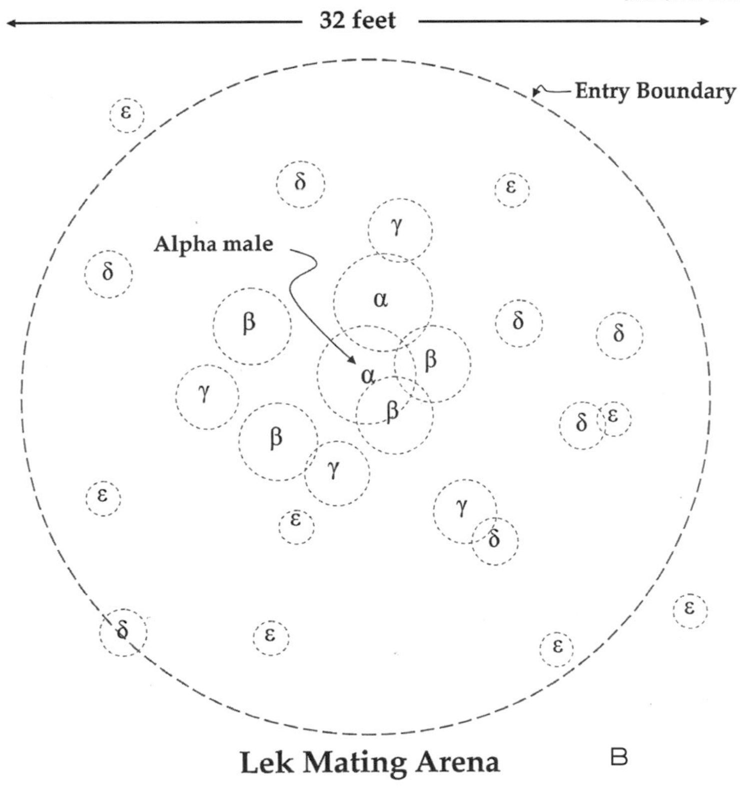

FIGURE 6.16 **IN LEKKING SPECIES, FEMALES CHOOSE MALES BASED ON STATUS AND PERFORMANCE**

Three male Guianan cock-of-the-rock (*Rupicola rupicola*) performing at a lek in tropical America during the breeding season (a). Each male defends a small displaying territory within the lek against neighboring males. Central positions are usually held by dominant males (the larger, central circles labeled with an α), with positions of reduced status found more peripherally (e.g. β, γ, etc.) as indicated in the model lek map (b). Females visit the lek and choose one or more males with whom to mate; all parental care is provided by the females in lekking species.

[6.16a] WILDLIFE GmbH / Alamy Stock Photo
[6.16b] Libb Thims / CC-BY-SA 3.0

CHAPTER 6 SUMMARY

- A mating system is the way a sexually reproducing population is structured with respect to who mates with whom, and it strongly influences patterns of parental care.
- There are five types of mating systems: *promiscuity* (no social bond), *monogamy* (socially bonded male and female), *polygyny* (mating group of one male and more than one female), *polyandry* (mating group of one female and more than one male), and *polygynandry* (mating group of at least two of each sex).
- Genetic monogamy is true monogamy, but genetic tests demonstrate that social monogamy, which is characterized by some degree of usually covert extra-pair mating outside the pair-bond, is widespread.
- There are several factors that can lead to monogamy, most notably the necessity or advantage of having two parents share the burden of raising offspring.
- Polygyny is far more widespread both in nature and among humans than polyandry; males tend to establish polygynous groups by either defending females directly from other males or by defending the resources that females require.
- Females reap numerous demonstrated benefits from non-monogamous mating, including fertility insurance, genetic benefits for offspring, and resources from multiple males.
- Polyandry is more likely to occur in species where sex roles are reversed, in which case there are reversed parallels with polygyny, such as the defense by females of resources important to males.
- Some social species have complicated mating patterns that can include cooperative breeding, reproductive skew (where sexual and breeding activity among members of the group is not evenly distributed), and alloparenting (where individuals other than biological parents provide some parental care).
- In lek polygyny, males gather for extended periods at a mating arena where they perform mating displays, and females attend to select partners.

FURTHER READING

Byrne, P.G., & Whiting, M.J. (2011). Effects of simultaneous polyandry on offspring fitness in an African tree frog. *Behavioral Ecology, 22*(2), 385–391.

DuVal, E.H., & Kempenaers, B. (2008). Sexual selection in a lekking bird: The relative opportunity for selection by female choice and male competition. *Proceedings of the Royal Society B: Biological Sciences, 275*(1646), 1995–2003. https://doi.org/10.1098/rspb.2008.0151

Ellsworth, R.M., Bailey, D.H., Hill, K.R., Hurtado, A.M., & Walker, R.S. (2014). Relatedness, co-residence, and shared fatherhood among Ache foragers of Paraguay. *Current Anthropology, 55*(5), 647–653. https://doi.org/10.1086/678324

Fisher, D.O., Double, M.C., Blomberg, S.P., Jennions, M.D., & Cockburn, A. (2006). Post-mating sexual selection increases lifetime fitness of polyandrous females in the wild. *Nature, 444*, 89–92.

Foerster, K., Delhey, K., Johnsen, A., Lifjeld, J.T., & Kempenaers, B. (2003). Females increase offspring heterozygosity and fitness through extra-pair matings. *Nature, 425*(6959), 714–717. https://doi.org/10.1038/nature01969

Friesen, C.R., Uhrig, E.J., & Mason, R.T. (2014). Females remate more frequently when mated with sperm-deficient males. *Journal of Experimental Zoology. Part A, Ecological Genetics and Physiology, 321*(10), 603–609. https://doi.org/10.1002/jez.1892

García-González, F., & Simmons, L.W. (2007). Paternal indirect genetic effects on offspring viability and the benefits of polyandry. *Current Biology, 17*(1), 32–36. https://doi.org/10.1016/j.cub.2006.10.054

Gerlach, N.M., McGlothlin, J.W., Parker, P.G., & Ketterson, E.D. (2012). Promiscuous mating produces offspring with higher lifetime fitness. *Proceedings of the Royal Society B: Biological Sciences, 279*, 860–866.

Henry, M.D., Hankerson, S.J., Siani, J.M., French, J.A., & Dietz, J.M. (2013). High rates of pregnancy loss by subordinates leads to high reproductive skew in wild golden lion tamarins (*Leontopithecus rosalia*). *Hormones and Behavior, 63*(5), 675–683. https://doi.org/10.1016/j.yhbeh.2013.02.009

Reid, J.M., Monaghan, P., & Ruxton, G.D. (2002). Males matter: The occurrence and consequences of male incubation in starlings (*Sturnus vulgaris*). *Behavioral Ecology and Sociobiology, 51*, 255–261.

Roberts, S.C., Gosling, L.M., Carter, V., & Petrie, M. (2008). MHC-correlated odor preferences in humans and the use of oral contraceptives. *Proceedings of the Royal Society B: Biological Sciences, 275*(1652), 2715–2722. https://doi.org/10.1098/rspb.2008.0825

Rovelli, V., Randi, E., Davoli, F., Macale, D., Bologna, M.A., & Vignoli, L. (2015). She gets many and she chooses the best: Polygynandry in *Salamandrina perspicillata* (Amphibia: Salamandridae). *Biological Journal of the Linnaean Society, 116*(3), 671–683. https://doi.org/10.1111/bij.12613

Scelza, B.A. (2011). Female choice and extra-pair paternity in a traditional human population. *Biology Letters, 7*(6), 889–891. https://doi.org/10.1098/rsbl.2011.0478
Schmitt, D.P., & Rohde, P.A. (2013). The human polygyny index and its ecological correlates: Testing sexual selection and life history theory at the cross-national level. *Social Science Quarterly, 94*(4), 1159–1184. https://doi.org/10.1111/ssqu.12030

7 Sexual Conflict

KEY THEMES

All organisms have finite time, materials, and energy to apportion between reproduction and other activities, leading to a diversity of reproductive patterns. But even within species, males and females have different ways of maximizing reproductive success. Very commonly, these do not align, leading to *sexual conflict*. Conflict occurs before mating, as well as during and following mating. Different mating strategies are influenced by genes, body condition, and opportunity, even leading to same-sex parenting.

On the surface at least, the production of offspring through sexual reproduction appears to be a beautifully cooperative venture. Male and female, sperm and egg, penis and vagina, resources and nurturing—all pairings seem to be complements designed for the undertaking. But when we look deeper, that is too simplistic. That there has always been some measure of conflict between the sexual and mating desires of men and women should have been the first clue that much about the affairs of the sexes is characterized by conflict. And, in addition to conflict with potential mating partners, there are also other real-world imperfections that can influence the mating patterns and options of both males and females, making things less than ideal.

Descriptions of sexual conflict and mating patterns reach far back in time, but the recognition of conflict and flexibility in mating patterns as a subject worthy of study and analysis is much more recent. In the past few decades, sexual conflict has come to be recognized as a useful explanatory tool for many patterns that are otherwise inscrutable. There were many impediments to its fair consideration in the past; recent social changes that have altered the way we think about sex and the sexes have doubtless been influential in allowing researchers to consider the subject more freely and dispassionately. Perhaps the most significant finding, documented in Chapter 6, is that the truly monogamous female is a relatively rare creature in nature—contrary to Victorian assumptions.

What is meant by sexual conflict? This is the difference between optimal strategies of males and females, as potential mates, in matters of copulation, fertilization, and parenting (Figure 7.1). Mismatched "best outcomes" for each sex result in conflict. Accordingly, the foundational idea in this chapter is this: given that reproductive success is the ultimate evolutionary measure of success (since that is how genes survive from generation to generation), natural selection will have constructed phenotypes designed to make the most of reproductive opportunities. Further, given that both sexes commonly engage or are at least prepared to engage with multiple partners to advance their reproductive success, much of the dance between the sexes must be rooted in conflict. This does not mean that cooperation is a mere illusion, but we do expect to find many adaptations and counter-adaptations in

FIGURE 7.1 **SEXUAL COERCION IS ONE MANIFESTATION OF SEXUAL CONFLICT**
These male mallards (*Anas platyrhynchos*) are mating with an unwilling female. *Sexual conflict* manifests in many forms, most of which are not so directly overt as *coercive* mating, shown here.

the sexes that have been selected to gain the upper hand in sexual matters. Mating systems reflect how this has played out at the level of breeding populations. But there is much individual nuance, and in this chapter we look more at the individual, and how he or she is adapted and uses those adaptations to navigate conflicts and constraints presented by his or her reproductive world.

7.1 STRATEGIES FOR SEXUAL SUCCESS

While mating and reproductive biology began experiencing a renaissance of attention from researchers, leading biologists were reminding the community that evolution requires that organisms behave in their own best interest or at least in the best interests of their kin. This was especially true for mating matters, with articulate reminders that the currency of success in biology is reproductive success. Any heritable character that improves reproductive success has the associated consequence of making it to the next generation—or at least having the genes responsible for the trait make it to the next generation. In contrast, individuals with heritable characters that lead to greater survival do not succeed more, unless that means they also reproduce more successfully. For all time, reproductive success has trumped survival, and that is the evolutionary legacy of all organisms in every generation, humans included.

7.1.1 A comment on terminology

When biologists speak of strategies, they usually mean an evolutionary strategy, and this invites confusion because it is significantly different from the normal use of the word. When humans devise a strategy, it is a prior plan of action designed to achieve a goal. It involves collecting information and then cognitively assessing how a desirable end can be achieved based on that information. An evolutionary strategy, on the other hand, is a genetically inherited pattern of behavior that determines an individual's actions. The strategy exists not because the organism figures out a plan but because in the past individuals that behaved in ways that resulted in survival and mating success were able to disproportionately populate subsequent generations through their successful genes. Genes that govern or enable those strategies spread in populations.

Given that sexual conflict is the difference between optimal male and female mating strategies, we expect to find it whenever the best outcome for the male does not match the best outcome for the female. Because it can easily be shown that these best outcomes often don't match, conflict is common, and it is expected that males and females are often engaged in a tug-of-war centered on different mating strategies. This tug-of-war has evolutionary consequences, because any adaptation in a male that furthers his agenda is likely to be countered by an adaptation in the female to prevent the male from obtaining control. This is known as *sexually antagonistic coevolution.*

7.1.2 Sexual strategies exist in the context of sexual conflict

In Chapter 5, the case was made that there is often a disparity in reproductive investment and commitment between males and females. Where these asymmetries persist, it is usually the female that bears the lion's share of reproductive costs. In such circumstances, males can easily increase reproductive output by more mating, but this is commonly not the case for females. This is an obvious source of potential conflict. But there are many additional and intersecting sources of conflict between males and females, with female non-monogamy being the most significant.

One reason so much attention is focused on female rather than male non-monogamy is the significance of the asymmetry in reproductive investment. Females usually invest more. But a second asymmetry also plays a role, which is paternity uncertainty. Where fertilization is internal, females can be certain their offspring are their own, but the uncertainty that males experience about whether "their" offspring are indeed their own influences the extent to which males are willing to provide parental care. Both of these asymmetries—greater female investment and greater male uncertainty—drive sexual conflict.

We've seen in Chapter 6 that there are several kinds of explanations for female non-monogamy. One is that female non-monogamy is not adaptive but is instead driven by male eagerness to mate, and that males either coerce females into mating or harass them sufficiently that the females give in. The other is a broad class of ideas that indicates that female non-monogamy is adaptive by serving the interests of the female, and Section 6.4 reviewed numerous hypotheses based on benefits to females. Technically, even the former can be construed as being adaptive for the female if she succeeds in minimizing harm to herself, but it is not adaptive in the sense of proactively optimizing her reproductive life. Where females are clearly seeking multiple matings, there must be a payoff for them because these non-monogamous strategies are pursued by females even though there are offsetting costs of multiple mating: physical harm from males, increased exposure to sexually transmitted disease, time and energy wastage that may reduce her reproductive output, and the risk of desertion by males whose paternity becomes uncertain or who otherwise move on to other mating opportunities.

In cases where females customarily have more than one partner, or at least where the possibility of additional partners is real, internal fertilization poses three specific challenges for males. First, to what extent does the female control what happens to his sperm and that of other males inside her reproductive tract? Second, if sperm and semen have some real cost associated with its production, how is wastage to be avoided given that the female can subsequently re-mate afterwards, potentially making that second male the successful sire? And third, if paternal care is expected, how can the male be sure of paternity so that providing care is worth his while?

Cuckoldry is the mating by a female with a male partner other than her social mate. Over time, males that have employed systems to cope with female infidelity have been the more successful ones, producing populations where males seek to minimize the risk of cuckoldry and associated confused paternity. But males do not hold all the cards. Females are not simply vessels in which male–male competition operates. As we will see, control of what occurs inside the female reproductive tract represents a battleground in many species, both between males competing to fertilize her eggs and between the female and those male partners. In this context, females of some species are known to employ a dual mating strategy, an idea that is sometimes applied to people (Box 7.1). This is really a different manner of describing social monogamy: bonding with a "committed" male partner but engaging with one or more other males selectively to increase her reproductive success, presumably through genetic benefits.

7.2 REALMS OF SEXUAL CONFLICT

Success for both sexes is measured by their fitness, so both male and female mating strategies must take into account the opportunities and constraints presented. Males commonly cope with two troublesome things. One is the biased operational sex ratio where females are in more limited supply due to gestation, lactation, maternal care, and fertility cycles, with the additional result that females are therefore discriminating in their choices. The other problem for males is paternity uncertainty in the context of male–male competition, especially where fertilization is internal and where there is frequent female non-monogamy. But, subject to these serious challenges, males can achieve ever-greater reproductive success by mating often.

Females also cope with two troublesome things. One is that it is much easier for females to be saddled with all or most of the reproductive costs, especially where production of young involves parenting or other substantial investment, like pregnancy and lactation or making large eggs. The other is male persistence, sometimes amounting to harassment or sexual coercion. But, subject to these serious challenges, females can achieve greater reproductive success by accessing the best males for either their genes or their resources, whether that involves choosing one male or numerous males.

Given these male and female reproductive profiles, one can easily see that while the two sexes seem made for each other with respect to reproduction (and their interests *are* sometimes aligned), there is great room for conflicting interests and numerous ways in which one sex might try to take advantage of the other. More formally, biologists say that the two sexes have conflicting optimal fitness strategies.

7.2.1 Sexual conflict before mating

Hanuman langurs (*Semnopithecus entellus*) live in groups that typically comprise numerous females and either one male or one dominant and one or more sub-dominant males (Figure 7.3). Male Hanuman langurs periodically and systematically kill infants, known as *infanticide*. At one time the untested explanation was that when the local langur population became too dense, the over-crowding was what triggered the killing. This, however, did not make much evolutionary sense. A closer analysis showed that periodically (often about every two to three years), groups get taken over through aggression by a new male or males.

BOX 7.1 DO HUMAN FEMALES SHOW CHARACTERISTICS OF A DUAL MATING STRATEGY?

Growing evidence for the widespread pattern of female non-monogamy in nature led many to predict that it should also occur at a significant level in humans. Accordingly, researchers have looked for evidence in humans, including direct evidence of "cheating" but also male and female behaviors and attributes that reflect the risks (for males) and opportunities (for females) of extra-pair mating in humans.

We all know that heterosexual human females do sometimes cheat on their partners. Although the frequency of such activity is difficult to quantify due to problems with self-reporting, estimated rates of female infidelity range from 20 to 50 per cent in Western societies. Medical evidence occasionally exposes these infidelities, such as where a mother's child has a blood group that is not possible given her partner's blood group, or when she becomes pregnant notwithstanding that her partner proves to be infertile. These reports have sometimes led to untrue claims that in Western societies, large numbers of children (10 per cent or more) are not sired by their supposed father (Figure 7.2).

What does a closer look at the evidence indicate about humans? Considerable evidence indicates that human females have succeeded by employing one of two strategies of behavioral adaptations or by employing them opportunistically depending upon the circumstances. One strategy is the familiar and more socially endorsed one: long-term committed and cooperative monogamy, with paternal care a part of it. The other, usually more covert, system is the dual mating strategy, where a woman is socially monogamous with a committed partner but engages with one or more other males selectively. The ultimate explanation (Section 2.1) of such activity is to increase her reproductive success, presumably through genetic benefits, but there will be additional proximate explanations.

As supporting evidence, advocates of this view look to shifting patterns in women's desires and inclinations across the monthly cycle and to patterns in cross-cultural studies, where non-monogamy is frequently documented. As for the first type of evidence, fertility is a measure of the probability of conceiving through sexual intercourse, and it peaks just before ovulation, which is the point during the woman's monthly cycle when an egg is released for fertilization. Human ovulation appears to be largely hidden. This is known as concealed ovulation, a pattern that has been speculated to increase the certainty of the pair-bond (see Section 10.2.2). In contrast, most other female mammals exhibit overt visual, olfactory, and behavioral indications of impending ovulation. Males rely upon these cues, exhibiting increased interest in fertile females.

Among pre-technological small-scale societies that have been looked at, there are relatively high estimates of extra-pair paternity (EPP). In one South American study from the 1960s, the estimated EPP rate for children was nine per cent. More recently, a study of a Namibian culture in Africa produced an EPP

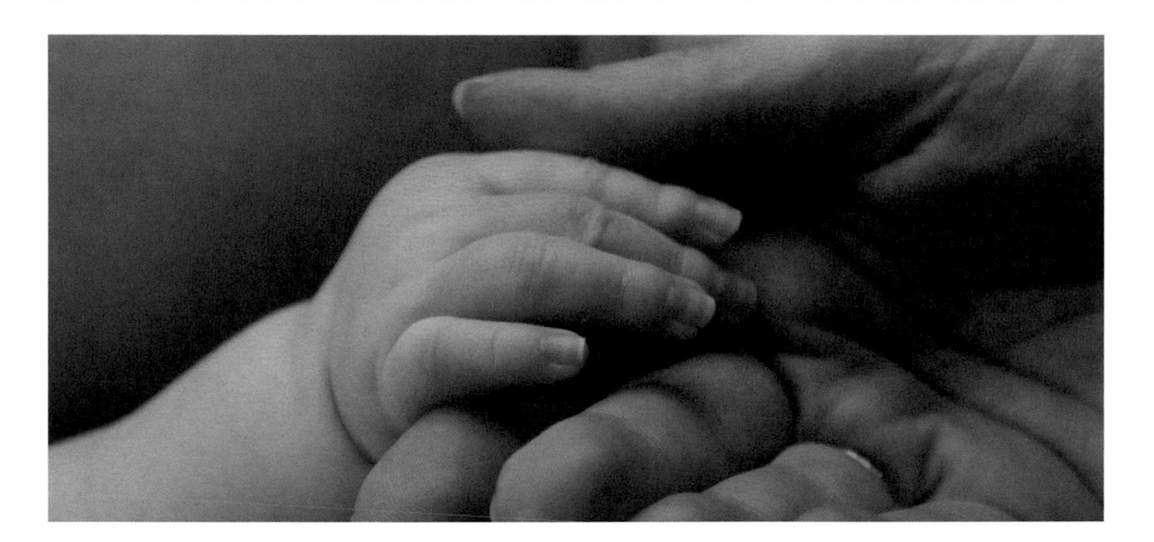

FIGURE 7.2 **EXTRA-PAIR OFFSPRING ARE RARE AMONG HUMANS**

The proportions of babies born to couples where the father is someone other than the mother's married or common law partner have been highly distorted in the popular press and the Internet. The true figure is close to one per cent, based on sophisticated genetic studies.

rate of 17.6 per cent. In the latter study, all such births occurred in arranged marriages, which was the more common type of marriage. In the rarer "love-match" marriages, there were no EPP cases. These data, however, were also based on self-reporting.

A 2016 study reviewed recent research using genetic methods to test for extra-pair paternity in modern societies. In addition to being more objective than self-reporting by mothers, Y-chromosome gene sequencing in men has allowed investigators to search back several generations to periods when there was a lack of reliable contraception and a more "natural" rate might be evident. Yet these more recent genetic studies have shown that EPP rates in humans have remained at about one per cent across several human societies (both Western and non-Western) for the past several centuries. This does not mean that there are no societies with higher rates of female non-monogamy, nor that such rates were equally low in the past, nor that there aren't valid arguments why extra-pair mating among humans might be advantageous for females. Whatever genetic benefits there may be for a woman to have extra-pair fertilizations may be offset by the severe costs of infidelity, including sexually transmitted infections, spousal retribution, and withdrawal of paternal and spousal care.

This one per cent EPP figure is not the same as the proportion of humans who admit to engaging in extra-pair sex; in modern Western societies, these rates tend to be considerably higher than ten per cent. Remember, contraception is likely to mean that extra-pair fertilizations now rarely result from extra-pair copulation, making cuckoldry rates low.

FIGURE 7.3 **WHEN HAREMS ARE TAKEN OVER, INFANTICIDE BY MALES OFTEN FOLLOWS**
In Hanuman langurs (*Semnopithecus entellus*), when new males take control of a social group, they evict the previous breeding males, kill the young, and the group starts a new breeding cycle, with the females mating with the new conquerors.

For the females, this is unfortunate, because they are often in the midst of raising young whose male parent has just been ousted. For the conquering males, this is less than ideal because the females are busy being mothers and are not ovulating. So, there is a conflict. It is "resolved" as follows: the new males kill the youngsters, the females respond by re-entering estrus, the new males inseminate them, and a new batch of young are born, which the males do their best to protect.

Lions (*Panthera leo*) behave in a similar manner. Interestingly, so does the sex-role-reversed African jacana (*Actophilornis africanus,*

Figure 5.10) but in a reversed pattern. If a female jacana is removed from her territory experimentally or dies a natural death, a new female moves in, kills the young that are being tended by the previous female's male partners, and then solicits those males to fertilize her so that she may lay them a new clutch of eggs to incubate and raise.

Other types of pre-mating sexual conflict include harassment and coercion (see Figure 7.1). Male-perpetrated harassment and coercion is widespread (and in species where sex roles are reversed, female-perpetrated harassment and coercion is likewise observed). Our own species is not immune, where it is disproportionately male-perpetrated and is known in extreme form as *rape*. There is a battle among intellectuals concerning rape motivation in humans. Evolutionary psychologists tend to interpret it as having psychobiological roots as a sexually motivated male mating strategy (although this does not mean it is genetically determined or is acceptable because of its "naturalness"). In contrast, feminist scholars tend to interpret rape as being socially constructed through cultures that almost universally endorse male inclinations to exercise power over women using violence.

There has been much experimental work on coercion among animals. Research using the blue-winged water strider (*Gerris buenoi*), an insect that "skates" on the surface of fresh water, was among the first to show that females do sometimes respond to male mating harassment by giving in (Figure 7.4). In these insects, females store sperm, and the last male to mate with her does most of the fertilizing. This is known as *last sperm precedence*. Thus, females that have capitulated to male harassment can still retain some control over who will fertilize her eggs by subsequently pursuing mating opportunities with a different male. In the experiment, females were housed with either one male or multiple males. Harassment rates toward females were almost three times as much when they were housed with multiple males. Although females in both scenarios attempted to avoid males, those in the multiple-male system resisted less and mated more than three times as much as females housed with a single male.

FIGURE 7.4 **MALE MATING INTENSITY CAN BE SUB-OPTIMAL FOR FEMALES**
Water striders (Family Gerridae) have been used in studying sexual conflict, including harassment. In their pursuit of mating opportunities, males can impair what is optimal for the female.
Sergei Mironenko / Shutterstock.com

In some species, anatomy or other sexual features prevent males from gaining the upper hand before mating. One of the oddest is the spotted hyena (*Crocuta crocuta*), an African social carnivore whose females are the only mammals that do not have an external vaginal opening. Both intercourse and birth occur through a canal that passes through a clitoris that matches the male's penis in size. Female hyenas are thus said to exhibit *virilization* or masculinization: the clitoris is described as penile, and there is even a pseudoscrotum. Both structures are strongly influenced by high doses of androgens, which are male hormones. Adult females are also larger than adult males, and they are dominant within the social group. How this system evolved is uncertain, but females are ascendant in courtship matters, controlling when, and with whom, copulation will occur. However, they appear to suffer substantial discomfort when giving birth. Male sexual advances are timid, and most are either ignored by the targeted female or replied to with aggression. Males desirous of mating appear to exhibit a strong motivational conflict between approaching females and running away from them.

Meanwhile, sexual conflict in hermaphrodites before mating and during mating exhibits interesting behaviors and weaponry and tends to support the argument that when it comes to individual reproductive success, it is preferable to be male (Box 7.2).

7.2.2 Sexual conflict during mating

For sperm competition to operate, females must mate with more than one male. When we made mention of sperm competition in Section 5.5.3, the focus was on male–male competition. Here, we add a vitally important component: sperm competition does not simply occur inside an indifferent female. Just as female pronghorn (*Antilocapra americanus*) assessing males as potential partners behave in a manner interpreted to incite them to compete, female non-monogamy can in some cases be interpreted as a means to inciting sperm to compete. In this way, given that choice is important to her, and given that she has access to multiple males, a female may exercise mate choice not just through the choice of a mating partner but also through operations in her own reproductive tract. Accordingly, the intra-sexual (competition) and inter-sexual (choice) branches of sexual selection that were considered in Chapter 5 play out here as sperm competition and cryptic female choice.

In some species of damselflies (Figure 5.11) such as the ebony jewelwing (*Calopteryx maculata*), the male has a dual-purpose intromittent organ. This "penis" not only serves as a device for delivering sperm, but it also has structural refinements that allow it to first deal with sperm already stored in the female's reproductive tract from a mating with a prior male. In some species whose strategy is to lower the likelihood of a prior male's sperm fertilizing the eggs, the sperm is removed altogether, but other species achieve as much success by repositioning the other male's sperm away from where they are likely to have access to eggs. This is known as *sperm displacement*, and it tends to result in a pattern of last sperm precedence for successful fertilization. Sperm displacement is thought to be used in numerous animal groups; it has even been suggested that the shape of the human glans penis and the foreskin are designed to displace sperm, although there is not strong evidence for this.

BOX 7.2 SEXUAL CONFLICT IN HERMAPHRODITES

Where individuals are both male and female, conflict can arise between mating partners if the preference is to mate in the same role. Accordingly, we expect to find evidence of conflict in hermaphrodites. Indeed, it is among hermaphrodites that the greatest variety and frequency of traumatic insemination occurs. Traumatic insemination is the delivery of sperm other than by the female reproductive tract, through a form of traumatic wounding.

Marine flatworms (Order Polycladida) are hermaphroditic, and some species engage in what has been termed *penis fencing* (Figure 7.5). These animals have extendable penises referred to as *stylets* that have dagger-like tips used to inject sperm into the body cavity of their "partner," provided they are able to pierce their partner's skin. Sometimes both partners succeed in this duel, which is known as mutual insemination or bilateral insemination.

However, in some species at least, such as *Pseudoceros bifurcus*, reciprocity is a possibility but is not the apparent goal. The partner that injects (or injects first) is favored, since "he" will suffer fewer wounds and is likely to sire more offspring through matings with additional individuals, without being burdened with egg production.

Traumatic insemination does occur in some non-hermaphrodites, the best known of which is the unsavory common bed bug (*Cimex lectularius*). Once mounted on the female, the male uses his dagger-like hooked penis to pierce the female's abdomen; the female reproductive tract is not used for intromission. The sperm travel by the insect equivalent of the blood system to the female's sperm storage areas and from there to the ovaries, where they fertilize the eggs. In a bizarre twist, males of some species in the bed bug group will also traumatically inseminate other males, the adaptive function of which remains uncertain.

FIGURE 7.5 **HERMAPHRODITIC FLATWORMS COMPETE FOR THE MALE ROLE**

These two hermaphroditic marine flatworms are engaged in penis fencing, where each seeks to win by inseminating the other.

Secret Sea Visions / The Image Bank / Getty Images

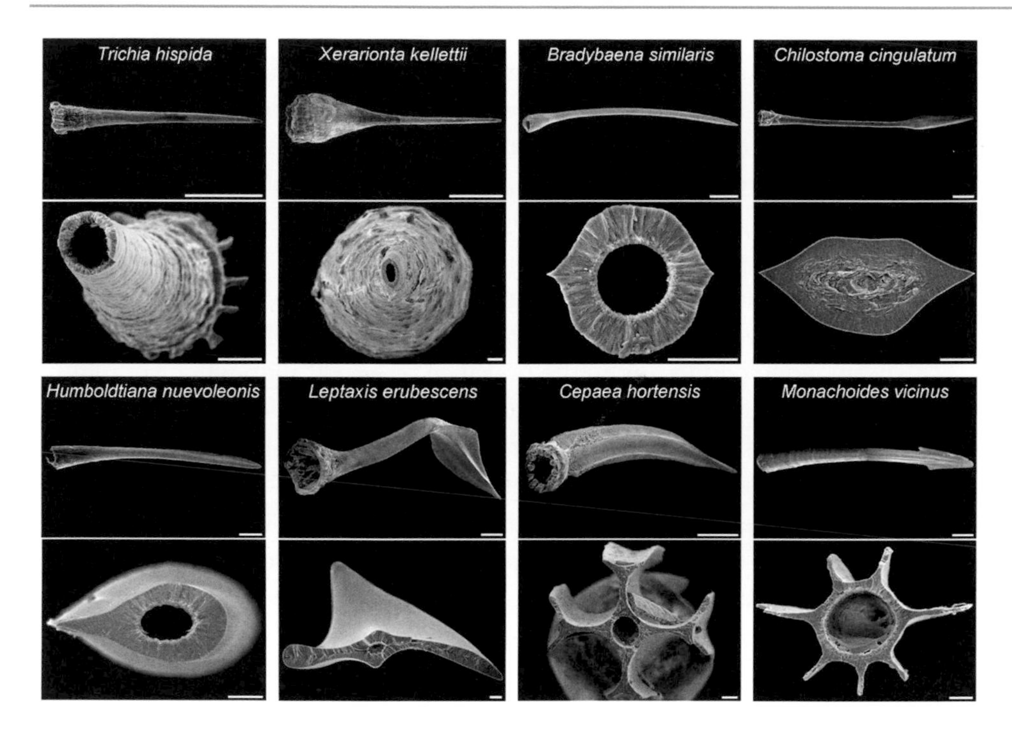

FIGURE 7.6 **HERMAPHRODITIC SNAILS USE LOVE DARTS TO MANIPULATE PARTNERS**

This image displays a diversity of love darts (side view and cross-sections) used by hermaphroditic snails (Superfamily Helicoidea). If the dart hits the target, the shooter then plays the less costly male role of breeding using sperm. The dart also promotes the storage of his sperm in the recipient, who plays the more costly female role.

In hermaphroditic land snails and slugs, there is also traumatic wounding, but it does not also involve traumatic insemination. During courtship in the brown garden snail (*Cantareus aspersus*), each party tries to achieve paternity by delivering a sharp *love dart* on contact (Figure 7.6). The darts do not deliver sperm; that only happens if copulation follows. However, a successful "stabbing" pays off for the stabber, because each individual is attempting to increase its reproductive success through its male function. About half the time there is a direct hit to the body, following which the number of donated sperm stored by the receiving party is more than doubled. These consequences demonstrate that paternity is enhanced when the dart hits its target, and research has shown that it is a bioactive mucous coating on the dart that is responsible for the shooter's success. This mucous positively alters the subsequent reception of the sperm package delivered by the shooter to "his" mate. Being a good shot appears to be important, since it is usually the first male's sperm that gets to fertilize the eggs, known as *first sperm precedence*.

Although it is true that sperm are relatively inexpensive to make compared to many of the costs females bear, they are not limitless, and males of many species have the capacity to adjust the sperm content of their ejaculates depending upon the apparent prospects for fertilizing success. This is known as sperm allocation. In a large review of 37 species, the prevailing response of males that faced a significant risk of sperm competition was to increase their sperm allocation. For instance, male Norway rats (*Rattus norvegicus*) ejaculate considerably more sperm when copulating in the presence of a rival than when copulating with no other males present. And in primate species where females commonly mate with multiple males, males have relatively larger testes (Figure 1.11) and dedicate more energy to sperm production and copulation than in primates that employ monogamous systems. The allocation of many sperm in an ejaculate is often coupled with a high rate of copulation. Extreme cases of such sperm allocation are known as *swamping*; some songbirds copulate hundreds of times per day during the egg-laying period.

On the other hand, there is also evidence that when it is not in a male's best interest to allocate a high number of sperm to an ejaculate, he has the capacity for adjustment. In one experiment, researchers coated virgin female Pacific field crickets (*Teleogryllus oceanicus*) with extracts from males, each of which had a distinct *olfactory* signature. New males given the opportunity to mate with her adjusted their sperm allocation downward in proportion to her apparent mating experience, as indicated by the number of extracts she carried. The interpretation was that these males would otherwise be wasting too much sperm where the prospects of fertilization were low. A well-publicized study using heterosexual human couples predicted that if members of a couple spent more time apart between instances of sexual intercourse, the ejaculate delivered upon their reunion would contain a higher sperm count. This is what was indeed found, the interpretation being that time apart correlated with the risk of cuckoldry, and large numbers of sperm can help to counter that risk. A control group provided ejaculates from male masturbation, which is sexual self-stimulation, often resulting in orgasm and ejaculation. The semen samples from this group showed no such pattern.

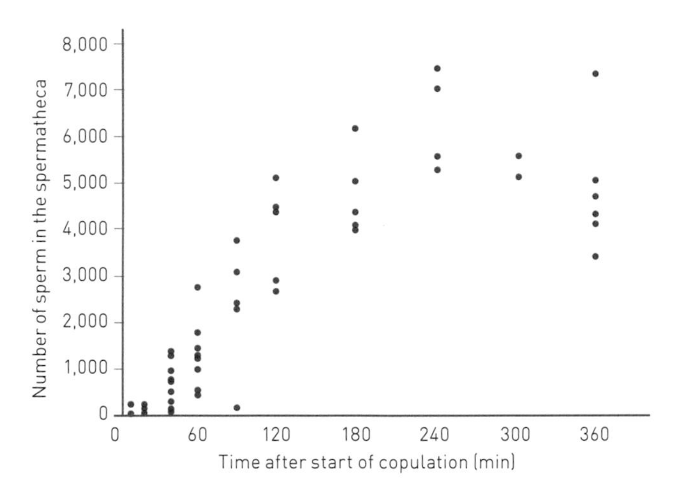

FIGURE 7.7 **LONGER COPULATION INCREASES SPERM TRANSFER IN SOME INVERTEBRATES** Male firebugs (*Pyrrhocoris apterus*) contend with the risk of sperm competition by engaging in very long copulations. This graph shows that the number of sperm transferred and stored in the female spermatheca reaches a maximum after about 3 hours, yet half of copulations last at least 12 hours.

[7.7a] From Schofl & Taborsky (2002). Copyright © 2002, Springer-Verlag. With permission of Springer

There is additional evidence for sperm allocation in humans. In a study of 52 heterosexual young men who had had no sexual activity for 48 hours and who were shown sexually explicit material, ejaculates contained significantly more sperm when the visual material depicted two males and a female (sperm competition) compared to three females. In a different study that investigated the reproductive consequences of mate guarding in humans (which is cross-cultural), men who had engaged in fewer behaviors consistent with guarding their partner allocated more sperm to their ejaculates when having sexual relations with that partner.

A third during-mating strategy is *prolonged* copulation. This is seen in numerous invertebrate species, such as firebugs (*Pyrrhocoris apterus*), where copulation lasts much longer than the time necessary to deliver the maximum number of sperm (Figure 7.7).

Another means by which sperm competition proceeds is through *chemical warfare*. At the start of her reproductive life, the queen western honeybee (*Apis mellifera*) can mate with several males for the purpose of storing the sperm that she will use for the rest of her life. Males mate once only, however, and so they compete strongly for the opportunity to do so. But having mated with the queen, the battle has not necessarily been won. Where the sperm of one male find themselves swimming among the sperm of another, their respective seminal fluids engage in battle; the seminal fluid of a particular male tends not only to enhance the survival of his own sperm, but it also can incapacitate or at least impair the sperm of his competitor.

The fact that conflict can occur during mating can also sometimes be seen by comparing the architecture of male and female reproductive parts. Although most birds do not have a penis, waterfowl do. Both the male penis—which in some species can achieve a human-length erection almost instantaneously—and the female vagina are helical (corkscrew-shaped). However, one is clockwise and the other is counterclockwise, and the vagina is also equipped with dead-end pouches. These attributes are consistent with an arms' race over control of insemination. In this case, the mismatch appears to favor female control over insemination, if not copulation. One study determined that only three per cent of successful inseminations come from forced copulations. This appears to be because the female can influence the shape of her vagina to accommodate a preferred male but can also direct the semen of unwanted suitors into the dead-end pouches away from her unfertilized eggs (Figure 7.8).

7.2.3 Sexual conflict after mating

Alas, sexual conflict never rests. Following mating, the race can continue, and the guarding by a male of a female he has mated with is a widespread strategy, which we've already described as mate guarding. Possessiveness in human males is thought by evolutionary psychologists to be an expression of mate guarding, which is itself a type of paternity protection.

In many species, the guarding is overt. We've already seen that the ebony jewelwing uses pre-mating sperm displacement, but like many insects it also uses mate guarding. The males often control egg-laying sites so that they will not only have access to females intent on laying eggs (an example of resource-defense polygyny; Section 6.3.1) but can also guard them while they lay eggs. Female jewelwings are not mere

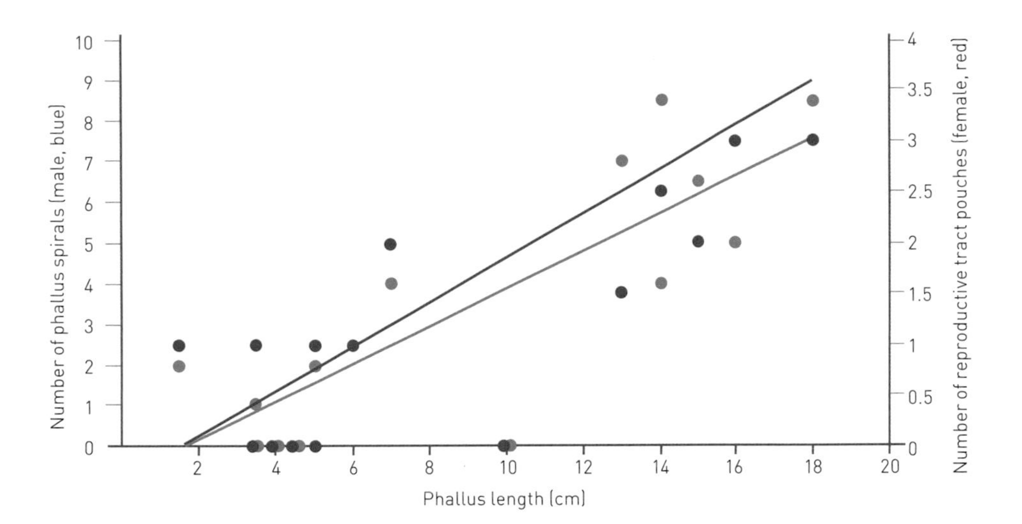

FIGURE 7.8 **REPRODUCTIVE ANATOMY SOMETIMES REVEALS A HISTORY OF SEXUAL CONFLICT**

In mallard ducks (*Anas platyrhynchos*), males harass females and coerce matings (Figure 7.1). Unlike most birds, male waterfowl have long penises and female waterfowl have correspondingly long vaginas. Here, each point on the graph represents one duck species. The longer the genital system, the more penis spirals and vaginal dead-end pouches there are, indicative of an arms race between the sexes to control insemination.

From Brennan, Prum, McCracken, Sorenson, Wilson, & Birkhead (2007)

victims of male interests, however. They actually store sperm in two different chambers, and the penis is really only effective in removing prior sperm from the larger of those chambers. Consequently, she can have sperm from previous males in the smaller chamber, and sperm from her most recent partner in the larger one. Now, when she is ready to deposit eggs, she can follow one of two strategies—an example of cryptic female choice. She can deposit eggs that mostly use sperm from the larger chamber (i.e., fertilized by her most recent partner) as he guards her. Or she can lay eggs that mostly use sperm from the smaller chamber—which are more diverse genetically, being from more than one male— by sheltering under a male already guarding another female. This latter strategy is known as *stealing a guard*. The latter option may be in her best interests because she already has diverse sperm and she can avoid costs of additional mating, which can include injury from rough encounters, by making use of her male guard.

Guarding has additional, different manifestations, many of which have been thoroughly studied in insects. One version of guarding is to delay dismounting, as firebugs do (Figure 7.7), especially where the operational sex ratio is male-biased. Under male-biased conditions, the New Zealand lesser spiny stick insect (*Micrarchus hystriculeus*) employs this behavior for several days, only briefly interrupting copulation to allow egg-laying. Other male insects continue to "ride" the female, even if copulation does not continue. *Cotesia* wasps add female mimicry as an interesting wrinkle. If a rival male approaches a couple that has just mated, the first-mating male will employ a female-mimicking solicitation behavior, which commonly attracts the rival male's courtship efforts.

It was once speculated that the "spiny" penises of many insects (and even of some mammals like that of the domestic cat, illustrated in Figure 9.6) were to serve as a kind of anchoring device to either make copulation last longer or delay the liberation of the female as a form of mate guarding. However, an elegant experiment that used a laser to remove such spines from cowpea seed beetle (*Callosobruchus maculatus*) penises rejected this hypothesis because they made no difference in mating duration. However, spined males did achieve more successful fertilizations. Previous work had shown that substances in the seminal fluid manipulated females by inducing egg-laying and by reducing receptivity to other mating-intent males. This latter function of the fluid is known as an *anti-aphrodisiac*. Evidently the spines abrade the walls of the female reproductive tract, increasing the invasion of her body cavity by this influential fluid.

The guppy penis also is adorned with "claws" (also illustrated in Figure 9.6). These claws have also been removed experimentally to reveal their utility. When females are receptive to mating, the penis claws seem to be of little consequence. However, male guppies also harass unreceptive females to mate, and those with the claws experimentally removed transferred considerably fewer sperm than the control group that still had claws (Figure 7.9). Evidently, the claws assist the males with the transfer of sperm when the female is not interested in mating.

Another male strategy to protect paternity is the mating plug (or copulatory plug). Although there are competing ideas about its purpose, a common explanation is that it serves as a "chastity belt." Mating plugs are reported in many invertebrate groups, as well as reptiles, birds, and mammals. After mating, male-produced fluids coagulate, thereby hardening and plugging the opening of the female reproductive tract. While the plug is usually not permanent, and females of many species can remove them, it does apparently give that male a timing advantage in achieving paternity. It is often a rival male that removes the plug, as has been shown in spiders and birds.

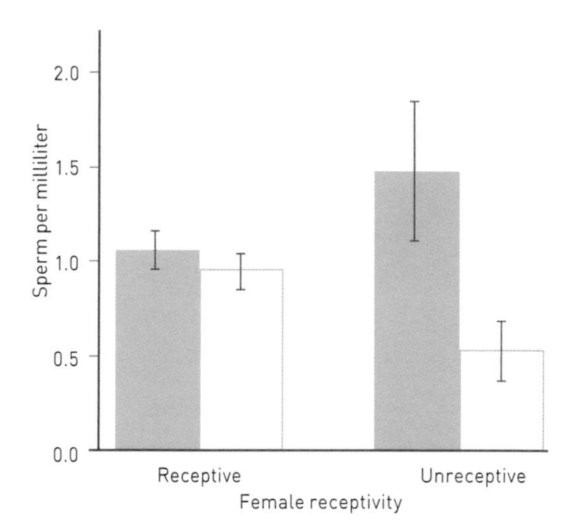

FIGURE 7.9 **THE MALE GUPPY'S CLAWED PENIS MANIPULATES UNRECEPTIVE FEMALES**

This graph contrasts amounts of sperm transferred by male guppies with a clawed penis (control group, filled columns; see Figure 9.6) compared to those with a surgically declawed penis (treatment group, open columns). When females are receptive, it makes no difference, but when they are not receptive to male advances, males achieve higher rates of insemination with the claw than without. The vertical bars in each measurement are known as standard error bars, which are determined through statistical analysis to represent the amount of uncertainty of each measure. The left pair of measures are not statistically different because the error bars overlap, but the right pair are significantly different.
From Kwan, Cheng, Rodd, & Rowe (2013).

A powerful female strategy in the post-mating sexual conflict arena is to eject sperm. Not all species can do this, and *sperm ejection* is not likely to be 100 per cent effective, but it does shift the balance back to the female. In the common fowl (*Gallus gallus*; Figure 6.1), most copulations are male-coerced. Despite the coercion, females still prefer dominant males, so if pursued by subordinate males, they use two strategies: one is to manipulate dominant males to reduce the likelihood of mating success in the subordinates, but if that fails, the second option is to forcibly eject the semen of the subordinate.

7.2.4 Sexual conflict during parenting

Most species do not provide parental care; once eggs are laid they are left to develop on their own and fend for themselves. In many species that do have parental care, it is often just one parent that provides it; this is most commonly the mother, although there are some exceptions,

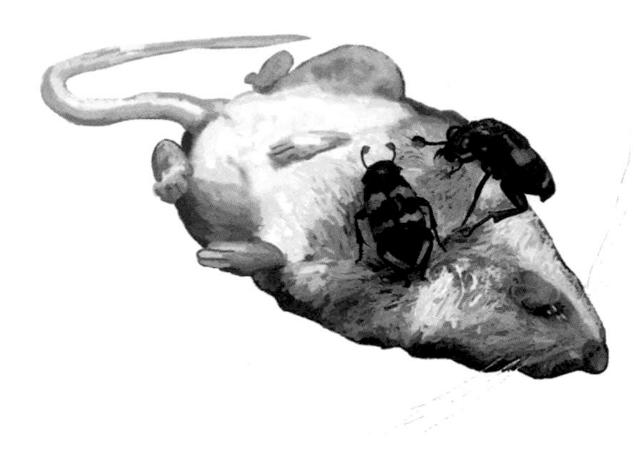

FIGURE 7.10 **CONFLICT CAN ARISE WHEN CURRENT AND FUTURE BREEDING INVESTMENT IS DIVIDED**
Burying beetles (*Nicrophorus* spp.) are rather unusual among insects in that they provide biparental care for their young. They find a carcass like the mouse shown here, mate on it, lay the fertilized eggs in it, and tend to the larvae during their first few days of life. Observations reveal conflict between the pair members over investment in the current brood and investment in future broods with other partners.
Drawing by Peter Mills

especially among fish and frogs. In those species with biparental care, alas, conflict between the sexes continues. The evolutionary root of the issue is this: one partner may have attractive opportunities for additional mating, providing he or she can get away with desertion or at least get away with splitting his or her attention between the current and future partners. If deserted, what should the now-single parent do? Desert as well? Or make the best of it? This is referred to as the *cruel bind*.

One well-studied system where *desertion* is common is another European songbird, the penduline tit (*Remiz pendulinus*). Incubation of the eggs and subsequent feeding of the chicks is uniparental, but it can be the undertaking of either the male or the female. Consequently, as many as a third of the nests are completely abandoned at the egg stage where first one parent, and then the other, leaves. Patterns of desertion suggest that each sex prefers the other sex to care for the nest so that the liberated parent can pursue additional matings with other partners. Females desert less often than males. Males are less likely to abandon a nest when the female has deserted him later in the season, when additional mating opportunities for him are unlikely. Other species have evolved even better ways to off-load parenting (Box 7.3).

A similar conflict is seen in burying beetles (genus *Nicrophorus*), but females seem to have the upper hand here. These insects have biparental care, where the parents feed the young by pre-digesting morsels from the carcass of a small vertebrate, such as a dead mouse, on which the eggs are laid (Figure 7.10). They even copulate on the carcass. In the case of arboreal burying beetles (*Nicrophorus defodiens*), if the carcass is relatively large, it is in the male's interest to attract a second female for egg laying. Males will attempt this by emitting a pheromone (an airborne hormone) whose purpose is to attract another female. This is not in his current partner's interest, however. Instead, she actively interferes with her partner's advertising, sometimes by biting him. When experimenters placed females on short tethers to restrict their capacity to rebuke their male partners, the males advertised significantly more. In a different species, the boreal burying beetle (*Nicrophorus vespilloides*), once the eggs hatch, the female produces a different sex-suppressing pheromone that has the effect of making her partner focus on the task at hand, which is to nourish the larvae. Clearly, it is not only males that have evolved the capacity to produce anti-aphrodisiacs for their arsenal.

7.3 MATING STRATEGIES CAN CHANGE WITH CIRCUMSTANCES

Life is complicated, so no single sexual or mating strategy is likely to be optimal in every situation, even within a single species. Accordingly, it is not surprising that individuals of behaviorally complex species show some flexibility in mating strategies, depending upon circumstances. When genetically influenced traits vary depending upon environmental factors, such species are said to manifest *plasticity*. Environmental factors, for instance, can impact mating opportunities and can even alter what constitutes an optimal mating strategy. An individual's age and social status are other factors that can play a role in influencing mating strategies. Moreover, producing offspring of a particular sex, whether sons or daughters, might be a good strategy depending upon circumstances.

7.3.1 Making the best of things

If you cannot win by the conventional rules of engagement, then perhaps you can win by some other method. Biologists use the term alternative mating strategies when there is more than one way to achieve reproductive success, although remember that use of the word *strategy* does not mean there is anything conscious about the choices

BOX 7.3 OFF-LOADING PARENTING IS ONE WAY TO REDUCE INTER-SEXUAL CONFLICT

Examination of the fates and fortunes of organisms shows that each individual seeks to achieve reproductive success, whether in monogamy, polygamy, or promiscuity. The tactics are many and diverse. Some species make direct use of other species to raise their offspring, the best known of which systems is brood parasitism. This is where an egg-laying female deposits her eggs in a nesting structure of another female or pair, known as a host, which then proceeds to raise her young. Some species only breed this way, and they are known as *obligate* brood parasites. Other species sometimes raise young on their own and sometimes parasitize others and are known as *facultative* brood parasites. Among facultative species, the brood parasitism often relies on nests of the same species, but obligate brood parasitism cannot, because no parents in the species ever raise their own offspring.

Brood parasitism has evolved in six groups of birds. The best-known examples are several of the New World cowbirds (Icteridae) and numerous Old World cuckoos (Cuculidae). In fact, the word *cuckold* is derived from the observation that cuckoos get others to raise their young. North America's brown-headed cowbird (*Molothrus ater*) is known to have used over 200 host species, but some brood parasites specialize in one or a few hosts. Almost any smallish songbird will do for the cowbird (Figure 7.11). What the female must do is find songbird nests at the stage where the host is assembling her own clutch by laying an egg per day. Then, when the host is absent, the female cowbird removes a host egg from the nest and shortly thereafter lays an egg of her own. Some hosts will recognize the egg and remove it or abandon the nest, but most incubate it and raise the hatched chick. Several North American ducks are facultative brood parasites, including the ruddy duck (*Oxyura jamaicensis*). In birds, the pattern of doing this intraspecifically is also known as *egg dumping*.

Brood parasitism is a successful means of reproduction for numerous fish species too. For instance, there are several species of "mouth-brooding" cichlid fish (Cichlidae) in Lake Tanganyika in east Africa. The fertilized eggs are taken by the female cichlid into her mouth for a time while they develop. A brood parasite known as the cuckoo catfish (*Synodontis multipunctatus*) spawns at the same time and in the proximity of the cichlids, and the smaller catfish eggs become part of the brood in the host cichlid's mouth. The catfish eggs hatch sooner and consume many of the cichlid eggs.

Some insects engage in similar behavior, but it usually does not exploit host parental care. Instead it involves exploiting food assembled by the host during its own egg-laying period, and therefore this reproductive behavior is more often known as *kleptoparasitism*. Kleptoparasitism is any form of feeding where one animal steals the food that another has caught or assembled. As a reproductive strategy in insects, cuckoo wasps (Chrysididae) lay their eggs on the pollen masses collected by other bee or wasp species, or they lay their eggs directly in the nests of those hosts. In some species the developing wasp larvae eat the host larvae, making such species of cuckoo

FIGURE 7.11 **BROOD PARASITES OFF-LOAD PARENTING TO UNSUSPECTING FOSTER PARENTS**

Many nests of North American songbirds, like this red-eyed vireo (*Vireo olivaceus*) nest, are parasitized by female brown-headed cowbirds (*Molothrus ater*). In this nest, there are two vireo eggs and one cowbird egg. Commonly, the host will accept the egg and will feed the hatched chick. Because the cowbird is often larger than the host young, it is likely to get more than its fair share of the food provided by the foster parents.

Scott Camazine / Alamy Stock Photo

wasps *parasitoids*, but in other species, the developing larvae are kleptoparasites, eating the food assembled by the host. Europe's ruby-tailed wasp (*Chrysis ignita*) is both.

made. In a scenario where a male is not going to succeed through direct competition, then success may be available via an alternative and perhaps more devious method. One widespread strategy that is employed by males known as satellite males. The term satellite is used because such males are likely to associate with territorial or dominant males because it will be in that vicinity where mating opportunities for them arise. Satellite males do not attempt to mate by establishing a territory, by being dominant, or by putting on a good display of one sort or another. Instead, they do not draw attention to themselves but instead mate quickly with females when opportunity allows.

What determines whether a male will opt for one strategy or another? A factor that is commonly influential is the physical condition of the individual. Sometimes alternative mating strategies are genetically determined (see Box 8.1), but in other cases alternative mating strategies are a result of condition dependence. Condition-dependent traits tend to produce one phenotype under one set of conditions and another phenotype under a different set. In the case of mating, males in better condition tend to manifest the dominant, territorial, displaying, and *parental* phenotype while males in poorer condition manifest the *satellite* phenotype, commonly called the *cuckolder* phenotype because they attempt to mate with females mated to a parental male (Figure 7.12).

FIGURE 7.12 **MALE MATING STRATEGIES CAN BE CONDITION-DEPENDENT**

This image of bluegill sunfish (*Lepomis macrochirus*) spawning shows alternative mating tactics that depend upon male condition. Here the female (shown in dark shadow) has moved to the nest of the parental male, with two cuckolder males nearby: a small young sneaker beside the female and a larger, older male mimicking a female a little to the side. When the female releases her eggs, all three males may simultaneously release their *milt* (fish semen) so that her eggs will be fertilized by multiple males, yet they will be parented by the non-cuckolder (parental) male. From Neff & Svensson (2013). Copyright © 2013 The Authors. By permission of the Royal Society

The chinook salmon (*Oncorhynchus tshawytscha*) is a commercially valuable fish species that is well known for its great migrations. Born in freshwater rivers, they spend several years in the ocean feeding before they return to their birth river, whereupon they breed and then die. During their final years, most males employ the same strategy: achieve a large size and then compete for and establish a small territory on the river bottom in which to build a nest so that one or more females will be attracted to it for egg laying. Fertilization is external, so that both the eggs and the *milt* (fish semen) are released at the same time in the river water at the male's nest.

But not all males do this. A minority of sexually mature males are called *jacks*, which are one or more years younger than the youngest females and which are also considerably smaller than the adult males. Jacks are condition-dependent satellite males that quickly exploit fertilizing opportunities. When the female sheds her eggs in the adult male's nest and the male simultaneously releases his milt to fertilize them, the small jack rushes in as a third party and also releases his milt, thereby fertilizing many of the eggs. Males that use this alternative mating strategy are called sneakers. Experimentation shows that by altering conditions, biologists can induce males to alter their strategy. For instance, when combinations of breeding adults (two males and a female) are manipulated in large aquaria, a jack sockeye salmon (*Oncorhynchus nerka*) will fight for access to that female, rather than sneak, if the second male is another jack.

Female mimicry is yet another ruse males use to improve their chances of reproductive success when more direct methods have low prospects for success. In fact, female mimicry by males can provide several benefits, including avoiding the costs of male aggression, gaining a surprise advantage in combat, and, of course, getting close to females for sneaky copulations. Some male flat lizards (*Platysaurus broadleyi*) retain a female-like appearance even as they mature sexually and so are sometimes called *she-males*. This too, is condition dependent, because the she-males ultimately develop into typical mature males that are bright-colored, territorial, and highly aggressive, known as *he-males*, which have greater reproductive success than she-males. When the he-male relies on visual cues, the she-male ruse works, reducing aggression from he-males. In fact, he-males will even court she-males for mating purposes. Based on molecular cues that can be smelled, however, the he-male is never tricked, so the she-males must keep their distance from the he-males if they wish to continue to gain the aggression-free advantage that their female mimicry provides. In the sand goby (*Pomatoschistus minutus*) fish, there are two types of satellite males, one of which is female-like. When the territorial nest-building males are removed, many individuals of both satellite types will build nests, and the female-like ones even develop male patterning.

Although being a satellite is likely to yield lower reproductive success than being a "macho" male in most cases, there is good evidence that sometimes it is an equally advantageous, if not better, strategy. Even if jack salmon don't usually produce as many fertilized eggs, they do have the advantage of being younger and so being less likely to die before getting to breed. And, in a study of South American fur seals (*Arctocephalus australis*), gene sequencing of females, satellite males, territorial males, and their offspring showed that some satellite males actually sire more offspring than some territorial males.

A related alternative strategy is known as prospecting. Meerkats (*Suricata suricatta*, Figure 7.13) are rather endearing and highly social African carnivores that are cooperative breeders, although most breeding in a group is done by the dominant pair. This poses a problem for subordinate males and females, who have the choice of either waiting for opportunities that might not arise in their short lives or seeking opportunities elsewhere. Genetic studies show that subordinate males achieve some reproductive success by prospecting. They

FIGURE 7.13 AMONG SOCIAL MAMMALS, SUBORDINATE INDIVIDUALS INCREASE MATING OPPORTUNITIES BY PROSPECTING

Meerkats (*Suricata suricatta*) are highly social. One mating tactic for subordinate individuals is mating using prospecting, where subordinate males leave the group and attempt to mate with females of other groups, commonly subordinate females.

accomplish this by temporarily leaving their home group to seek copulations with females in other groups. Although they may not achieve mating success in their own group, these males achieve some success by mating with both subordinate and dominant females of other groups. This is also the most immediate way for subordinate females in these other groups to achieve some reproductive success, because these females often have no access within their own group to males who are not relatives. Instead of being called extra-pair paternity (EPP), this is known as *extra-group paternity* (EGP).

7.3.2 Life history theory argues for plastic mating strategies

Life history theory is that branch of biology that deals with things like life expectancy, age at first reproduction, how and when to balance investing in reproduction versus in growth and survival, etc. Some species are described as manifesting an r-selection set of life history traits. Put very simply, these species tend to live in more uncertain environments, and they grow fast, breed young, produce many offspring but invest less in each, and have a short life expectancy. The "r" is a mathematical term from equations describing the growth of such populations. In contrast, other species are described as having a K-selection set of traits, where the "K" is another term in a growth equation. These species live in more stable environments, and they grow slower (but to a larger size), they breed at a later age, they produce fewer offspring but invest more in each one, and they have a longer life expectancy. Whereas r-selected species live "fast," K-selected species live "slow."

K-selection and r-selection are usually applied to species, but occasionally the terms are also applied to different groups within species. Thus, if a species exhibits some plasticity in life history characteristics, we will expect to find some variation across the r- and K- spectrum. This way of viewing things has been applied to human sexual behavior. Humans tend to be a K-selected species, although there is a spectrum. K-type humans, who have access to greater resources and therefore have greater health and longer life expectancy, tend to reproduce later and less often, and they invest more in their offspring than r-type humans, who live under more precarious conditions. It is environmental conditions and opportunities, not genetics, that determine whether human individuals are r-type or K-type.

One especially relevant component of life history theory is known as *reproductive value*. This is the potential number of offspring an individual has during its lifetime. More interesting for our purposes is the residual reproductive value (RRV). This is how much reproductive value is left at any given point in an individual's life. Species that breed once, like Pacific salmon (*Oncorhynchus* spp.), have zero RRV following breeding. In other species, once individuals reach maturity their RRV declines as they age. Individuals with high RRV are more likely to invest in maintenance and self-preservation because they can expect considerable future reproduction. Individuals with low RRV are more likely to invest in reproduction as their time runs out, often at the expense of self-preservation.

7.3.3 Sex allocation

There is one other significant variable that can affect sexual and mating strategies that merits attention: sex allocation, which is not to be confused with sperm allocation. Over the past four or five decades, it has been discovered that many organisms, even some birds and mammals that use chromosomal sex determination, manipulate the sex of their offspring to achieve greater reproductive success. This has been analyzed in the context of a profound rule in reproductive biology known as Fisher's principle: the reproductive effort expended by a population is equally divided between male and female offspring. If males and females are equally expensive to produce, the offspring sex ratio will be 50:50.

Why is this? In a different world, one could imagine having more females than males, since females are more limiting and one male can "service" multiple females. Imagine, for instance, starting with a population that was 20 per cent male, where a single male could service four females. However, this would not be stable, since any genetic change that caused parents to produce more males would be rewarded, because a male produces four times as many offspring as a female. So, male production would increase, and this would bring the ratio back to 50:50. In other words, the rarer sex always has the reproductive advantage.

But what if the two sexes are not equally costly to produce? For instance, what if males are twice as costly to produce because they are bigger or because they require a longer development period? Parents should then invest more in daughters. But, once daughters are twice as common in the population, it makes sense to invest equally in male and female offspring again, because although a male is twice as costly to produce, he now has twice the value because he is less common.

Biologists continue to accept the Fisher principle, but a related idea known as the Trivers-Willard hypothesis has generated considerable research. This states that a mother's general condition will influence whether she will invest more in daughters or sons, assuming that her condition will influence the condition of her offspring. Trivers and Willard argued that in most cases a male in good condition will out-reproduce his sister in similar condition, but when both are in poor condition, the sister will out-reproduce her brother. They suggested that

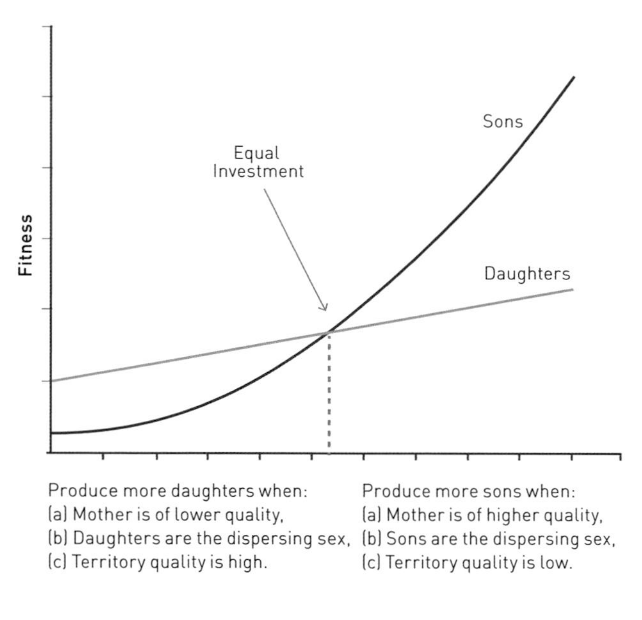

FIGURE 7.14 **SEX ALLOCATION IS ADAPTIVE WHEN HAVING OFFSPRING OF A PARTICULAR SEX IS ADVANTAGEOUS**
In some circumstances mothers preferentially produce daughters, while in others they preferentially produce sons. This is known as sex allocation. The preferred allocation of sex to offspring reflects which sex is likely to have an advantage. The Trivers-Willard pattern relates to mother quality (a); if she is healthy she tends to produce sons (e.g., red deer). If sons disperse away more than daughters (b), there is a tendency to produce sons (e.g., some primates). If the territory quality is high (c), the tendency is to produce daughters, although this is related to having daughters remain to assist in future nests (e.g., Seychelles warbler [*Acrocephalus sechellensis*]).
Adapted from Clutton-Brock, Albon, & Guinness (1984); Silk & Brown (2008); Komdeur, Daan, Tinbergen, & Mateman (1997)

what follows is that females should be able to adjust the sex ratio of their offspring, citing several examples of mammals where this pattern prevails, such as red deer and Barbary macaques (*Macaca sylvanus*), a primate from North Africa. Since then, there are numerous additional ecological factors that have been shown to influence whether a mother will invest more in daughters or sons (Figure 7.14).

Additional research has revealed that some mothers can allocate sex according to characteristics of the father. Florida's brown anoles (*Anolis sagrei*) are non-monogamous lizards. Females produce more sons using the sperm of large males and more daughters using the sperm of smaller males (Figure 7.15). Some birds also produce more sons when the father rates high in sexually selected ornaments; otherwise, they produce more daughters. In another study of bighorn sheep (*Ovis canadensis*), females invested more in lactation for a daughter when the father was a relatively unsuccessful sire but more for a son when the father was high status and more successful. In humans, there tends to be about 105 males born for every 100 females, but whether this is adaptive or the by-product of some other mechanism is not clear.

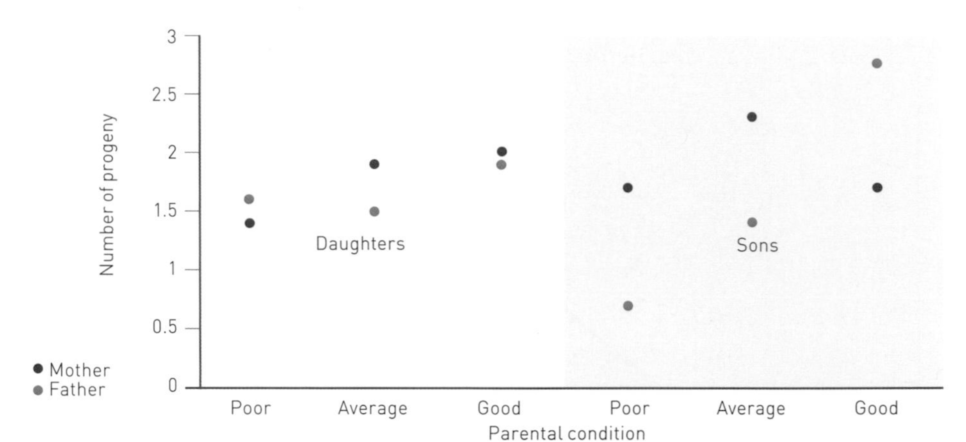

FIGURE 7.15 **SEX ALLOCATION CAN REFLECT THE BODY CONDITION OF THE FATHER**

Florida's brown anoles [*Anolis sagrei*] are non-monogamous. Maternal body condition has no significant effect on the number of daughters or sons produced [red dots]. Paternal body condition has no significant effect on the number of daughters produced [blue dots, left half], but paternal body condition has a profound effect on the number of sons produced [blue dots, right half], with fathers in superior condition producing significantly more sons. Some birds also produce more sons when the father rates high in sexually selected ornaments, and daughters otherwise.

[7.15a] Adapted from Cox, Duryea, Najarro, & Calsbeek [2011]
[7.15b] iStock.com / sdbower

7.4 SAME-SEX PARENTING

Another twist on mating strategies is same-sex parenting. Among Western cultures at least, same-sex partnerships represent a newly endorsed family construct that has been facilitated by social change at many levels: progressive legislation, changes to adoption rules, surrogate parenting, sperm banks, and short-term heterosexual liaisons intended to result in pregnancy. It appears that a significant minority of people in all cultures, usually estimated in the range of two to eight per cent, have a non-heterosexual orientation, and there is no reason to think that this rate hasn't been true through our history.

Recall from Chapter 2 that strict homosexuality, more so than bisexuality, presents a theoretical challenge. After all, it is obvious there are strong reproductive rewards for heterosexual behaviors and no such rewards for strict homosexuality. In other words, heterosexual behavior contributes to the gene pool of the next generation, but homosexual behavior does not. So, from a biological viewpoint, it is difficult to imagine how forces of natural selection could preserve alleles associated with homosexuality, since they are unlikely to be passed on.

There have been numerous biological hypotheses to respond to this apparent paradox. They fall into two different categories. One category is that homosexuality is somehow adaptive, with the prevailing idea being that homosexuals effectively assist heterosexual family members in their parenting efforts, consistent with the concept of inclusive fitness and kin selection (Section 6.6.2). That homosexuality is adaptive, however, does not have strong support, and so the more popular idea is that homosexuality is influenced by biology but is non-adaptive. This does not mean it is not a valid orientation, but it means that it arises recurrently and is not a trait that has been favored by natural selection.

The search for alleles that lead to homosexual orientation has mostly come up short. It is true that there are alleles on several chromosomes that have been reported to correlate with (but not dictate) homosexual orientation, but the results have not always been repeatable, and there is hardly a consensus that there is a gene or genes where alleles are correlated with homosexuality. All the same, many studies of twins, especially those that are *monozygotic* (meaning "one zygote," hence identical twins), indicate there is a biological component to homosexual orientation for both gay women and gay men. One proposal has been that certain alleles contribute highly to reproductive success in women (i.e., they are adaptive in females), but those same alleles in males tend to contribute to a homosexual orientation. This is a system known as *balancing selection* because the allele(s) are favored by natural selection in one case but disfavored in another.

Recent research has focused on what is known as epigenetics, which essentially means "in addition to genetics." In *epigenetic inheritance*, there are inherited influences on the development of an offspring that are not encoded in the offspring's genotype. One category of epigenetics is a *maternal effect*, which is the influence of any of the maternal genotype, phenotype, or prenatal environment on the development of an offspring's phenotype, and this, of course, could include sexual orientation. For instance, for a male, the likelihood of developing a homosexual orientation has correlated in more than one study with the number of older brothers he has through the same mother. Known as the *older brother effect*, if true, this is almost certainly a maternal effect. Female offspring developing in a uterine environment that is relatively higher in testosterone levels have been shown to exhibit behavioral traits after birth that are associated with males, but so far these results do not include sexual attraction to women. Biological contributors to lesbianism are more intractable to study because females are more plastic in their orientation than are males—who less often identify other than as homosexual or heterosexual.

With respect to mating, direct homosexual parenting has been documented in nature, especially in birds. Female–female pairings that produce young by EPCs (Section 6.2.1) have been demonstrated in more than one gull species (*Larus* spp.). These pairings can be female–female from the outset, especially where there is a female-biased sex ratio, or they can be created when a polygynous trio loses the male partner. Laysan albatrosses (*Phoebastria immutabilis*) are monogamous and have one-egg nests. In a Hawaiian population that was female-biased at the time of study, 31 per cent of nests were attended by pairs of females (Figure 7.16). These female–female pairings produce fewer offspring than the male–female pairings, but they produced more offspring than unpaired females. Albatrosses often mate for life, and some of these female–female pairings have proved to be stable over many years. Whether these pairs are truly homosexual in the sense of sexual behavior or orientation cannot be assumed. This is especially so given that the eggs are fertile and the population has a female-biased sex ratio, perhaps supporting the interpretation that these pairings are making the best of a less than ideal situation. Indeed, some females will switch to male partners in a subsequent breeding season but are able to do so only if they had been a successful parent in their previous same-sex attempt.

FIGURE 7.16 **SAME-SEX PARENTING IS KNOWN FROM A VARIETY OF BIRD SPECIES**

About 30 per cent of Laysan albatross (*Phoebastria immutabilis*) pairs on the Hawaiian island of Oahu are both females. This may be partly due to the fact that the population is about 60 per cent female. The same-sex pairs can raise young successfully from an egg fertilized by a male in the colony, but success is at a much lower rate than male–female pairs.
David Patte / US Fish and Wildlife Service

Homosexuality is defined by sexual behavior and orientation, not by social bonds and shared parenting, and so this pattern of parenting is more often referred to as *homosociality*. All the same, homosexual behavior has been directly observed in numerous bird species. A large review of the evidence reveals that such individuals often also engage in heterosexual behavior, making them bisexual. Another pattern is revealed by the evidence: male–male homosexual behavior is more common among polygynous birds, while female–female homosexual behavior is more common among monogamous birds, especially when males do a disproportionately large amount of parenting. These findings indicate that emancipation from parental care for either males or females correlates with increasing levels of homosexual behavior.

CHAPTER 7 SUMMARY

- An evolutionary strategy is a genetically inherited pattern of behavior; when applied to mating, it is a suite of behaviors that optimize mating activity for a given set of circumstances.
- Sexual conflict commonly results because the optimal mating strategy for the male is not aligned with the optimal mating strategy for the female.
- Sexual conflict can occur before mating; for instance, males of some species will commit infanticide to gain mating opportunities, and males of many species engage in sexual coercion.
- Sexual conflict can occur during mating; for instance, males of some species remove or dislocate sperm from prior males or employ "chemical warfare" that subjugates female interests.
- Sexual conflict also occurs after mating; males of some species will guard the female or delay "dismounting" to prevent her mating with other males; females may select sperm through cryptic mate choice or by ejecting sperm; and either parent may desert, leaving their mate with the task of raising the young.
- Sexual conflict occurs in hermaphroditic species, commonly because the goal of each is to play the male role because it has the more immediate payoff for reproductive success.
- Mating strategies can vary with circumstances; some strategies are condition-dependent: they are only used when the individual is in good health, and the individual switches to alternative strategies when health is sub-optimal.
- Life history theory considers traits like age at first breeding, frequency of breeding, longevity, investments in offspring versus investments in survival, and whether to have many small offspring or few large offspring.
- Some species can allocate sex to their young; this is beneficial because sometimes it is a better strategy to produce daughters and other times to produce sons.

- Homosexual behavior is common in the animal world, although it is usually in the context of a bisexual interest in sex; there are cases of female–female bird pairs successfully raising young, although they experience lower success rates than heterosexual pairs.

FURTHER READING

Arnqvist, G., & Rowe, L. (2005). *Sexual conflict.* Princeton, NJ: Princeton University Press. https://doi.org/10.1515/9781400850600

Bell, M.B.V., Cant, M.A., Borgeaud, C., Thavarajah, N., Samson, J., & Clutton-Brock, T.H. (2014). Suppressing subordinate reproduction provides benefits to dominants in cooperative societies of meerkats. *Nature Communications, 5,* 4499. https://doi.org/10.1038/ncomms5499

Brennan, P.L.R., Prum, R.O., McCracken, K.G., Sorenson, M.D., Wilson, R.E., & Birkhead, T.R. (2007). Coevolution of male and female genital morphology in waterfowl. *PLOS One, 2* (5), e418.

Chase, R., & Blanchard, K.C. (2006). The snail's love-dart delivers mucus to increase paternity. *Proceedings of the Royal Society B: Biological Sciences, 273*(1593), 1471–1475. https://doi.org/10.1098/rspb.2006.3474

Clutton-Brock, T.H., Albon, S.D., & Guinness, F.E. (1984). Maternal dominance, breeding success and birth sex ratios in red deer. *Nature, 308,* 358–360.

Cox, R.M., Duryea, M.C., Najarro, M., & Calsbeek, R. (2011). Paternal condition drives progeny sex-ratio bias in a lizard that lacks parental care. *Evolution, 65*(1), 220–230. https://doi.org/10.1111/j.1558-5646.2010.01111.x

Engel, K.C., Stökl, J., Schweizer, R., Vogel, H., Ayasse, M., Ruther, J., & Steiger, S. (2016). A hormone-related female anti-aphrodisiac signals temporary infertility and causes sexual abstinence to synchronize parental care. *Nature Communications, 7,* 11035. https://doi.org/10.1038/ncomms11035

Komdeur, J., Daan, S., Tinbergen, J., & Mateman, C. (1997). Extreme adaptive modification in sex ratio of the Seychelles warbler's eggs. *Nature, 385,* 522–525.

Kwan, L., Cheng, Y.Y., Rodd, F.H., & Rowe, L. (2013). Sexual conflict and the function of genitalic claws in guppies (*Poecilia reticulata*). *Biology Letters, 9* (5).

Larmuseau, M.H.D., Matthijs, K., & Wenseleers, T. (2016). Cuckolded fathers rare in human populations. *Trends in Ecology & Evolution, 31*(5), 327–329. https://doi.org/10.1016/j.tree.2016.03.004

Leivers, S., Rhodes, G., & Simmons, L.W. (2014). Sperm competition in humans: Mate guarding behavior negatively correlates with ejaculate quality. *PLOS One, 9*(9), e108099. https://doi.org/10.1371/journal.pone.0108099

Neff, B.D., & Svensson, E.I. (2013). Polyandry and alternative mating tactics. *Philosophical Transactions of the Royal Society B: Biological Sciences, 368*(1613), 20120045. https://doi.org/10.1098/rstb.2012.0045

Perry, J.C., Sirot, L., & Wigby, S. (2013). The seminal symphony: How to compose an ejaculate. *Trends in Ecology & Evolution, 28*(7), 414–422. https://doi.org/10.1016/j.tree.2013.03.005

Schofl, G., & Taborsky, M. (2002). Prolonged tandem formation in firebugs (*Pyrrhocoris apterus*) serves mate-guarding. *Behavioral Ecology and Sociobiology, 52*, 426-433.

Shackelford, T.K., & Goetz, A.T. (Eds.) (2012). *The Oxford handbook of sexual conflict in humans.* Oxford: Oxford University Press. https://doi.org/10.1093/oxfordhb/9780195396706.001.0001

Silk, J.B., & Brown, G.R. (2008). Local resource competition and local resource enhancement shape primate birth sex ratios. *Proceedings of the Royal Society, 275*(1644), 1761–1765.

Thompson, M.E. (2014). Sexual conflict: Nice guys finish last. *Current Biology, 24*(23), R1125–R1127. https://doi.org/10.1016/j.cub.2014.10.056

Zietsch, B.P., Morley, K.I., Shekar, S.N., Verweij, K.J.H., Keller, M.C., ... , (2008). Genetic factors predisposing to homosexuality may increase mating success in heterosexuals. *Evolution and Human Behavior, 29*(6), 424–433. https://doi.org/10.1016/j.evolhumbehav.2008.07.002

8 Sex Determination and Differentiation

KEY THEMES

Sex determination refers to the factors that govern whether an individual will be male or female, whereas *sex differentiation* is the complex pattern of sexual development that flows afterward. Sexual development is regulated by a complicated interplay of genes and hormones. Sex in most mammals is determined by a gene on the Y-chromosome, setting in play male, rather than female, development. There are different types of genetic sex determination that don't rely on this *SRY* gene, and other species rely on non-genetic, environmental factors to determine sex.

In many societies both past and present, patriarchal systems have often put more value on the birth of male children. King Henry VIII (reign: 1509–1547) wanted a male heir to consolidate the Tudor family stronghold on English political power (Figure 8.1). His first marriage to Catherine of Aragon lasted more than two decades and produced a healthy daughter, but no sons that survived infancy. Henry's increasing desperation for sons partly fueled his notorious penchant for periodically replacing a current wife with a new one. Henry terminated his six marriages in diverse ways—annulment (twice), execution (twice), being widowed (once), and his own death (once!). His desire to terminate his first marriage had large social consequences: when the Roman Catholic pope would not sanction the annulment of Henry's first marriage, Henry rejected papal authority. He made the Church of England independent of the political authority of the Roman Catholic Church, a significant chapter in the development of Protestantism in Europe.

It is now well recognized that what determines the sex of a child resides mostly with the father, because offspring sex is determined by the genetic content of sperm, not by the genetic content of the egg. It wouldn't be until the late 1600s, using the relatively new invention of the microscope, that people saw for the first time that semen included "swimming cells" we now know as spermatozoa, or more succinctly, sperm. Still, the content of that sperm remained mysterious for more than 200 additional years. It wouldn't be until the early twentieth century when it was demonstrated that sex is determined by X and Y sex chromosomes, the combination of which is determined by the sex chromosome carried by sperm (see Figure 3.10 for the human case). Although we now know that a variety of factors can influence whether male-determining or female-determining sperm fertilize the egg, several of Henry's wives suffered the consequences of not producing sons, notwithstanding that it was Henry's sperm that gave him mostly daughters.

It turns out that not all organisms have a sex-determination system dependent upon the genetic content of sperm. In butterflies and birds, we'll see that the determining factor is the genetic content of the egg. And in many groups, there are non-genetic systems that determine offspring sex—some of which are surprising and often perplexing. This

FIGURE 8.1 **HENRY VIII WAS DETERMINED TO HAVE SONS**
Much of the dramatic narrative surrounding Henry VIII and his many wives was driven by his desire to have a male heir in patriarchal sixteenth-century England. Oblivious to the mechanisms of sex determination, his six marriages were at least partly driven by this obsession.

does not mean that genes in these other groups are not involved in producing male versus female phenotypes. Genes are *always* involved in sex differentiation. But they are not always involved in deciding whether a male or female phenotype will be produced, which is sex determination. On top of this complexity, hermaphroditism, where individuals manifest both sexes in one body, is not uncommon in nature. Hermaphroditism can be expressed simultaneously or sequentially—in the latter case, sometimes reversibly.

When Henry VIII was married to Jane Seymour, he did beget one son, Edward, who survived infancy. Upon Henry's death, Edward succeeded to the throne, but he was only nine years old, and he died some six years later. However, two of Henry's daughters, Mary and Elizabeth, were highly influential characters in history, becoming the first women to sit on the English throne. Elizabeth I was queen for 45 years, during which period she strongly influenced English policy and contributed to England's rise to international prominence during the Elizabethan era. Alas, Henry's aspiration for a son is not a merely antiquated aspiration; in many cultures, both mothers and fathers prefer to have male offspring (Figure 8.2).

8.1 ARE MALE AND FEMALE BODIES THE ONLY TWO OPTIONS IN SEX DETERMINATION?

Before we look at systems of sex determination, we should first ask: does nature have systems where male and female are not the only options? Most familiar animal species are sexual reproducers where individuals are either male or female for life. This familiarity can bias our expectations in several ways, not least of which is an expectation that that is how all species do things. Although we retain the working definition that males produce sperm and females produce eggs, there are quite a few deviations from the one-male-phenotype, one-female-phenotype model.

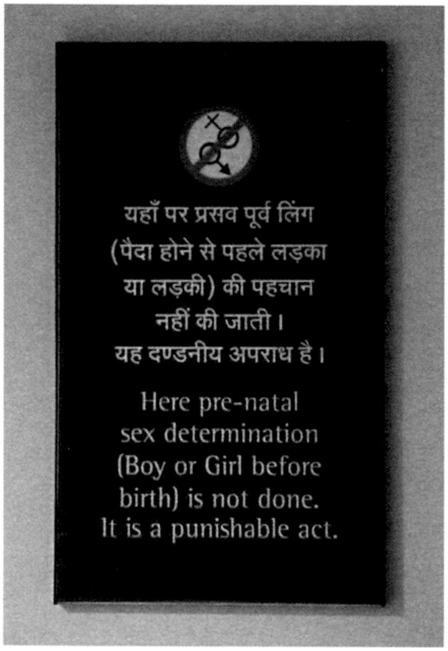

FIGURE 8.2 MODERN PRENATAL MEDICAL TECHNIQUES CAN DETERMINE FETAL SEX Coupled with the capacity to abort unwanted pregnancies, this is a controversial procedure, as there are several cultures that continue to value sons over daughters, especially for subsequent children where the first child is a daughter. This is a sign from a clinic in New Delhi, India.

Certainly, among most vertebrate animals, individuals are either male or female. In contrast, most plants are *monoecious* ("one house"), where individual plants either have perfect flowers (both male and female parts) or imperfect flowers (some blooms are male, others are female). There are many exceptions, however. Willow trees (*Salix* spp.) are one, where an individual tree produces only male, or only female, flowers. They are said to be *dioecious*, meaning "two houses."

Monoecious is a term that we only apply to plants, although the condition is found in some animals too. The concept of having both sexes in one body has long fascinated people. The Greek mythological character Hermaphroditus was born male. The nymph Salmacis loved him so deeply that she pleaded with the gods to never have them part. The solution provided by the gods was to fuse them together, and so *hermaphrodite* has come to be the term applied to animals that contain both sexes in one body. It is not the proper term for humans who have attributes intermediate between the typical male and female phenotypes; for that, we use the term *intersex*.

Earthworms are hermaphrodites, as are many other invertebrate animals. If you pay attention to the ground after a good rain, it is not uncommon to find two worms engaged in sex, whereby each individual is both receiving and donating sperm, and each of which then goes off to produce worm offspring. Such organisms are said to be simultaneous hermaphrodites, since they are male and female at the same time. A few vertebrates are simultaneous hermaphrodites. The colorful hamlet fish (*Hypoplectrus* spp.) is one example (Figure 8.3). Unlike worms, these fish do not donate and receive sperm simultaneously. In one sexual transaction an individual will be the male, while in another—even moments later and perhaps with the same partner—that same individual will be the female.

Simultaneous hermaphroditism contrasts with the sequential hermaphroditism that we introduced when we examined clownfish in Chapter 1. Sequential hermaphroditism is in fact rather common among reef fish. What we haven't noted until now is that in some species, the individuals are male first (as in clownfish) while in others, such as in the group known as the fairy basslets (Family Serranidae), they are female first, with the pinnacle of success being the super-male in a group. We say male-first species exhibit protandry (meaning "first male") and female-first species exhibit protogyny (meaning "first female").

Although the existence of both protogynous and protandrous fish species makes it look like they are equally effective or successful reproductive systems, the pattern appears very much to be related to mating systems, which we examined in Chapter 6. In species that tend to form pairs (monogamy), the female benefits from being larger so that she can produce more eggs, and so these species are protandrous. For other species where a large male can control a group of breeding females (polygyny), he benefits from being large, and these species are protogynous. This is known as the *size-advantage model* of sequential hermaphroditism, where gonadal transformation occurs when the reproductive potential of an individual is greater as the opposite sex than its current sex.

FIGURE 8.3 **SIMULTANEOUS HERMAPHRODITES HAVE FUNCTIONAL TESTES AND OVARIES**

Some species of hamlet fish (*Hypoplectrus* spp.) are simultaneous hermaphrodites, making both sperm and eggs and taking turns as male and female during sexual transactions.

Luiz A. Rocha / Shutterstock.com

Hermaphroditism is one set of patterns that deviates from the familiar model of separate bodies that are either male or female. Box 8.1 presents another deviating pattern, where there can be more than one type of each sex.

8.2 THE FAMILIAR METHOD OF SEX DETERMINATION RELIES ON X- AND Y-CHROMOSOMES

The sex chromosomes are a unique homologous pair that differs in males and females, unlike the 22 pairs of autosomal chromosomes. In the nuclei of human cells, we find 22 pairs of autosomes that do not differ between men and women. As for the 23rd pair, in women it is two X-chromosomes and in men it is one X-chromosome and one much smaller Y-chromosome (see Figure 3.10). Here we should underscore one further similarity between the chromosomes of men and women: an X-chromosome from a woman is no different from the X-chromosome from a man. In fact, one of the woman's X-chromosomes came from her father.

We also noted in Chapter 3 the rapid solidifying of the new field of genetics in the first decade of the twentieth century. This is also the decade that saw the discovery of the XY system of sex determination. The workings of the XY system were revealed ultimately by research in several different labs, but the first confident assertion was published in 1905 by Nettie Stevens, a female researcher in the United States who managed to contribute during a very male-dominated period.

8.2.1 Using insects to discover the role of sex chromosomes

Nettie Stevens was single, without significant financial means. This slowed her rise to a position of status. In 1903, she applied for money to the Carnegie Institute, a new foundation funded by Andrew Carnegie, the immensely wealthy Scottish-American industrialist. In her letter of application, this is how she expressed the question she was interested in tackling: "I am especially interested in the histological side [i.e., cell anatomy] of the problems in heredity connected with Mendel's Law, and know that there is need of a great deal of painstaking work along that line." In her application she was supported, not thwarted, by male colleagues, and she was successful in getting the money.

Researchers often have to choose a study organism when they seek to answer biological questions. Mendel chose the pea plant, for example, which we saw was a wise, or at least fortuitous, decision. Because Nettie Stevens was keen to test Mendel's freshly revealed ideas against patterns that could be seen in actual cells, she sought species whose

BOX 8.1 CAN THERE BE MULTIPLE TYPES OF ONE SEX?

Some species have two or more types of females or two or more types of males. Most social insects are in the insect order Hymenoptera that includes bees, ants, and wasps. Western honeybees (*Apis mellifera*) have two female phenotypes, queens and workers (Figure 8.4), and some ant species have three types because they also have large-jawed female soldiers whose bodies carry chemical arsenals. In all cases of large colonies, the males are little more than breeding machines and are of one type only.

In contrast, some fish and crustacean species have one female phenotype and multiple male phenotypes. The marine isopod (*Paracerceis sculpta*), found naturally along the Pacific coasts of California and Mexico, is such an example, with three male phenotypes (also Figure 8.4). Whereas the multiple female phenotypes in ants can be explained by a division of labor in a complex social structure, the multiple male phenotypes in marine isopods are genetic versions of what we describe as alternative mating strategies in Section 7.3.1. There, we linked alternative matings strategies with body condition, but in this case it is linked to genes.

Alpha male marine isopods fit a stereotype: they are large and they overtly and aggressively defend a sponge in which the females they attract are likely to lay eggs. Beta males are female-sized. In fact, their strategy is one where they mimic females, allowing them to infiltrate the harem of females that are attracted to the alpha male or to his sponge. Beta males will even sexually solicit the alpha male, to add to the ruse. When opportunity arises, such as when the alpha male is busy being territorial or mating with other females, beta males quickly inseminate female members of the harem. Finally there are gamma males, which are tiny and infiltrate a harem using a sneaking mating strategy. Notwithstanding their small size, they can produce prodigious quantities of sperm, which they use when they interrupt pairs that are mating.

But what determines what type of male each individual turns into? It is a one-gene/three-allele system (like the human ABO blood group system) located on one of the autosome pairs. This gene is appropriately named *AMS*, for alternative mating strategy. In populations there are three possible alleles at this locus, but of course a male can only have two, since individuals are diploid. The *beta* allele is dominant to both alpha and gamma, so beta males are beta-beta, beta-alpha or beta-gamma. The *gamma* allele is dominant to the alpha allele, so gamma males are gamma-gamma or gamma-alpha. And, therefore, *alpha* males must be homozygous alpha-alpha. Although one might imagine that one male type with its associated mating strategy might be reproductively superior to the others, over the long term and considering different densities of females on different sponges, they work out to be about equally successful. Alpha males do no better overall than beta or gamma males.

Not to be outdone, the female sex in some dragonflies and damselflies (Odonata) uses

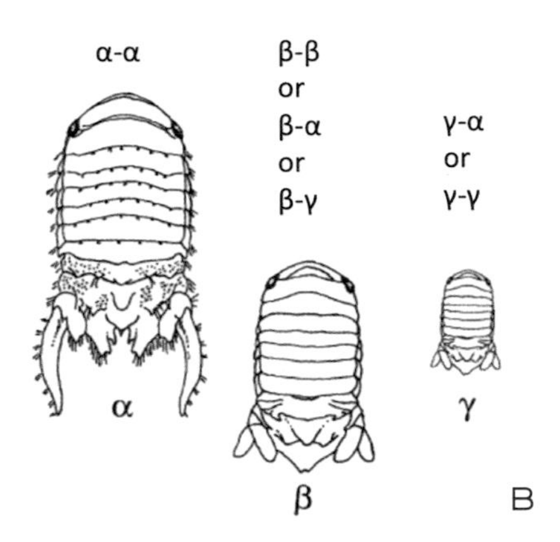

FIGURE 8.4 **SOME SPECIES HAVE MORE THAN ONE TYPE OF EACH SEX**

Some species have more than one type of each sex. Among members of a western honeybee (*Apis mellifera*) colony (a), mating is done by the queen (left) and the relatively uncommon males, known as drones (middle). Most of the colony individuals are the non-reproductive female workers (right). In the marine isopod (*Paracerceis sculpta*) (b), a male develops into one of three types, depending upon the alleles he has for the *alternative mating strategy* gene which is a one-gene, three-allele (α, β, γ) autosomal system for determining which male body type (and associated mating strategy) will develop. There is one combination for the *alpha* male, two for the *gamma* male, and three for the *beta* male.

[8.4a] WILDLIFE GmbH / Alamy Stock Photo

[8.4b] From Shuster (1992). Copyright © 1992 Koninklijke Brill NV. By permission of Brill

male imitation to suit its own ends. In some species there are "female" females and "male" females. The latter are called *androchromes* ("male colors") because their colors and patterns mimic that of males. The advantage is not well understood, but one observation is that androchromes benefit by being free from male harassment. Both ordinary females, which are more common, and androchromic females are reproductively female, mating with males and laying eggs in water. This, too, appears to be a one-gene, three-allele system.

cells would be easier to use than human cells. She worked with several species, but it was her work with the mealworm beetle (*Tenebrio molitor*) that yielded exciting results. Fortunately, mealworm chromosome behavior, like those of Mendel's peas, proved to be applicable to humans and to many other sexually reproducing species.

FIGURE 8.5 **THE XY SEX-DETERMINATION SYSTEM IS FOUND IN DIVERSE ANIMALS** Species with the XY sex-determination system include Nettie Stevens' mealworms, in which the system was first fully characterized, as well as mammals like humans. This Punnett square shows why it is the sperm that determines the sex of the offspring.

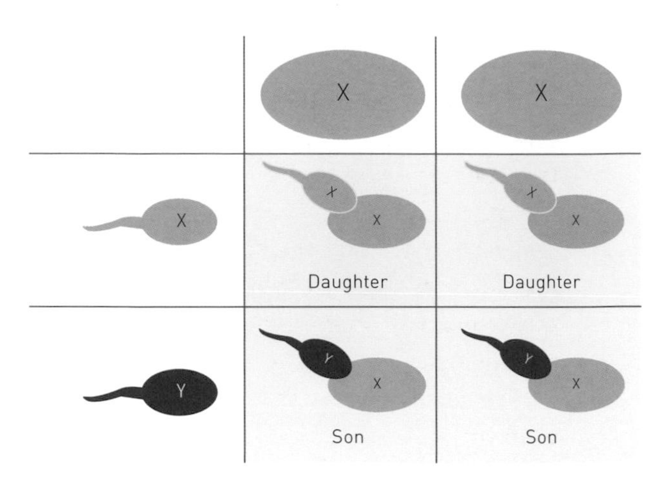

Mealworms are easy to raise, as is evident from the monumental volume of mealworms sold as food in the reptile trade. They have ten pairs of homologous chromosomes. Stevens' research confirmed three things. The first was that in the nuclei of the body cells there were 20 chromosomes. In females, all 20 were about the same size, but in males there were 19 about the same size and 1 much smaller chromosome. The second thing was that in the production of eggs and sperm, the eggs all contained 10 chromosomes (which were all about the same size), but the sperm were of 2 types. About half the sperm were like the eggs in that they contained 10 chromosomes of about the same size. But in the other half, there were nine chromosomes of the same size, as well as the much smaller chromosome.

Third, in her 1905 paper, she joined these two observations by concluding that the sperm carried the sex-determining factor. If it carried the much smaller chromosome, the resultant offspring was a male; if it carried one of the normal-sized chromosomes instead as the tenth chromosome, then the resultant offspring was a female, as shown in the Punnett square in Figure 8.5. Stevens called this small chromosome the Y-chromosome, as the X-chromosome had already been named and it was therefore a logical choice.

8.2.2 The X- and Y-chromosomes in humans

Humans are not unlike mealworms in the behavior of their chromosomes. We have 23 chromosome pairs, not 10, but as in beetles, men have an X- and a Y-chromosome in each cell nucleus and women have two X-chromosomes. So in humans, as in beetles, the sperm determines the offspring sex. This is because the egg always contains an X-chromosome and the sperm carries either an X (in which case the fertilized zygote will have two X-chromosomes, producing a daughter) or a Y (in which case the fertilized zygote will have one X and one Y, producing a son), as shown in Figure 8.5.

We already know that the Y-chromosome is considerably smaller than the X-chromosome. Here, we'll add some numbers to convey the difference. Chromosomes are numbered according to their physical size; so human chromosome 1 is the biggest and chromosome 22 the smallest (although we now know that chromosome 21 is actually slightly smaller). Chromosome length is not a perfect predictor of its informational content, but we have now confirmed that chromosome 1 indeed codes for the most proteins. The X-chromosome is about 62 per cent the length of chromosome 1, but the Y is only about 24 per cent. More significantly, in terms of coding information for the making of proteins, the X-chromosome is about 41 per cent that of chromosome 1, but the Y is only about 2.5 per cent. Considering the sometimes-dramatic differences in male and female phenotypes, a persisting mystery through the balance of the twentieth century was how the Y-chromosome—an apparently inconsequential chromosome, based on size—could generate such dramatic phenotypic differences.

8.2.3 Little genetic differences between men and women, but big phenotypic differences

It would be easy to imagine that the genetic content of male and female zygotes contained dramatically different packages of genes. After all, there are some dramatic phenotypic differences between males and females that must be accounted for. But there are in fact vanishingly small differences in the genetic content between them. Recall that in humans the 22 pairs of autosomes are not different between males and

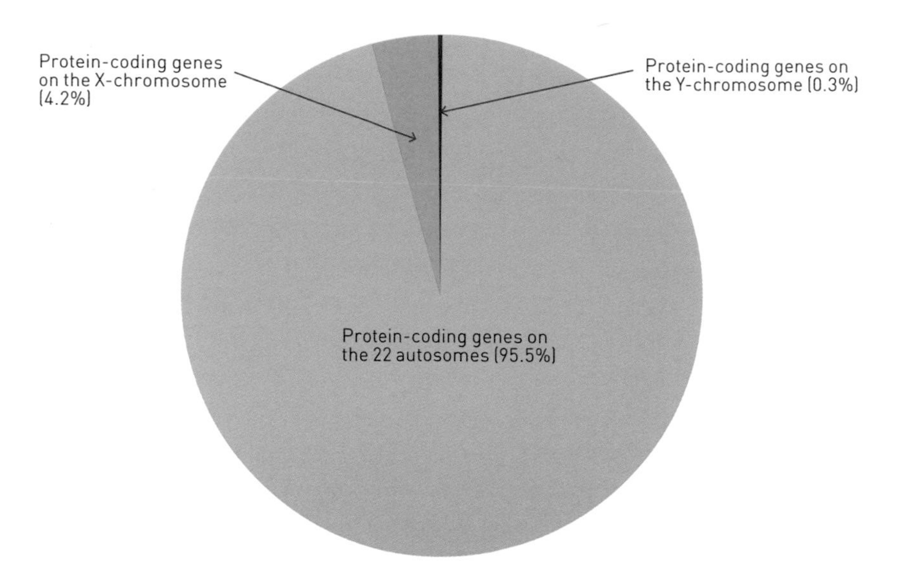

FIGURE 8.6 **THE X-CHROMOSOME HAS FAR MORE GENES THAN THE Y-CHROMOSOME**
This pie chart shows the distribution of genes in humans among the 22 autosomes, the X-chromosome, and the Y-chromosome. Males and females both have X-chromosomes, and although females have two, generally only one is active in any particular cell.

females. That amounts to more than 95 per cent of the genes (Figure 8.6). This does not, however, mean that the five per cent of genes found in the sex chromosomes represent the genetic differences between males and females. Far from it! Even though the X-chromosome is called a sex chromosome because it partners with the Y-chromosome, unlike the Y-chromosome it is not found only in females. You might be inclined to object by saying that there are two X-chromosome copies in females and one in males—which is true—but in each cell in female humans and other mammals one X-chromosome is almost entirely inactivated, known as X-inactivation or *lyonization*. This means the genetic content of the second X-chromosome is not being transcribed and translated into proteins (Section 3.6.3), although which X-chromosome is inactivated varies from place to place in the female body. So, while female cells do indeed have two X-chromosomes and male cells have only one, the cells of males and females differ rather little in that they both have only one active X-chromosome.

The Y-chromosome carries considerably fewer genes than even the smallest of the autosomes. So, since almost all the genetic information carried by the sex chromosomes is on the X-chromosome (about 1,100 genes, as opposed to about 80 genes on the Y-chromosome), more than 99 per cent of the genetic content of males and females is sex-neutral (i.e., found in both male and female cells) when one tallies up the content of the autosomes and the X-chromosome. The most significant difference, then, between the genetic content of human males and females is the presence of the Y-chromosome in males and absence of it in females. How then, can the smallest chromosome in most mammalian species generate such dramatic phenotypic differences?

8.2.4 The *SRY* gene and transcription

The short answer is that the Y-chromosome contains a single gene that can be likened to a binary switch. If this gene is present (which occurs in males, since the gene is on the Y-chromosome), it activates a program that sets the developing individual on a path of sexual differentiation into a male. Alternatively, if the gene is absent (as in females), this activation does not occur; instead, a different program proceeds that sets the developing individual on a path of sexual differentiation into a female.

In almost all mammals, this switch is what is called the *SRY* gene. This acronym derives from the gene's location: it is found on the S*ex-determining* R*egion of the* Y*-chromosome*. During the 1990s, there was considerable competition among research labs to find this factor. The gene and its location were ultimately localized in the following way. In mice, small pieces of DNA from the Y-chromosome were introduced into embryonic mice. Once those mice were born, the researchers looked for genetic females (i.e., XX individuals) that were phenotypically male, since these females must carry a fragment of Y-chromosome with the male-determining code. Once these mice were found, the researchers then characterized the piece of the Y-chromosome they carried. In humans, the *SRY* gene becomes active at about week seven of pregnancy. So the embryo, which is the term for the developing offspring during the first eight weeks of development, is sex-neutral until the *SRY* gene is turned on.

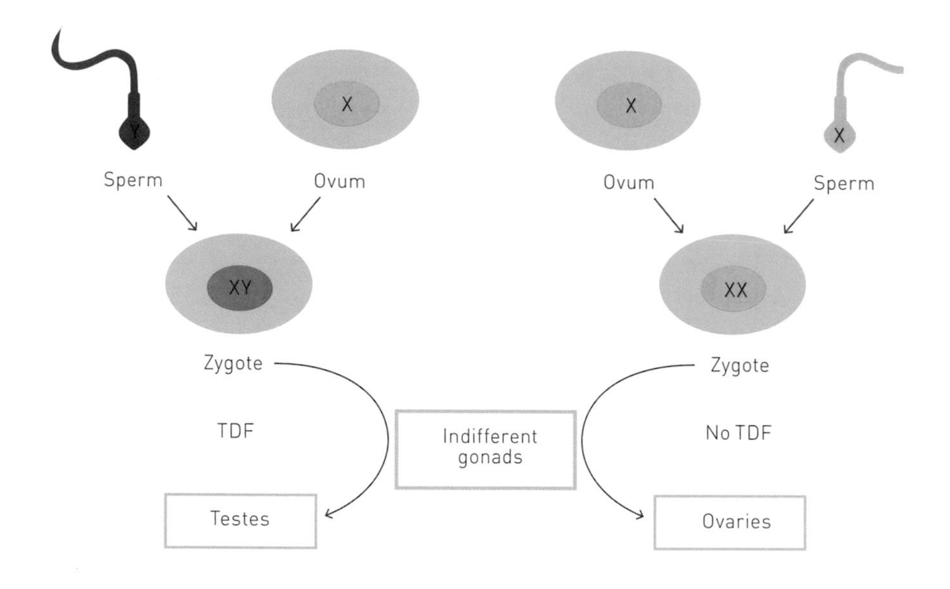

FIGURE 8.7 **TESTIS-DETERMINING FACTOR (TDF) IS A DECISIVE GENE PRODUCT**
This is a simple schematic of sex determination in placental and marsupial mammals during the first trimester. Y-chromosomes carry a gene that produces testis-determining factor (TDF); otherwise, ovaries develop.

If you recall that genes constitute information that code for gene products, which are usually proteins, it is natural to ask what protein the *SRY* gene codes for. The short answer is another acronym: TDF, which stands for testis-determining factor. TDF is indeed a protein, but it is a protein that is in a class known as transcription factors. Recall that *transcription* is that part of protein synthesis whereby the DNA of a gene is transcribed into mRNA in preparation for making proteins. Transcription factors are a large class of proteins that regulate whether transcription of another gene (or genes) is undertaken. It is in this sense that they also act like switches that turn genes "on" or leave them "off." (Perhaps 10 per cent of all human genes code for transcription factors.) A schematic of how TDF works in mammals is shown in Figure 8.7.

SRY is a gene for the testis-determining factor, but that shouldn't lull one into thinking there is a simple correlation between that gene and the testis. It would be incorrect to say that the *SRY* gene is a "gene for" the testis. There is no gene for the testis anymore than there is one for the ovary. Many genes are involved in making testes and ovaries. Instead, *SRY* is a gene that initiates male development. In fact, although *SRY* does code for TDF, which ultimately leads to the production of the testis, it is accomplished through what is known as a developmental cascade. TDF activates another gene found on chromosome 17, which in turn has an impact on other genes that ultimately lead to the development of the primordial gonad into a testis instead of an ovary. The cascade metaphor is used in biology whenever there is a key event that leads inexorably to other key events—similar to water falling along a particular path. In this case, the initial event is either the production of TDF or no production of TDF. That fork in the road leads to different cascades that are increasingly dissimilar, resulting in either male development or female development.

You may have guessed that females have many autosomal genes that contribute to maleness that are never expressed or are expressed in other, sex-neutral circumstances. This is true, and the converse is true for males too. More generally, many genes within an individual don't produce proteins at all times, and some may never produce proteins. In some cases, a gene's proteins are produced only in one sex, or only during fetal development or early infancy, or not until adolescence and thereafter.

8.3 NOT ALL GENETIC SEX DETERMINATION RELIES ON THE XY SYSTEM

There are two major systems of sex determination: genetic sex determination (GSD), which relies on a genetic attribute, and environmental sex determination (ESD), which relies on an environmental factor. Humans employ GSD, as do all other mammals and many other animal species, such as Nettie Stevens' mealworm beetles. During that same first decade of the twentieth century, the colleague of Nettie Stevens who is co-credited with discovering genetic sex determination, Edmund Beecher Wilson, found a different GSD system that was not a good model for humans because it had no Y-chromosome.

8.3.1 Genetic sex determination without Y-chromosomes

Wilson was working with broad-headed bugs (*Protenor* spp.). Female broad-headed bugs have 14 chromosomes in 7 pairs, including a pair of X-chromosomes. But males have 13 chromosomes in 6 homologous pairs, plus a single X-chromosome. Therefore, eggs contain seven chromosomes including an X, but sperm contain either seven chromosomes (including an X) or six (in which case there is no X). As you may have anticipated, the X-carrying sperm fertilize eggs that will accordingly become females, and the "O"-carrying sperm fertilize eggs that will accordingly become males.

Stevens' XY system and Wilson's XO system are commonly denoted by the genus names of two insect species, each of which has six pairs of autosomes. The XX/XO system is named the *Protenor system* after the broad-headed bugs (*Protenor* spp.) where it was first discovered, and the XX/XY system so familiar to humans is named the *Lygaeus system* after the milkweed bugs (*Lygaeus* spp.) The sex-determination systems for these two insects are contrasted in Figure 8.8.

One can see why the discovery of the *Protenor* system could have set back the scientific community in their quest to resolve the sex-determination challenge. Here, there is no such thing as a Y-chromosome, making generalization difficult!

In order to discuss both these systems, there is one more pair of terms that should be introduced. The sex that produces two types of gametes (i.e., those that determine whether the fertilized zygote will be a male or a female) is known as the heterogametic sex, as it generates two different types of gametes, one carrying the X-chromosome and the other not. In all cases we've looked at so far, males are the heterogametic sex. The homogametic sex has gametes that do not determine offspring sex; so far, we've seen that females are the homogametic sex because their eggs all contain a single X-chromosome.

8.3.2 Are females ever the heterogametic sex?

Certainly. All birds have chromosomal sex determination of the *Lygaeus* variety (like mammals), but the heterogametic sex is the female. In birds, the sex chromosomes are not referred to as the X- and

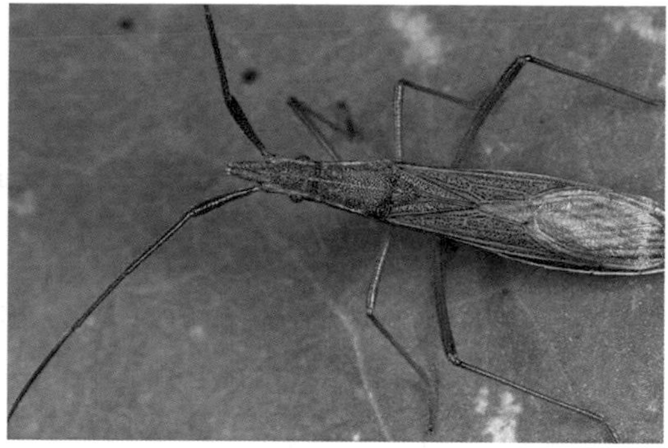

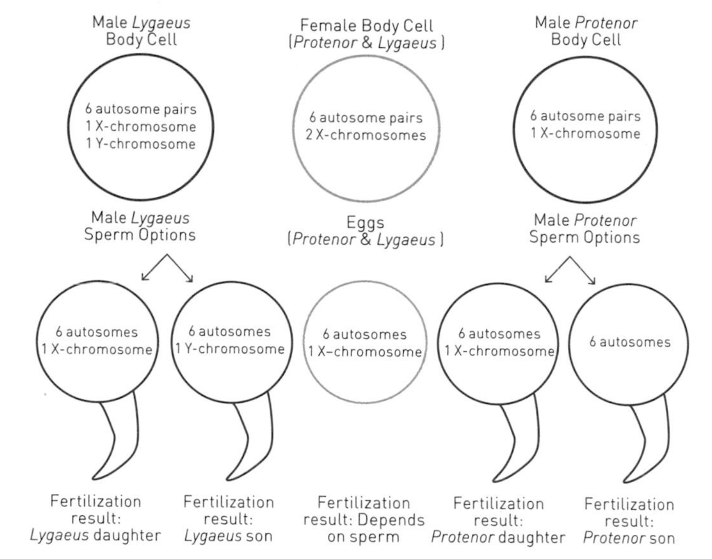

FIGURE 8.8 **TWO KEY SEX-DETERMINATION SYSTEMS ARE NAMED AFTER RESEARCH INSECTS**
Milkweed bugs (*Lygaeus*, top left) employ the familiar XX/XY system and the broad-headed bugs (*Protenor*, top right) employ the XX/XO system. At this level of analysis, the difference is that male *Lygaeus* have both an X- and a Y-chromosome (meaning that sperm carry one or the other), and male *Protenor* have only a single X-chromosome (meaning that sperm either carry, or do not carry, the X-chromosome). In humans, chromosomal sex is determined by the *Lygaeus* mode.

Y-chromosomes; instead, we use Z- and W-chromosomes, where the W-chromosome is smaller than the Z-chromosome. But that is just terminology that was instituted to distinguish the two systems functionally and evolutionarily. The significant point is that female birds are the heterogametic sex (ZW) and male birds are the homogametic sex (ZZ). Therefore, every sperm contains a Z-chromosome, but eggs contain either a Z- or a W-chromosome. Among other organisms, butterflies and snakes also have ZW sex chromosomes, making females the heterogametic sex.

Interestingly, birds do not have the equivalent of the *SRY* gene. So, what is it about the avian chromosomal sex-determination system, then, that triggers a male or a female program of development? There is a Z-chromosome gene known by the moniker *DMRT1* that appears to be the best candidate for the determinative factor. You might be puzzled by this, however. Because it is at a locus on the Z- instead of the W-chromosome, both males (ZZ) and females (ZW) have the gene—very different from the mammalian *SRY*, which is found only in the male. Like *SRY*, *DMRT1* codes for a transcription factor involved in initiating sex determination. The difference between *SRY* (mammals) and *DMRT1* (birds) is this: in mammals, males have the transcription factor and females don't, and that starts the male differentiation program. In birds, both sexes have the transcription factor but they have it in different doses (approximately twice the amount in the ZZ males), and the dosage difference is what sets off sex specific differentiation in development programs. *DMRT1* is known as a *dose-sensitive transcription factor*, and in fact the *D* in its acronym stands for *doublesex*. This is not the whole story of course; birds do have two unique genes on the W-chromosome (which are therefore found only in females) that are necessary for female development, but they are not determinative from the outset.

8.3.3 Chromosomal sex determination where hermaphroditism is a phenotype

A favorite study animal of geneticists is the transparent roundworm (*Caenorhabditis elegans*), a small soil organism introduced in Section 2.2.4. Recall that there are two sexual phenotypes: hermaphrodites and males. Hermaphrodites have five pairs of autosomes and one pair of sex chromosomes ($2n = 12$), both of which are X-chromosomes. When an egg is fertilized by a sperm that carries no X-chromosome (like in the *Protenor* system), the result is a zygote that is XO, which is a male. On the other hand, when an egg is fertilized by a sperm that carries an X-chromosome, the result is a zygote that is XX but that is a hermaphrodite (Figure 8.9).

Unlike typical male–female systems, the male-hermaphrodite system accommodates two very different types of sexual reproduction we've already encountered in Chapter 2. One of these is *crossing*, where

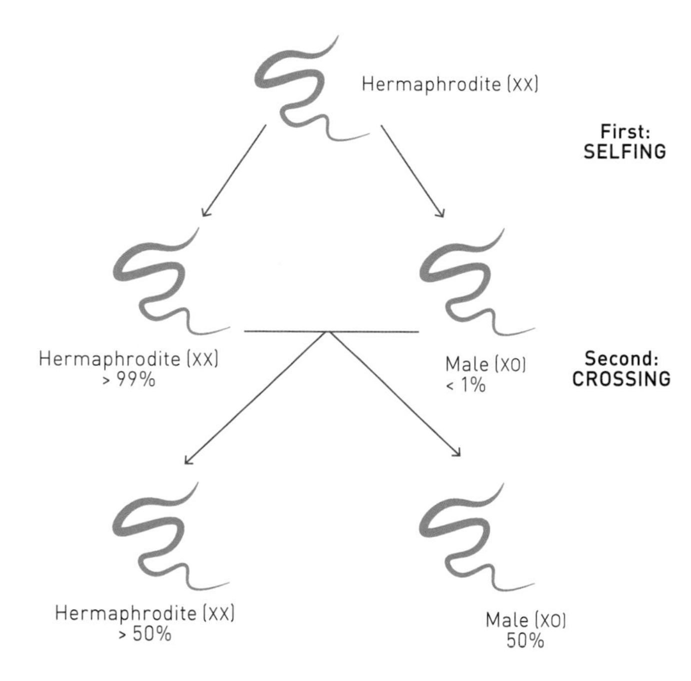

FIGURE 8.9 **SOME SPECIES HAVE NO FEMALES**
Transparent roundworms (*Caenorhabditis elegans*) have two sex phenotypes, males and hermaphrodites. Individuals are one or the other depending upon how many X-chromosomes are in cell nuclei. Males have one X (XO) and hermaphrodites have two (XX). Hermaphrodites commonly engage in *selfing*, a sexual system that uses their own sperm to fertilize their own eggs. This produces hermaphrodite offspring. Occasionally selfing produces an individual with only one X-chromosome because of imperfect meiosis; these individuals are male. Males reproduce only by *crossing* with hermaphrodites; the hermaphrodites in that role are functionally female, and their offspring are approximately 50:50 hermaphrodites–males.

gametes from two individuals—in this species, one male and one hermaphrodite—fuse to form a zygote. The other type is *selfing,* where sperm and eggs from one individual—in this species, a hermaphrodite—fuse to form zygotes. This is technically sexual reproduction because it is the merger of a sperm and egg, but it occurs within one individual. Although all offspring from selfing are like the single hermaphroditic parent in that all genes come from that parent, parent and offspring are not genetically identical to each other because of recombination that shuffles the proportions of the parent's alleles during the formation of the gametes. Self-fertilization of this sort is more formally known as *autogamy* ("self-marriage").

For males (XO) to achieve reproductive success, they must mate with a hermaphrodite. Males produce sperm, of course, half of which contain an X-chromosome and half of which do not. When a male mates with a hermaphrodite, she is a functional female (except for the fact that she may still have her own sperm stored in her body), and there is nothing different between this mating and a more usual male–female mating with which we are conceptually familiar. Because the sperm carry an X-chromosome or no sex chromosome, sperm determines the offspring sex, and about 50 per cent will be males and about 50 per cent will be hermaphrodites. Crossing produces more eggs than does selfing, but selfers can breed without having to encounter the relatively rare males.

8.3.4 Genetic sex determination in honeybees

Male honeybees have no fathers, and therefore they can have no sons. Curiously, these drones do have a grandfather and grandsons though, through the maternal side. This sounds strange indeed! As a system of sex determination, the method employed by the Hymenoptera is quite dramatically different from what we have seen so far. They employ a system called haplodiploidy (Figure 8.10), and to understand it, one needs to keep in mind the definitions of haploid and diploid. Recall that a haploid cell has one of each chromosome, but a diploid cell has pairs of each. In humans, body cells are diploid but gametes are haploid because meiosis (Section 3.7) reduces the set of 46 chromosomes to 23—one from each pair.

Put simply, in honeybees and most other social insects, females are diploid and males are haploid. That is, males have one of each chromosome in all their cells but females have two, like humans. Both males and females emerge from eggs, but males emerge from unfertilized eggs (which are haploid, making the males haploid), and females emerge from fertilized eggs (which are diploid, making females diploid). This arises because the queen's eggs are viable whether they are fertilized or not. Most of her eggs are fertilized, resulting in most of her offspring being daughters. But a small number are not fertilized, and those produce males. In this way, you can see how males can only have daughters, not sons. However, the daughters can have sons, meaning males can indeed have grandsons.

Like the female transparent roundworm, western honeybee queens can store sperm in a *spermatheca*. Recall that mating is a big event for the drone bee, because the mating process rips open his abdomen, and he quickly dies. The queen appears to be able to control the release of sperm from her spermatheca, so that she can thereby

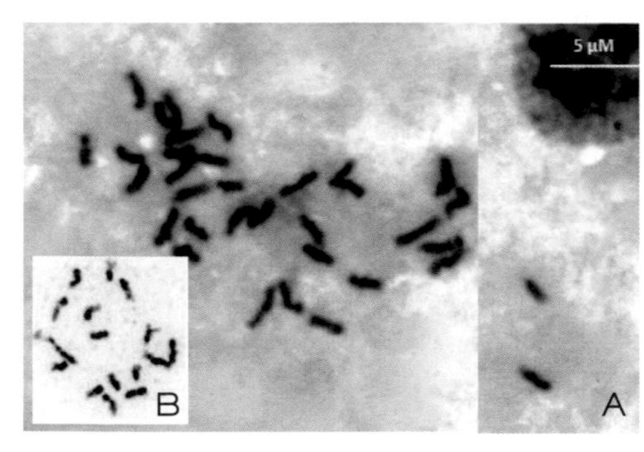

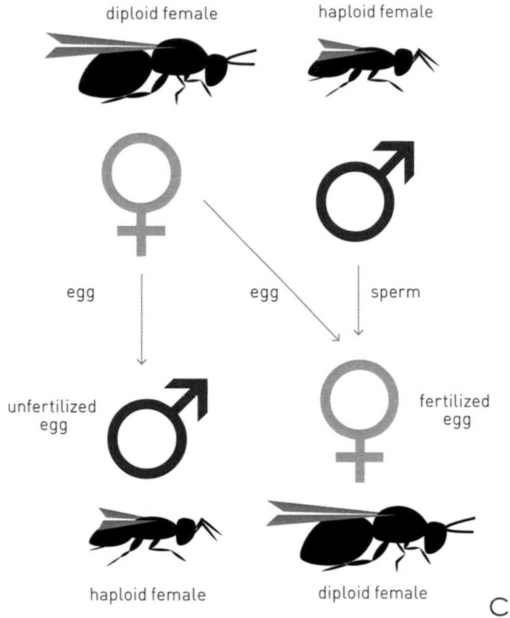

FIGURE 8.10 **SEX DETERMINATION IN SOCIAL INSECTS IS KNOWN AS HAPLODIPLOIDY**
The social insects (Hymenoptera: bees, ants, wasps) employ an unusual sex-determination system. All cells in male bodies are haploid, whereas female body cells, like those of human bodies, are diploid. The set of chromosomes in western honeybee (*Apis mellifera*) (a) differs by sex: a female (worker or queen) has 32 chromosomes (16 pairs), while a drone has 16 chromosomes. In all hymenopterans, such as the *Nasonia* wasps (b), drones hatch from unfertilized eggs, and so have genetic material just from the mother, but females hatch from fertilized eggs, with genetic material from both parents.
[8.10a] Adapted from Honeybee Genome Sequencing Consortium (2006) and Brito & Oldroyd (2010)

control the sex of her offspring. In practical terms, queens can control the phenotypic makeup of the colony by producing more reproductive males (which can leave the colony to mate elsewhere) or by producing more worker females.

So, honeybees have no sex chromosomes. Queens and workers have 16 chromosome pairs ($2n = 32$), and drones simply have 16 chromosomes. (The thing that distinguishes queens from workers is not genes, but diet during the developmental period.) In any event, this haplodiploid pattern makes the system quite different from other sex-determination systems. But in another sense it is not so dissimilar. Just as most mammals have an *SRY* gene, social insects also have a switch that initiates either a male differentiation or a female differentiation cascade. This is the complementary sex determiner (*CSD*) gene. If the individual has two different alleles for this gene (i.e., if the developing bee is heterozygous at this locus), an active protein is produced leading to a cascade that produces a female phenotype. On the other hand, if there is only one copy (as there must be in haploid males), or, more rarely if there are two copies but in homozygous form, an inactive form of the protein is produced and this defaults to male development through a different cascade. So, diploid males are possible (if the *CSD* locus is homozygous), but they are relatively rare. If they do hatch, they are quickly dispatched by their sisters acting in the role of worker nurses!

8.4 ENVIRONMENTAL SEX DETERMINATION

The statement that "XY zygotes turn into males and XX zygotes turn into females" applies to humans and to quite a few other species, but as we've seen, it is a very incomplete summary of what is a remarkable diversity of genetic sex-determination systems. That's why it is more meaningful to generalize genetic sex determination in species with two sexual phenotypes without reference to chromosomes, which we do as follows. During a key stage in development, there is commonly a gene product that—by its presence or absence (e.g., the mammalian *SRY* gene) or by its dosage (e.g., the avian *DMRT1* gene) or by whether the individual has two different alleles for the gene (as in the Hymenopteran social insects)—constitutes a master switch that sets the individual on a male-differentiated or a female-differentiated development path.

The other class of sex-determination system is environmental sex determination. It would be fundamentally incorrect, however, to think that genes are not involved in this type of system. In fact, when we consider male and female differentiation, genes are no less involved in environmental sex determination than in genetic sex determination. Although the master switch is an environmental factor that triggers a male or female development program, the whole pattern of differentia-

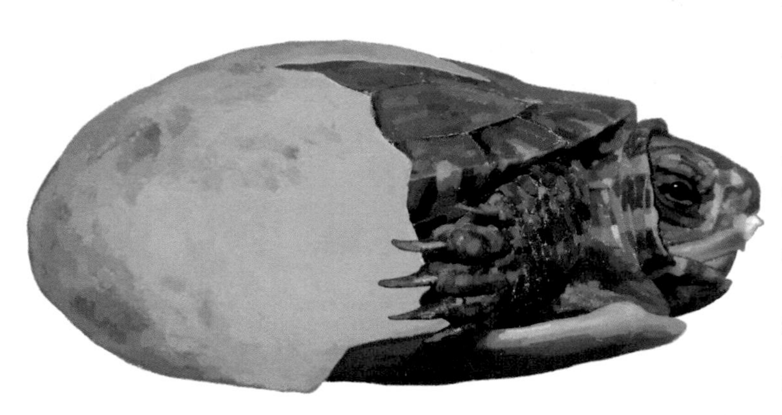

FIGURE 8.11 **ENVIRONMENTAL TEMPERATURE DETERMINES SEX IN MANY DEVELOPING REPTILES** This hatchling painted turtle (*Chrysemys picta*) hatches following many weeks of environmental, rather than maternal, incubation. During its early development, the embryo was sex-neutral until it reached the *thermosensitive period*, when the incubation temperature triggered either a male or a female developmental cascade.
Drawing by Peter Mills

tion is just as influenced by genes, their protein products, and hormones. Environmental sex determination is itself broadly divisible into two categories, non-social and social.

8.4.1 Temperature is a non-social sex-determination system

The painted turtle (*Chrysemys picta*) is the most widespread and perhaps most familiar freshwater turtle in North America. In spring, each reproductive female seeks a nesting site for laying her eggs. Turtles lay shelled eggs like those of birds, although they're leathery. Most female birds are well suited to incubating their eggs, having a patch of bare skin through which their body heat is transferred to the egg. Turtles are not so suited, since their body temperature is set by the environment rather than by an internal thermostat, and they have a hard shell both top and bottom, unlike the malleable belly of the mother bird. Instead, female turtles lay eggs in a shallow hole each mother excavates in sand, gravel, or loose soil. Nest sites are usually in relatively unshaded areas so that the heat from the summer sun will incubate the eggs, allowing for their development and eventual hatching.

Like many turtles, painted turtles do not have genetic sex determination. Offspring sex is determined by incubation temperatures that occur during a key period of embryonic development. This period is

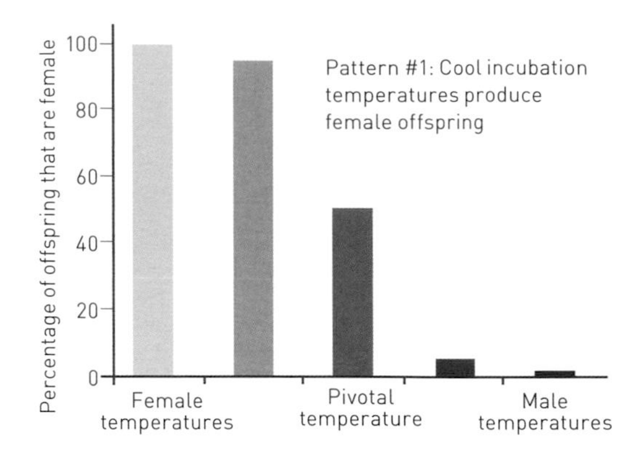

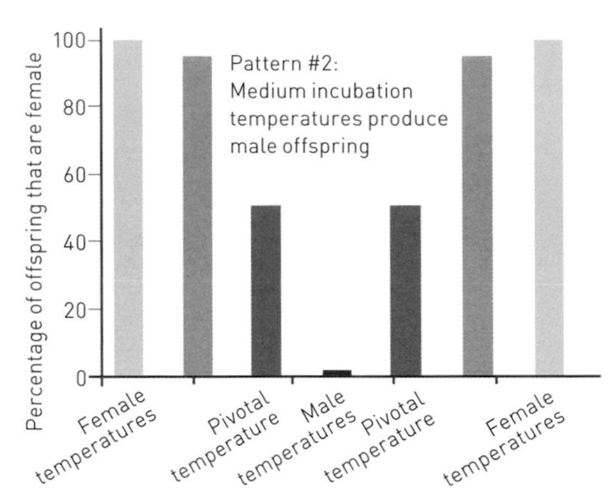

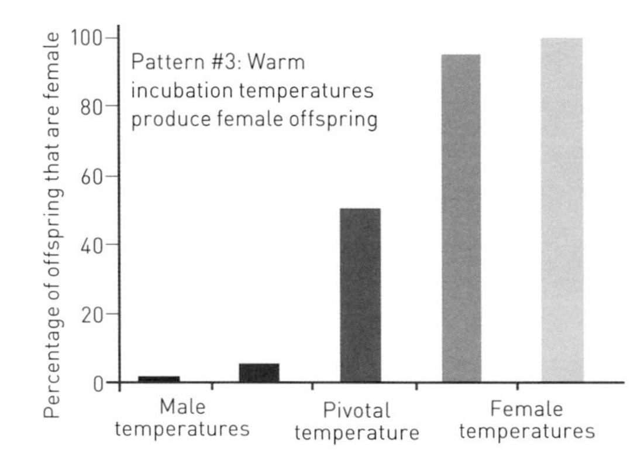

FIGURE 8.12 **REPTILES EXHIBIT THREE PATTERNS OF TEMPERATURE-DEPENDENT SEX DETERMINATION**

In these graphs, column height shows the corresponding percentage of female offspring for particular incubation temperatures during the early *thermosensitive period*. Low temperatures are to the left and high temperatures to the right. Temperatures at which the sex ratio is 50:50 are known as pivotal temperatures.

known as the *thermosensitive period*, and the system is known as *temperature-dependent sex determination* (TSD). Among reptiles, all crocodilians (e.g., American alligator, Nile crocodile), most turtles (Figure 8.11), and a variety of lizards exhibit TSD. Many fish also exhibit TSD.

In reptiles, there are three basic patterns (Figure 8.12). In the FM system, exemplified by American alligators (*Alligator mississippiensis*), females are produced at cooler temperatures and males at warmer temperatures. The FMF system, found in the popular pet the leopard gecko (*Eublepharis macularius*), produces males at intermediate temperatures and females at the extremes. Finally, the MF system produces males at cooler temperatures and females at warmer temperatures; most painted turtle populations are of the MF types, but in cooler locations like southern Canada, successful hatchlings from the lowest temperature extremes are often female, making it technically FMF. The MFM pattern is unknown in reptiles.

When natural turtle nests are examined in the field, it is easily seen that the sex ratios of the hatchlings vary by nest location; some nests produce both sexes, but the majority produce only one. Manipulated nests where the temperature is controlled during the thermosensitive period shows that there are temperatures, known as *pivotal temperatures*, where the sex ratios are 50:50. In the FMF system, there are two pivotal temperatures. Pivotal temperatures vary among species and also vary geographically within species. It is not surprising that the highest pivotal temperatures are found in equatorial turtles, where temperatures are on average higher.

It is one thing to say that temperature determines sex, but it says nothing about the actual mechanism. One well-documented cause of TSD involves the enzyme *aromatase*. Like all enzymes, aromatase is a protein, and there are therefore one or more genes that code for it. Unless you've taken organic chemistry, the word aromatase probably makes you think of odors. In fact, aromatase is responsible for the *aromatization* (a chemical transformation) of androgens ("male" hormones) into estrogens ("female" hormones). Not surprisingly, aromatase also goes by the name *estrogen synthase*, as in "assembling estrogen." Aromatase itself is temperature-sensitive during the thermosensitive period; because aromatase activity varies with temperature, so too does the conversion of androgens into estrogens, leading to sex differentiation.

Research has demonstrated other environmental influences on sex determination in particular groups of species. For instance, nutrition level and photoperiod (the proportion of the day that is sunlit) can also constitute triggers that determine sex during development, just as temperature does. Why would this be advantageous? Theoretically at least, it might be advantageous for a food-stressed mother to produce the less costly (perhaps smaller) sex. As far as photoperiod is concerned, that is a useful seasonal signal, and it may be advantageous to produce one sex during one season and the other sex during a different season. Another odd example is found in the green spoon worm (Box 8.2).

BOX 8.2 ONE LICK SEALS HIS FATE

In one bizarre marine species, hormones again determine the sex of a juvenile, but not through aggressiveness, sex ratios, or other similar factors. In this case the hormone is delivered by a lick from an adult female. The green spoon worm (*Bonellia viridis*) has exceptional sexual dimorphism (Figure 8.13). Females, which are approximately 40 times as long as the tiny males, are burrowers. They occasionally change burrows, but generally spend their adult lives in the same small patch of rocks. To feed, they use a tongue-like structure, usually referred to as a *proboscis*, that extends from their head. This can protrude from the body for more than a meter (more than 10 times the female body length). This structure is used primarily for feeding, but it has a significant additional function.

Larval spoon worms are sexually undifferentiated. For their first week or so, they are free-swimming, but they then settle to the bottom. Unless another female intervenes, a larva is destined to develop a female phenotype (the default genetic program), in which case it can take several years to mature to full size. Most larvae become male, however, because they do indeed encounter the female, or more exactly, encounter her wandering proboscis and a chemical compound found there called *bonellin*. Bonellin constitutes a masculinizing stimulus. Upon contact, the larva's sexual fate is sealed, and it quickly differentiates into the male phenotype.

FIGURE 8.13 **FEMALE SPOON WORMS ENSURE NEARBY LARVAE DEVELOP INTO MALES**

The green spoon worm (*Bonellia viridis*) is a marine dweller that is dramatically sexually dimorphic. The female shown here is relatively large and free-living, whereas males are relatively tiny and dependent upon the female. Sexually undifferentiated larvae are destined to become female unless touched by *bonellin*, a molecule found on the long *proboscis* of mature females. Upon being touched by a mature female's tongue, a larva's sexual fate is male development. Males remain small, they enter the nearest female's body cavity, and they reside there for the rest of their lives in a female's *androecium*.

Solvin Zankl / Visuals Unlimited, Inc. / Science Photo Library

The life of these dwarf males is a small one, both physically and behaviorally. Upon contacting bonellin, they attach themselves to a female, not uncommonly the one that has touched them, and enter her body, eventually residing in a special structure near the ovary called the *androecium* ("male house"). There, with an average of about three other males, they remain for life—sustained by the female and with no set of tasks other than to supply her with sperm for her eggs.

8.4.2 Environmental sex determination can override genetic sex determination

Continuing research in sex determination has found cases of vertebrate species where either temperature *or* chromosomes can be the sex-determination trigger. Skinks (Scincidae) are lizards with a worldwide distribution. One Australian species, the eastern three-lined skink (*Bassiana duperreyi*, Figure 8.14) has XX/XY genetic sex determination. This is not absolute, however. Although an XX individual *tends* to be female, researchers have found that two things can change this. First, XX eggs subjected to low incubation temperatures during embryonic development can sometimes develop into males. Second, it was found that if the researchers reduced the amount of yolk (which is essentially food for the developing embryo), they could significantly increase the proportion of males among the hatchlings—providing the temperatures were relatively low. Somehow, this "environmental" factor of yolk volume also influences sex determination. So, here is a species whose sex determination is influenced by genes, temperature, and yolk size.

FIGURE 8.14 **CHROMOSOMAL AND TEMPERATURE-DEPENDENT SEX-DETERMINATION SYSTEMS CAN CLASH**
Australia's eastern three-lined skink (*Bassiana duperreyi*) normally has genetic sex determination where males, like the animal with the bright orange throat shown here, are XY. However, this male could be a genetic female (XX) whose chromosomal sex-determination system was overridden during incubation by low temperatures, especially where the egg has impoverished yolk, producing a functional (but XX) male.

8.4.3 Can mothers use temperature to select their offspring's sex?

Can females exploit temperatures to choose the sex of their offspring? You might reflexively respond, Why would a female wish to do that? But as we saw when we considered mating strategies in Section 7.3.3, there are circumstances where sons and daughters do not have the same prospect for success, and selecting offspring sex might be an advantage for a mother.

In a study of yet another Australian lizard, there is some evidence females actively use temperature to influence offspring sex. The viviparous skink (*Eulamprus tympanum*) is relatively unusual among lizards in that it does not lay eggs. Instead, it bears live young. Unlike egg-layers whose eggs are stuck in one place (the excavated nest) for their whole development, live-bearing females can alter the temperature of their developing embryos by choosing where to be—basking in sunlight, hiding in shade, or various options in between. Research in natural circumstances found that when adult males are scarce, females disproportionately produce sons, and when adult females are scarce, females disproportionately produce daughters.

8.4.4 Social systems of environmental sex determination

Social interactions are environmental factors, and they can strongly influence hormonal responses in organisms. As a result, it should not be surprising that social interactions can constitute important environmental triggers for sex determination and differentiation. In this way, the brain, which processes social information, is a key player in mediating social cues into hormonal signals that influence sexual phenotype. In fact, our review of anemone clownfish in Chapter 1, with its sexually flexible adult phenotype, has already demonstrated this. We also noted in Section 8.1 that among reef fish that tend to form pairs, large individuals fare better reproductively as females and so are protandrous, but in species that form groups with a dominant male, large individuals fare better as males and so are protogynous. These are additional cases that demonstrate that social circumstances can trigger sex changes, and hormones are always at play.

One of the early observations on fish changing sex was reported for the bluestreak cleaner wrasse (*Labroides dimidiatus*, Figure 8.15), a species that makes its living by eating dead tissue and parasites off the skin of other fish. When the male controlling a harem of breeding females was removed by the experimenters, the largest female rapidly changed sex to take over as the new male master of the harem. In less

FIGURE 8.15 **SEQUENTIAL HERMAPHRODITISM IS INFLUENCED BY SOCIAL CIRCUMSTANCES**
Cleaner wrasses (*Labroides* spp.) are well known for the mutualistic relationship they have with larger fish, obtaining food by cleaning them of parasites, especially in the region of the mouth and gills. The bluestreak cleaner wrasse (*Labroides dimidiatus*) is protogynous, becoming sexually mature as a female first, and then becoming a male if there is a "male space" to occupy. This is experimentally reversible if two males are placed together; the smaller one will revert to a female.
iStock.com / ifish

than a day, she will have adopted male behaviors, including spawning (mating) motions. The time taken to produce active sperm is longer, but is achieved within 10 days. This is reversible; when experimenters place two males in the same tank, the smaller male switches back to a female.

Socially induced sex changes in the fairy basslet fish are also hormone-mediated, but the trigger is different. Here, the sex change relates to the sex ratio of a population. Populations tend to be most stable when they are about one-third male. If that proportion drops, then one or more of the larger females in the population converts to being a male.

In the case of the protandrous anemone clownfish, when the female is removed or dies, the male destined to replace her (commonly the largest male) immediately starts to shed its male phenotype through the operation of hormones. Over a period of six or seven weeks, these hormones regulate the change in the gonads from testes to ovaries. The hormonal changes she experiences also reinforce behavioral changes in her that result in aggressiveness. In turn, this influences the internal physiology of the males in her group; remarkably, her aggressiveness regulates hormone levels in the other males, maintaining their subordinate phenotypic maleness.

Socially induced sex changing in reef fish is a diverse area of study, yielding insights that vary from species to species and circumstance to circumstance. We make one further reference to one of these subtleties. In the protogynous bluehead wrasse (*Thalassoma bifasciatum*) of the Caribbean Sea, the sex ratio that results from the continual transformation of females into males as males die varies depending upon the size of the social group. The adult's whole life is spent on a single reef. On small reefs, all mating occurs in male territories, so reproductive males are relatively rare, meaning most of the breeding population is female (more than 80 per cent). But as reefs become larger, with more attendant complexity, there are more reproductive opportunities for males (including smaller, less territorial ones), and the male:female ratio drops, to as low as 50:50.

8.4.5 Anomalous sex determination caused by a member of a different species

Many organisms make their living by being *parasitic*. Parasites obtain a variety of benefits from their hosts—food, shelter, stable environments, etc. The main goal of some parasites is to hijack a host's reproductive system, converting it to a system that serves the reproductive interests of the parasite, rather than the host. Any organism that disables the host's reproduction is known as parasitic castrator. Parasitic castrators sometimes "steal" the reproductive energy and hardware of either or both sexes, including the energy associated with host reproduction.

FIGURE 8.16 **PARASITIC CASTRATION OCCURS WHEN A HOST'S REPRODUCTIVE SYSTEM IS TAKEN OVER**
This green shore crab (*Carcinus maenas*) has been taken over by the parasitic castrator *Sacculina*, a barnacle. The crab's reproductive system is hijacked by the parasite for its own purposes, and the control is so effective that the crab "mothers" the *Sacculina* eggs. If the *Sacculina* has entered a male crab, she re-develops him into a female crab phenotype that will serve her reproductive interests equally well, and he will mother her eggs.
Nature Photographers Ltd / Alamy Stock Photo

In some cases, when the parasite finds itself in a male as opposed to a female body, it suits the parasite to convert that male into a female. One such case is the *Sacculina* barnacle, which is a parasitic castrator of crabs, especially green crabs (*Carcinus maenas*). If a larval female *Sacculina* barnacle invades a female crab, the *Sacculina* takes control of the crab's internal processes, including both the flow of nourishment and her reproductive system. The female *Sacculina* larva exploits this nourishment, growing and then emerging as an adult at the spot where the crab's eggs would normally be. The male *Sacculina* seeks her out there, and he implants himself into her body (as in the green spoon worm) for the purposes of *Sacculina* reproduction. The now-infertile crab then cares for the *Sacculina* eggs, her behavior having been modified by the parasite (Figure 8.16).

If the female *Sacculina* enters a male crab instead, not only does she take control of the male crab body as she would have done for the female crab body, she also alters the hormonal "soup" of the male crab, sterilizing it and changing the body layout, so that the male crab re-develops as a female phenotype in both body and behavior. Among the female behaviors exhibited by the sex-changed male is the set of female egg-nurturing, egg-protecting, and birthing behaviors that benefit the *Sacculina* barnacle and her offspring. So, it matters little whether the female *Sacculina* invades a male or a female crab—a "castrated" crab whose mothering instincts are applied to the benefit of the *Sacculina* offspring is the result.

8.4.6 Anomalous sex determination by ecotoxins

Any system in the body that is regulated by hormones can be derailed by hormone disruptors. Sexual differentiation is highly regulated by hormones, and it is well documented that it can be inadvertently disturbed by *anthropogenic* (human-produced) *endocrine disruptors* in a host of species, including fish, amphibians, birds, and mammals. Endocrine disruptors enter the environment in multiple ways—inadequate waste management, industrial effluents, pesticide application to crops, and even normal household wastewater. Because endocrine disruptors can (a) increase or decrease the concentrations of other hormones, (b) convert one hormone into another, or (c) can imitate estrogens (female hormones), disruptions are varied in their consequences, but feminization of males is a widely reported consequence.

Bisphenol A (BPA) is an abundantly produced synthetic compound in use since the 1960s in the production of polycarbonate plastics. Among other purposes, these plastics are used to line food containers (e.g., plastic bottles and cans), and their consequent abundance has resulted in significant and widespread environmental contamination. BPA is an endocrine disruptor, and an unintentional consequence of its widespread use is that it is a widely distributed *ecotoxin* found in detectable quantities in many North American habitats, as well as in human bodies. (A Canadian survey that ended in 2011 found detectable amounts in more than 90 per cent of those people sampled.) When painted turtle eggs were exposed to BPA while being incubated in a lab set at temperatures that produce males (26°C; Section 8.4.1), 89 per cent of them developed female gonads, compared to zero per cent of a control group of unexposed eggs, also incubated at 26°C. As a result of public concern, BPA has been banned from use in baby bottles in most developed nations. Today many plastic containers are advertised as "BPA-free," although BPA continues to be widely used otherwise.

CHAPTER 8 SUMMARY

- Sex determination is the factor that governs whether an individual will become a male or a female, whereas sex differentiation is the pattern following determination that generates a male or female phenotype.
- Regardless of the mode of determination, differentiation into a male or female phenotype involves a complex interplay between genes and hormones.
- Most, but not all, sexual species are phenotypically binary; in hermaphrodites individuals are functionally both male and female, either sequentially or simultaneously, and in some species there is more than one male phenotype or female phenotype.
- The genetic constitutions of males and females are extremely similar, yet the end results can be dramatically different due to different developmental pathways.
- In most mammals, the *SRY* gene located on the Y-chromosome codes for a "testis determining factor" that sets the individual on the path of developing into a male phenotype; otherwise, the absence of the *SRY* gene sets the individual on the path of developing into a female phenotype.
- Not all genetic sex-determination systems rely on a Y-chromosome and the *SRY* gene. For example, in birds it is the female that has the two different sex chromosomes, and sex determination relies on two different "doses" of a genetic factor. Meanwhile, in social insects males are haploid and females are diploid.
- Many species rely on environmental sex determination, such as the temperature at which eggs are incubated or social factors that trigger hormones that influence development.
- Some parasites can cause sex changes in the host organisms, and some ecotoxins can disrupt sexual development.

FURTHER READING

Bachtrog, D., Kirkpatrick, M., Mank, J.E., McDaniel, S.F., Pires, J.C., Rice, W., & Valenzuela, N. (2011). Are all sex chromosomes created equal? *Trends in Genetics*, *27*(9), 350–357. https://doi.org/10.1016/j.tig.2011.05.005

Brito, R.M., & Oldroyd, B.P. (2010). A scientific note on a simple method for karyotyping honey bee (Apis mellifera) eggs. *Apidologie, 41*(2), 178–180.

Brush, S.G. (1978). Nettie M. Stevens and the discovery of sex determination by chromosomes. *Isis, 69*(2), 163–172. https://doi.org/10.1086/352001

Bull, J.J. (2015). Evolution: Reptile sex determination goes wild. *Nature, 523*(7558), 43–44. https://doi.org/10.1038/523043a

Charlesworth, B. (1991). The evolution of sex chromosomes. *Science, 251*(4997), 1030–1033. https://doi.org/10.1126/science.1998119

Dreger, A.D., Chase, C., Sousa, A., Gruppuso, P.A., & Frader, J. (2005). Changing the nomenclature/taxonomy for Intersex: A scientific and clinical rationale. *Journal of Pediatric Endocrinology & Metabolism, 18*(8), 729–733. https://doi.org/10.1515/JPEM.2005.18.8.729

Göth, A., & Booth, D.T. (2005). Temperature-dependent sex ratio in a bird. *Biology Letters, 1*(1), 31–33. https://doi.org/10.1098/rsbl.2004.0247

Hayes, T.B., Anderson, L.L., Beasley, V.R., de Solla, S.R., Iguchi, T., Ingraham, H., ... , & Willingham, E. (2011). Demasculinization and feminization of male gonads by atrazine: Consistent effects across vertebrate classes. *The Journal of Steroid Biochemistry and Molecular Biology, 127*(1–2), 64–73. https://doi.org/10.1016/j.jsbmb.2011.03.015

Honeybee Genome Sequencing Consortium. (2006). Insights into social insects from the genome of the honeybee *Apis mellifera. Nature, 443*(7114).

Lafferty, K.D., & Kuris, A.M. (2009). Parasitic castration: The evolution and ecology of body snatchers. *Trends in Parasitology, 25*(12), 564–572. https://doi.org/10.1016/j.pt.2009.09.003

Graves, J.A.M. (2006). Sex chromosome specialization and degeneration in mammals. *Cell, 124*(5), 901–914. https://doi.org/10.1016/j.cell.2006.02.024

Shuster, S.M. (1992). The reproductive behavior of α-, β-, and γ-male morphs in *Paracerceis sculpta,* a marine isopod crustacean. *Behavior, 121*(3/4), 231–258.

Warner, R.R. (1984). Mating behavior and hermaphroditism in coral reef fishes. *American Scientist, 72*(2), 128–136.

Werren, J.H., Baldo, L., & Clark, M.E. (2008). *Wolbachia*: Master manipulators of invertebrate biology. *Nature Reviews Microbiology, 6*(10), 741–751. https://doi.org/10.1038/nrmicro1969

Yasseen, A.S., & Lacaze-Masmonteil, T. (2016). Male-biased infant sex ratios and patterns of induced abortion. *Canadian Medical Association Journal, 188*(9), 640–641. https://doi.org/10.1503/cmaj.160183

9 Human Sexual Anatomy and Regulation

KEY THEMES

The endocrine system comprises several organs and associated hormones that travel the bloodstream and influence target cells. Sexual development and regulation, including behavior, are strongly influenced by hormones, which vary with sex, age, and reproductive condition. Human male reproductive anatomy is designed to deliver sperm, whereas human female reproductive anatomy is designed to receive sperm and to nurture and deliver offspring. Some genetic disorders in genes on the X-chromosome are found disproportionately in males, and some chromosomal abnormalities lead to atypical sexual phenotypes. Sexually differentiated organs are subject to sex-specific cancers.

This chapter is about the "birds and the bees," the idiomatic euphemism commonly employed for the fundamentals of sex education. Sex education for children has long been a contentious matter. The current prevailing view is that it should be early, especially so that young people may avoid feelings of shame and so that they are prepared for the risks associated with abuse, unwanted pregnancy, and sexually transmitted infections. More conservative elements within our society take the position that too much information too young can be disturbing to children or encourage precocious sexual activity. Regardless, studies show that significant numbers of adults, even those who are sexually active, often suffer from having absorbed misinformation along the way, especially about the opposite sex. Parents frequently admit to learning things by reviewing the sex education provided to their children in elementary school!

We will see in this chapter and the next that sexual differentiation during development as well as sexual regulation in adults is intimately connected to the interplay of numerous hormones. Only some of the great diversity of hormones in the body are directly related to sexual function, but among the best known are sex steroids like testosterone and various estrogens. Supplements of both are used to manipulate the body in various ways. In many cases, these supplements have legitimate medical purposes that are carefully prescribed, but some people obtain hormone supplements outside of the medical system, a risky undertaking that can have undesirable side effects (Figure 9.1).

9.1 HUMAN SEXUAL DIFFERENTIATION AND FUNCTION IS HIGHLY DEPENDENT ON HORMONES

We know that a fertilized human egg has genes from its mother and father, in the form of 23 chromosomes from each. We know that as the single-celled zygote divides and divides again on its journey to becoming a complex multicellular organism, those genes are translated into proteins. But we also know that in each individual, not all genes are translated, or they are only translated at certain times. Of utmost importance for the development of males and females, we

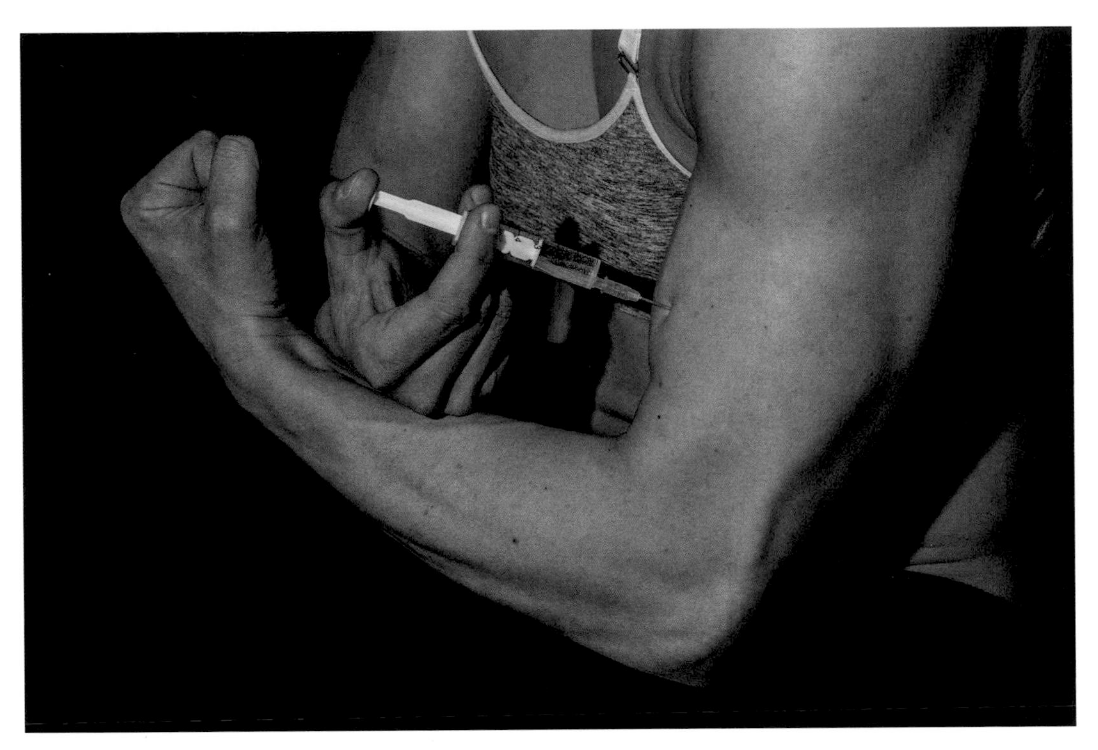

FIGURE 9.1 **INJECTIONS OF ANDROGEN STEROIDS CAN HAVE SERIOUS HEALTH CONSEQUENCES**
Athletes like this woman who are keen to increase muscle mass and enhance performance sometimes supplement natural hormone levels by injecting male-associated steroid hormones known as androgens, including real and synthetic versions of testosterone. In addition to life-threatening health risks and undesirable behavioral effects, the approach is illegal in competitive sports.
Srdjan Randjelovic / Shutterstock.com

know that two different developmental cascades follow from either having or not having the *SRY* gene, even though males and females are otherwise genetically very similar. These cascades do not manifest themselves immediately. For the first six weeks of pregnancy, the embryo is sexually undifferentiated. But from that point on, males and females develop along separate paths, and sexual anatomy becomes increasingly different.

We also have seen that many of the differences in males and females derive not so much from having different genes but from having different genes either "turned on" or "turned off," or perhaps expressed more or less, in one of the sexes (Chapter 8). So far, we have paid little attention to a very significant component of development and regulation, which is the cocktail of hormones moving around in the bloodstreams of individuals. Most of these hormones are found in both males and females but usually in different—sometimes dramatically different—concentrations. In addition, the influence of hormones upon cells depends upon *receptors* on the membranes of those cells, and this is another way that hormones can influence males and females differently.

9.1.1 The endocrine system regulates hormones in the bloodstream

The hormones of interest for the biology of sex are part of the endocrine system. Endocrine hormones are produced by endocrine glands from which they are released into the bloodstream. In this way, they can reach cells in all parts of the body, at least where those cells are equipped with hormone receptors. *Exocrine glands* do not rely on the bloodstream and therefore have local effects, and we will not be dealing with them further except when we consider lactation in the female breast.

When endocrine hormones are received by the cells of the targeted organ, they serve to regulate physiology and behavior. The influence of the hormone is experienced by the target cell once the hormone binds to the target cell's receptors. When that happens, it results in the activation of a signal pathway known as *signal transduction*, and cell responses can amount to dramatic impacts on metabolic and gene activity in the target cells.

Endocrine hormones that travel distances vary in molecular structure and origin and can be divided into three major classes. *Amines* such as adrenalin are small and are derived from amino acids, the building blocks of proteins. Peptide hormones such as oxytocin are also amino-acid based, but they are larger molecules made of chains of amino acids, and they are manufactured through the modification of gene products. The third class of hormones is a group of lipids known as steroid hormones, which are physiologically manufactured through the modification of cholesterol. Peptides and steroids are especially important when considering the biology of sex. For steroids, the cell

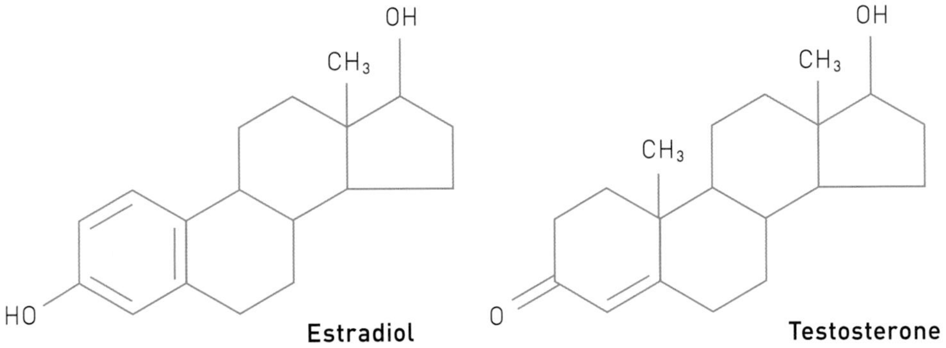

OH

CH_3

HO

Estradiol

OH

CH_3

CH_3

O

Testosterone

FIGURE 9.2 TWO PRIMARY HUMAN STEROIDS ARE ESTRADIOL AND TESTOSTERONE

Each is constructed of four fused rings where each ring "corner" is made of carbon and hydrogen atoms. The two molecules are very similar but have one key difference: where the estradiol has a hydroxyl group (–OH), the testosterone has a double-bonded oxygen atom known as a carbonyl group (=O), and this changes the shape and reactive behavior of the steroid.

receptor site is inside the target cell or even inside the nucleus of the target cell, where the chromosomes reside. This is because steroids can move directly through cell and nuclear membranes. For peptides, the cell receptor site is on the surface of the target cell, and the activation involves secondary molecules that influence activity inside the target cell.

The best-known steroid hormones are the *androgens,* associated with males, and the *estrogens,* associated with females. The most familiar androgen is testosterone, and the most familiar estrogen is estradiol. Both sexes, in fact, have both of these hormones. Testosterone and estradiol are very similar molecules (Figure 9.2). The key difference is in their *functional groups,* where at one particular carbon estradiol has a *hydroxyl group* (–OH) and testosterone has a *carbonyl group* (=O). These different functional groups make a difference to the three-dimensional shape and the reactive behavior of the two different steroids. Some receptors on cell surfaces accept testosterone molecules and others accept estradiol molecules.

There are four components of the endocrine system that are of significance for sexual function and differentiation—the hypothalamus found in the base of the brain, the pituitary gland adjacent to it, the *adrenal glands* found adjacent to the kidneys, and the *gonads* (Table 9.1). Located deep in part of the vertebrate forebrain, the hypothalamus is also part of the nervous system, but by sending hormones to the pituitary system and by sensing hormone levels in the bloodstream, it constitutes the main communication structure between the nervous and the endocrine systems, known as the *neuroendocrine pathway.*

9.1.2 Major hormones that regulate sexual development, function, and behavior

We've already referred to the hormones in the bloodstream as a cocktail, a term used to indicate that there is always a mix of hormones being moved through the body in this manner. To give a fairly complete account of those involved in sexual development and function, we refer to nine of them (Table 9.1). They are characterized by whether they are steroids or peptides, where they are produced, where they are targeted, and what their principle effects are, especially as they may influence males and females differently.

Hormone levels vary depending upon four primary things: feedback, age, sex, and in females, the menstrual cycle. Negative hormonal feedback occurs when a high concentration of a substance causes a second part of the system to reduce production of that substance, so it tends to stabilize a system. Positive hormonal feedback occurs when a high concentration causes a second part of the system to increase even more production of that substance. Testosterone levels in males are subject to a classic negative feedback system (Figure 9.3a). Female hormonal activity is regulated by both negative and positive feedback, depending upon menstrual cycle stage and upon whether the egg has been fertilized. In preparation of fertilization, high concentrations of estradiol positively feed back to produce more estradiol. However, if fertilization has not occurred, the production of progesterone ultimately has a negative impact on estradiol concentrations by influencing activity in the hypothalamus (Figure 9.3b).

Puberty is the stage of development in which a sexually immature human transitions to an adult capable of sexual reproduction. Prior to puberty, the hormones in Table 9.1 are produced in relatively low concentrations for the most part, in contrast to adulthood. In males, testosterone levels are relatively high in the first six months of life, but

Hormone	Hormone type	Gland of release	Target	Principal effect
Gonadotropin-releasing hormone (GnRH)	Peptide	Hypothalamus	Anterior pituitary gland	Regulate the release of gonadotropic hormones (those that target the gonads)
Luteinizing hormone (LH) (a gonadotropin)	Peptide	Anterior pituitary	Gonads	Males: Production of androgens (testosterone)
				Females: Regulation of ovulation (egg-release)
Follicle-stimulating hormone (FSH) (a gonadotropin)	Peptide	Anterior pituitary	Gonads	Males: Development of mobile spermatozoa
				Females: Production of estrogens (estradiol) and role in menstrual cycling
Testosterone	Steroid	Adrenal gland and gonads	Gonads, muscles, and other body tissues	Males: Stimulates testes and generates secondary sex characteristics
				Both sexes: Precursor of estradiol, and mediates sexual excitement and pleasure
Estradiol	Steroid	Gonads, adrenal gland, and placenta	Gonads, muscles, and other body tissues	Males: Involved in generation of sperm and erectile function
				Females: Stimulates ovaries and generates secondary sex characteristics; regulates sexual function and pregnancy
Progesterone	Steroid	Gonads, adrenal gland, and placenta	Uterus	Females: Regulates the condition of the uterine lining (the endometrium)
Oxytocin	Peptide	Posterior pituitary gland	Uterus	Females: Muscle contraction
			Mammary Glands	Females: Milk release
Prolactin	Peptide	Anterior pituitary gland	Mammary glands	Females: Milk production
Anti-Müllerian hormone (AMH)	Protein-type	Prenatal testis; ovarian cells during reproductive years	Müllerian ducts	Males: Dismantling of the Müllerian ducts during pre-natal development
				Females: Involved in regulation of egg production

TABLE 9.1 Nine key endocrine hormones that play roles in the context of sexual differentiation and function. There are additional hormones of importance in sexual development and regulation that exert their influence more locally, such as dihydrotestosterone in the male prostate and genitals.

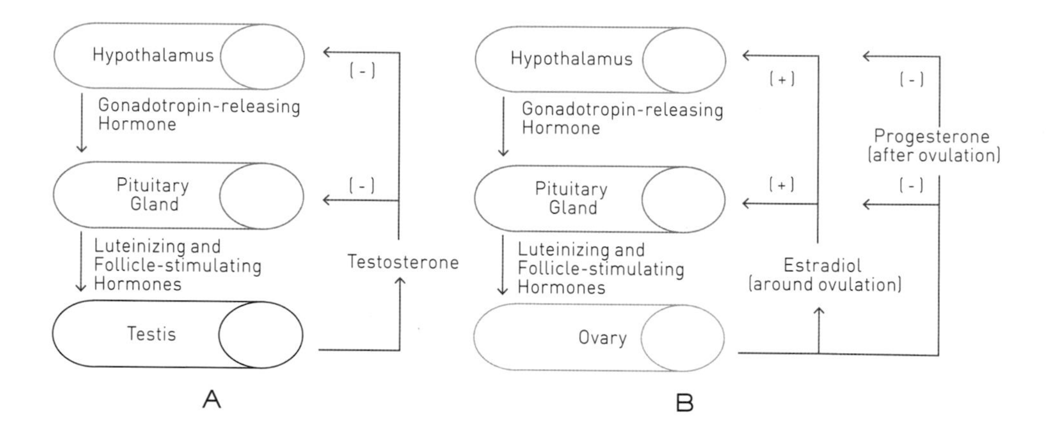

FIGURE 9.3 **HORMONE REGULATION RELIES ON FEEDBACK SYSTEMS**

Feedback control of hormones in the human hypothalamic-pituitary-gonadal axis differs between males and females. The male system (a) is classic negative feedback since elevated testosterone generated in the testis reduces the secretion of luteinizing and follicle-stimulating hormones, which in turn reduces testosterone production. The female system (b) is more complicated and varies with the menstrual cycle. Around ovulation, estradiol positively feeds back, but eventually progesterone levels produce a negative feedback if pregnancy does not occur.

they then drop until puberty, at which point testosterone levels surge, peaking in the late teens. Although production wanes moderately as men age, males tend to maintain relatively high levels of testosterone throughout life. Much more than in men, circulating hormone levels in adult women vary dramatically in accordance with three factors: the menstrual cycle, whether the woman is pregnant, and menopause. We will consider these in more detail.

Testosterone is considerably more common than estrogen in the bloodstreams of both men and women. Adult males have approximately 20 times more testosterone than women, although this is a very rough figure as it varies among individuals. Estradiol levels vary considerably in women in accordance with the stage of the menstrual cycle, being highest around ovulation, which is that part of the cycle in which the egg is released for possible fertilization. At that point, estradiol blood levels exceed those of males by about eight times, although this varies too. Women cease ovulating when their supply of eggs is depleted, and this also means that reproductive cycles terminate, known as menopause (Section 10.7). After menopause, estradiol levels more or less match those of men.

9.2 SEX DIFFERENTIATION IS PART OF DEVELOPMENT

When one considers that organisms that reproduce sexually start as single-celled zygotes but end up a short time later as individuals of, in some cases, trillions of cells of many different varieties organized into different tissue types and organs, it is not surprising that developmental biology has long been a rich area of biological research.

Developmental biology comprises all the changes that occur between the initial single-celled zygote and the ultimate phenotype. In animals, it includes not only the pre-birth *gestation* or *prenatal period* (which is about nine months in humans), but also the subsequent two decades or so that represent further developmental maturation, including the period of sexual maturation during puberty.

We have already distinguished *primary* and *secondary* sex characteristics. Primary sex differentiation refers to the gonads—testes and ovaries—and the "hardware" that is immediately associated with them in terms of sexual function. Secondary sex differentiation refers to all the accessory differences between males and females, both anatomic and behavioral, that are not directly part of the reproductive system—things like size, proportion, voice, and certain sex-influenced behaviors. In humans, female breasts and male facial hair are examples of secondary sex characteristics.

With respect to primary sex characteristics, early in development there is a sex-neutral primordial gonad that at about seven weeks begins to differentiate into the male or female sex organs under the influence of hormones produced during development. Although the result is dramatically different, male and female gonads have the same origin.

The "boy or girl?" question posed to new parents reveals the profound cultural importance attached to the sex of the newborn. It is so deep that we must remind ourselves that in some organisms there is no pre-birth sexual development—or at least none that can be considered male or female (as in clownfish). But in most vertebrate animals, as well as many other organisms, considerable sex differentiation does occur before birth, and sexual maturation is part of the post-birth development process.

9.2.1 Human sexual differentiation before birth

The different male and female developmental cascades that begin with, or without, the *SRY* gene do not diverge until the seventh week of pregnancy. Consequently, for the first six weeks the embryo is sexually undifferentiated. Beginning in the seventh week, differentiation of the gonads and associated ducts into a male or a female system begins. Since the gonads of both sexes are within the body at this stage, these changes are internal. The associated differences between male and female external genitalia don't begin to appear until several weeks later.

Prior to differentiation, the six-week-old embryo has a pair of generic gonads (neither testes nor ovaries) and two sets of ducts, one associated with each sex (Figure 9.4). The *Müllerian ducts* will, if retained, develop into the internal female reproductive system related to the monthly release of eggs and pregnancy. The *Wolffian ducts* will, if retained, develop into the male apparatus for delivering sperm. As for the genitalia, at six weeks a small bud appears that will become either the penis or the clitoris. Near the end of the second month, the genitals are distinguished only by having a *urogenital groove* (where *uro* refers to urinary), from which will develop either the male or female system.

In the seventh week, the male embryo (which is less than 2 cm long) begins to produce testosterone. Testosterone acts on the unisexual gonads by making them rapidly develop into testes, even to the extent that the tubules that make sperm are already evident before the end of the second month of pregnancy. Testosterone also triggers the production of two other hormones by the developing testes.

Concurrently, anti-Müllerian hormone (AMH) causes the dismantling of the Müllerian ducts, those that would otherwise be destined to become the female reproductive system, and instead the Wolffian ducts survive and develop into the male reproductive system. In the absence of testosterone, AMH is not produced, the Wolffian ducts degenerate, the Müllerian ducts persist so they can develop into the female reproductive system, and the gonad undergoes a slower transformation into an ovary. In reproductive-age women, AMH also plays a role in the ovulation schedule by regulating the release of eggs.

The other key hormone in early male development is dihydrotestosterone (DHT). This is synthesized from testosterone, and it is like an extra-potent testosterone, interacting more strongly with steroid receptor sites in target cells and so more able to influence the behavior of those target cells. DHT is essential for the prenatal development of male genitalia. Without it, the testes remain underdeveloped inside the body, rather than in the scrotum.

9.3 THE MALE REPRODUCTIVE SYSTEM

It's probably fair to say that the male reproductive system is simpler than the female system in the sense that it has one purpose: to produce and deliver lots of sperm to the female reproductive system. The female system is designed to produce an egg monthly and to receive sperm, but it is also designed to house the developing fetus when pregnancy results. In addition, the female system is complicated by the menstrual cycle and by menopause (the cessation of ovulation), whereas male reproductive function does not cycle and merely slows with age.

9.3.1 Further male differentiation during the prenatal period

When we considered the beginning of the male developmental

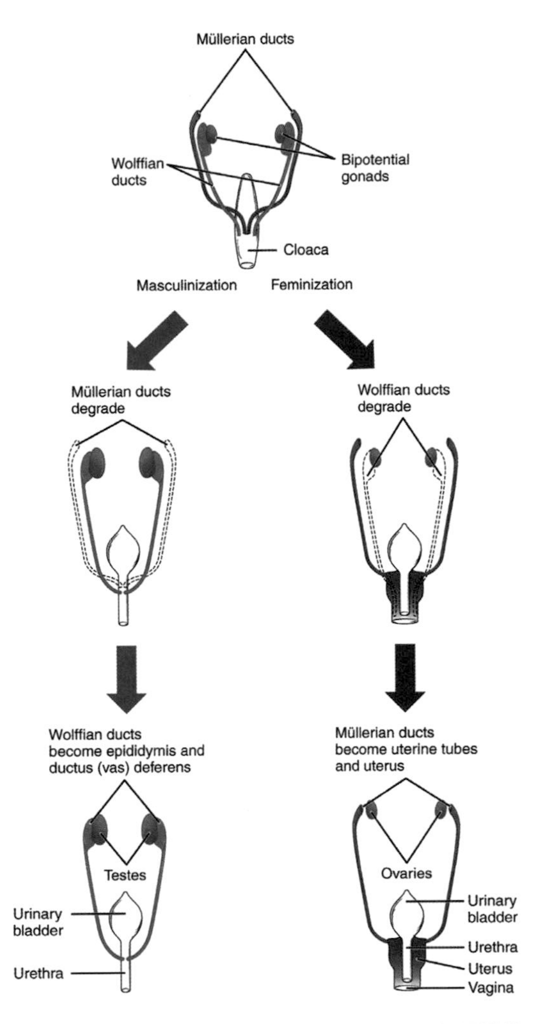

FIGURE 9.4 SEXUAL DIFFERENTIATION IN HUMANS BEGINS IN THE SECOND MONTH

By the end of the sixth week of development (top image), the human embryo has a pair of developing bipotential gonads, which means that they can become either ovaries or testes. The embryo also has two sets of ducts, one associated with the female system (Müllerian ducts) and one with the male system (Wolffian ducts). Under the action of hormones, the gonads develop as ovaries or testes, and only one duct system is retained and developed thereafter, ultimately leading to the birth of a girl or a boy.

cascade in the previous section, we saw that in the second month of pregnancy the female Müllerian ducts were dismantled under the action of AMH, the unisexual gonad was rapidly transformed into the testis under the action of testosterone, and a unisexual "bud" destined to be the penis had appeared.

By the end of pregnancy the XY baby is born as a boy, equipped with testes in a scrotum, a penis that is capable of erection, and a latent system for producing and delivering sperm once puberty arrives. Much of this anatomy we will consider in the adult, at which point it becomes reproductively functional, but we refer to two key prenatal events here.

By week 14 the urogenital groove has disappeared in males, and the swellings on either side have joined to become the scrotum. At this stage, however, the testes are still in their original position, deep in the abdominal cavity near the embryonic kidneys. Changes to the ligament that connects the testes to the body wall bring about a change in this position. As the body grows, this ligament becomes relatively shorter. By the fourth month of pregnancy, this has the effect of pulling the testes downward to a position still within the body but close to the future scrotum. Between the seventh month and full-term, the testes descend further into the scrotum, where they remain. In this sense, the testes are "outside" the body. This is thought to be advantageous on account of being about 2°C cooler, since the enzymes involved in sperm production operate better at slightly below normal body temperature.

Boy babies born prematurely commonly have undescended testes, but they usually descend into the scrotum shortly after birth. Full-term babies occasionally have one or both testes in an undescended state, but again, the problem usually corrects itself within the first year. If not, medical intervention is advised, since testes not inside the scrotum are prone to dysfunction and cancer.

9.3.2 Male sexual development at puberty

Puberty is the period during which the individual becomes sexually mature, changing from a child to an adult. In boys, it usually occurs between ages 12 and 16, although further maturation occurs until the end of adolescence. The central event in the onset of puberty in both sexes is the release of gonadotropin-releasing hormone (GnRH) from the hypothalamus, which stimulates the release of the two gonadotropin hormones: follicle-stimulating hormone (FSH) and luteinizing hormone (LH). FSH and LH both stimulate the gonads: in the case of FSH, to produce gametes and in the case of LH, to produce steroids. In males, LH stimulates a dramatic increase in testosterone levels in the bloodstream, which triggers multiple changes. The testes enlarge with the attendant production of mature spermatozoa and semen. The penis enlarges too, with the capacity for ejaculation of semen that contains spermatozoa. Body hair increases, especially in the pubic area and armpits, and facial hair develops. Height increases through a "growth spurt," and muscles grow. The cords of the larynx lengthen, allowing the voice to produce deeper pitches.

9.3.3 Sexual anatomy of the human adult male

The primary male sex organs are the testes, since they produce sperm and testosterone, and the accessory organs are the associated structures involved in delivering sperm during intercourse (Figure 9.5). Sperm are stored in the *epididymis,* which lies alongside the testis. The sperm leave the scrotum via a pair of tubes called the *vas deferens,* transporting them to an *ejaculatory duct.* The ejaculatory ducts join the *urethra,* a tube that exits through the penis and which does double duty by transporting urine most of the time but sperm during ejaculation.

The tip of the penis in mammals is known as the *glans* (Latin for "acorn") or glans penis. In the non-excited state, when the penis is flaccid, the glans is covered by the foreskin, which is loose skin near the tip of the penis. During sexual excitement, however, when the penis becomes erect, the glans is exposed. Male circumcision is the removal at a very young age of much of the foreskin so that the glans is thereafter exposed, even in a flaccid state. Most older men in North America have undergone circumcision, and it is still near universal among the world's traditional Jews and Muslims. It has become considerably less common in the North America since the 1980s, although it is still far from rare. This is partly because although circumcision has mostly been driven by religious tradition, it has long had medical apologists who argued the hygiene

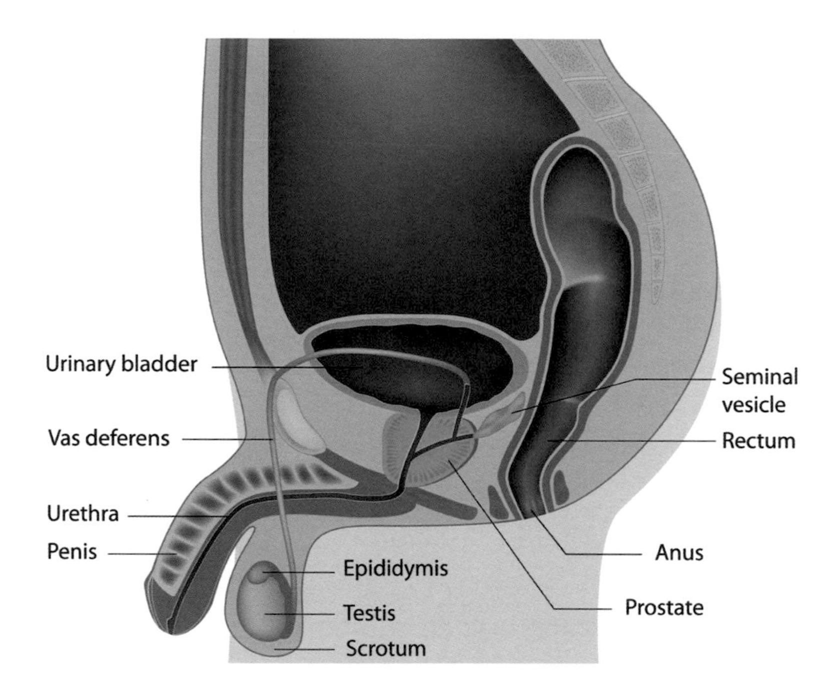

FIGURE 9.5 THE MALE REPRODUCTIVE SYSTEM

To compare and contrast with the female system, see Figure 9.7.

Alila Medical Media / Shutterstock.com

benefits that circumcision creates. It is still sometimes prescribed by the medical community to reduce the incidence of chronic infection, and there is now evidence that circumcision reduces the transmission of HIV in heterosexual men.

Sperm are delivered through the erect penis in semen (also called seminal fluid). Several structures near the base of the penis add components to the semen. One of these is the prostate gland, which is a non-endocrine part of the male reproductive system located at the base of the genitals. Because the female reproductive tract is acidic, the components of semen are collectively *basic* (the opposite of acidic), allowing the sperm to live longer once delivered. Semen also contains sugars, which serve as energy sources for sperm, and substances known as prostaglandins. These are hormone-like substances that impact cells in the female reproductive tract, in particular by inducing uterine contractions that help propel sperm deeper into the female system. Intromittent organs, of which the human penis is one, are highly diverse in structure across the animal kingdom (Box 9.1).

BOX 9.1 INTROMITTENT ORGANS ARE HIGHLY VARIABLE IN NATURE

Not all animals with internal fertilization have males equipped with an intromittent organ, but such devices are the most common solution to the challenge of getting sperm inside the female reproductive tract. Among these organs there is much diversity in morphology, size, mechanical operation, and origin (Figure 9.6). Most fish have external fertilization, but there are exceptions, as any aquarist who keeps live-bearing guppies will realize. The guppy intromittent organ is a modified fin known as a *gonopodium* that grows at the terminus of the male reproductive tract. This is a good example of an exaptation (Section 2.1.1), where a body part originally used for mobility has been co-opted for a sexual purpose.

Birds and reptiles produce shelled eggs, although some reptiles like rattlesnakes never lay the eggs and instead give birth to live young. Because young develop inside these shelled eggs, fertilization must occur before shell formation, and so these two classes of organisms also have internal fertilization. In birds and reptiles, the orifice that delivers liquid and solid waste, as well as semen in males and fertilized eggs in females, is called the *cloaca*. In males of most bird species during breeding periods, the cloaca swells, forming a stout structure called a cloacal protuberance, and this becomes the intromittent organ. Sex is fast in most birds, sometimes described as a "cloacal kiss," and it can even occur while the pair is in flight. Some waterfowl instead have a large phallus; that of the Muscovy duck is about as long as a human's, it is corkscrew-shaped, and its erection is achieved in less than a second.

Among reptiles, snakes and lizards actually have a pair of penises, each of which is called a *hemipenis*. These remain hidden at the base of the tail under most circumstances. During fertilization, however, the hemipenes are everted, meaning they becomes erect and cease to be covered by scales, and one or the other is then inserted into the female cloaca.

Unlike humans, most mammalian penises contain a bone called a baculum that assists in maintaining penile stiffness (Figure 9.6c), as in dogs and walruses. In contrast, the human erection, characterized by a significant stiffening and increase in length and volume of the penis, relies largely upon hydraulic pressure from blood. Penis size is not merely predictable by the size of the animal. That of the elephant (*Elephas* spp.) is very large, even considering the animal's size, while the gorilla (*Gorilla gorilla*) penis is tiny, being only about a third as long as a human's when erect. Mammalian penis anatomy is particularly varied at the tip. As already noted, many are equipped with a swollen terminus known as the glans, as in the human, while other species have tapering tips. The pig (*Sus scrofa*) penis has a corkscrew tip. Some species such as the domestic cat (*Felis catus*) have a penis with rings of backwards-pointing spines of keratin (Figure 9.6e)—difficult to associate with pleasure! Similarly, the guppy gonopodium has spines (Figure 9.6f), and the tip of each snake hemipenis is often elaborately embellished with spines and other projections, particular in pattern to each species. Why a

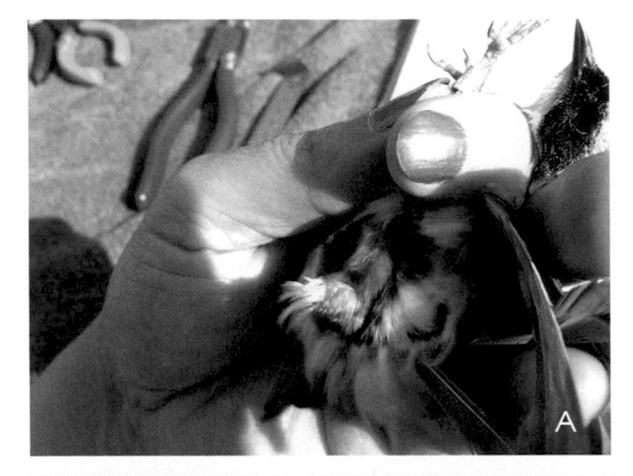

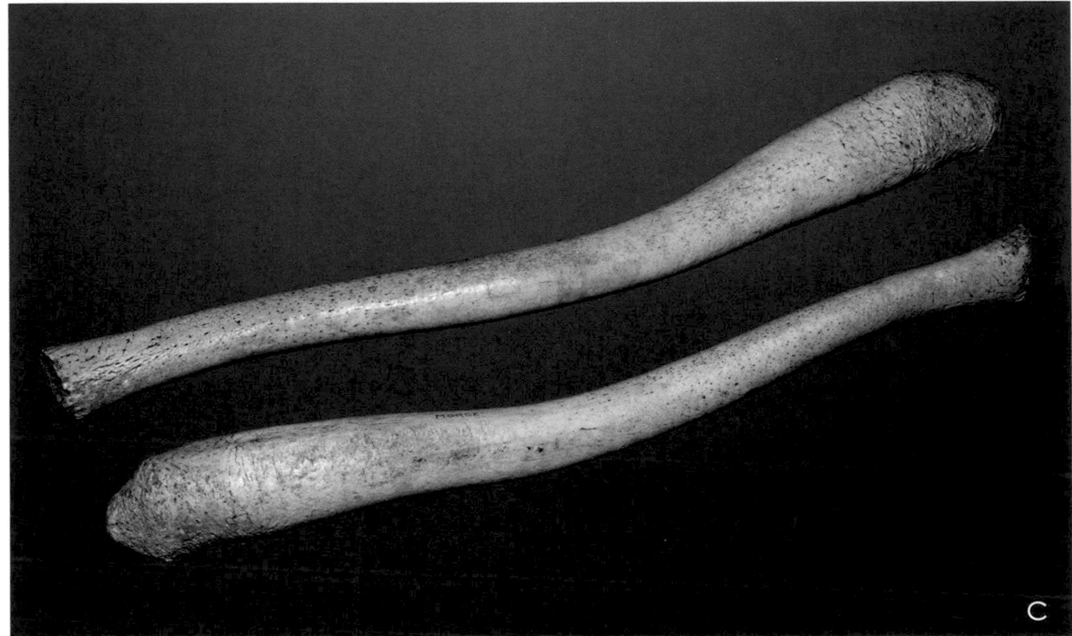

penis would be equipped with spines is a current area of research; they may prolong copulation so that fertilization is more likely, or they may manipulate the female by inducing ovulation or reducing her interest in mating with another male.

The mammalian penis and clitoris originate from the same embryonic tissue; in a few species, most notably the spotted hyena (*Crocuta crocuta*), the clitoris is as large as the penis, and it is capable of erection and is used in mounting behavior (Figure 9.6g).

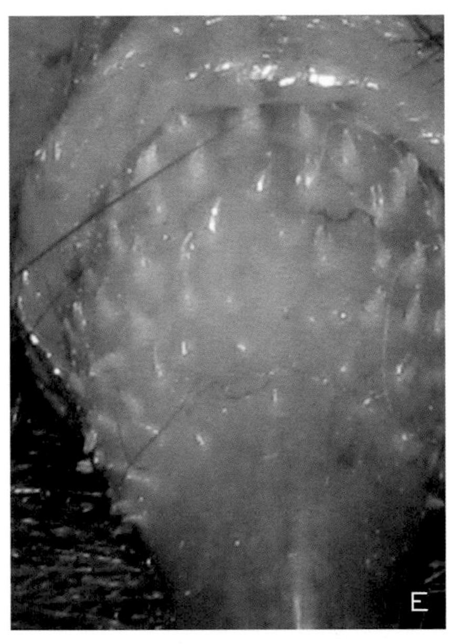

FIGURE 9.6 **INTROMITTENT ORGANS ARE HIGHLY DIVERSE IN NATURE**

The intromittent organ of songbirds like the common yellowthroat (*Geothlypis trichas*, a) is a mere swelling called a cloacal protuberance that exists only during the breeding season. Some birds like the mallard (*Anas platyrhynchos*, b) have larger and more complex organs. Penis diversity among mammals is great in terms of size and structure. Most mammals, like the walrus (*Odobenus rosmarus*, c) have a penis equipped with a bone called a *baculum*, which in the case of that species can reach 56 cm. Like humans, whales like the gray whale (*Eschrichtius robustus*, d) do not have a baculum. Some mammals like the domestic cat (*Felis catus*, e) have spiny projections, especially at the tip. The *gonopodium* of the guppy (*Poecilia reticulata*, f) is an intromittent organ that is really a modified fin, equipped with a claw and spines. The female spotted hyena (*Crocuta crocuta*, g) has a giant clitoris that equals the size of the male penis, and it can equally become erect and be used in mounting behaviors.

9.4 THE FEMALE REPRODUCTIVE SYSTEM

The human female reproductive system accommodates sexual intercourse, by which sperm is received in seminal fluid, but it also accommodates the fertilized egg by providing a supportive environment for nine months of prenatal development. The changes in the female body begin while she herself is still in the womb, beginning in the seven-week-old embryo. In the absence of the *SRY* gene and the consequent relative absence of testosterone, the Wolffian ducts degenerate and the Müllerian ducts develop into the female reproductive system.

9.4.1 Further female differentiation during the prenatal period

By the end of pregnancy, in the normal course of things, the XX baby is born as a girl. She will have internal ovaries, a set of fallopian tubes through which eggs will later travel, a uterus (commonly called a *womb*) that will later be able to accept a fertilized egg and nourish the developing fetus, and a vagina that will grow such that it is capable of accepting an erect penis during sexual intercourse and of delivering a baby following nine months of pregnancy. In addition, she will have a vulva (external genitals) that includes the outer lips (labia majora), the inner lips (labia minora), the clitoris, and the vaginal opening.

The urogenital groove with bordering swellings appears near the end of the second month in both sexes. In the male, this groove had disappeared by week 14 by transforming into the scrotum, but in the female fetus it persists to become the vaginal opening, and the swellings develop into the labia majora and minora. In the male, we saw a dramatic migration of the testes into the scrotum. The ovaries do not move so dramatically, but they do move to their adult position during fetal development.

9.4.2 Female sexual differentiation at puberty

Puberty in females usually arrives one to two years earlier than in males. As with males, it is triggered by the gonadotropins (LH and FSH) that have the effect of causing the ovary to generate and secrete estradiol that travels through the bloodstream, causing a suite of changes.

The first visible change is usually breast development, the onset of which is known as *thelarche*. Pubic and armpit hair begin to grow. As with boys, there is a growth spurt and a change in body proportions, but in females this also includes widening of the hips. Menstruation (the start of which is menarche) is usually a later achievement of puberty, and it can be several years after the start of the process, averaging at about 12.5 years in North America. Although girls reach puberty sooner than boys, it takes a bit longer for girls to reach the stage where they can actually become pregnant. In contrast, boys often experience their first ejaculation within a year of the onset of puberty, well before they have achieved a full complement of adult male features.

With a worldwide advance in the age of menarche in girls (see Figure 1.7), researchers are seeking to understand more about what triggers the start of puberty. In both sexes, growth factors related to body weight and body fat play a role, as does the system that regulates the transcription of genes. Notably, a gene with the label *KISS1* is a central player. Alarmingly, some research shows that blood levels of pollution-generated endocrine disruptors (Section 8.4.6) correlate with age of onset, making it later for boys and earlier for girls.

9.4.3 Sexual anatomy of the human adult female

The primary female sex organs are the ovaries, since they produce eggs as well as estrogens and progesterone. The accessory female sex organs are the associated structures involved in sexual activity, including intercourse, and pregnancy (Figure 9.7). We have already introduced the components of the adult vulva in Section 2.3.1—vaginal opening, clitoris, and outer and inner labia. These are the female genitals, and it is they that respond in an obvious manner during sexual excitement. Circumcision of the penis foreskin does not appear to cause any sexual dysfunction in males since the glans is not removed or scarred. However, female circumcision, which is widely practiced by certain religious communities, is a full or partial excision of the clitoris, resulting in the loss of its capacity for generating sexual pleasure. This is a culturally-entrenched practice in some places that is not only painful but also reduces a woman's quality of life and represents a severe restriction of her sexual autonomy.

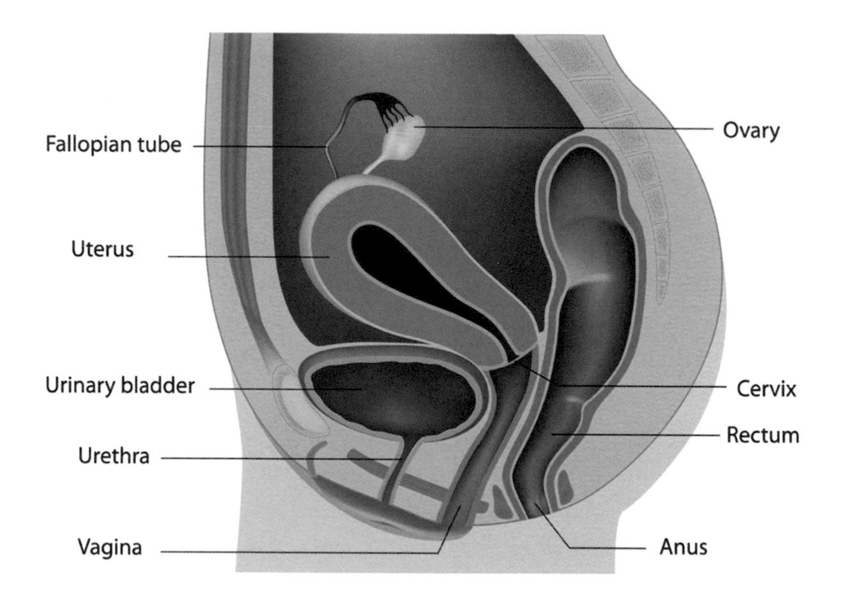

FIGURE 9.7 THE FEMALE REPRODUCTIVE SYSTEM
To compare and contrast with the male system, see Figure 9.5.
Alila Medical Media / Shutterstock.com

Once per menstrual cycle, the egg (or more rarely, eggs) leaves one of the ovaries and is swept into the corresponding fallopian tube through which the egg travels toward the uterus. Collectively, this constitutes ovulation. If it is a cycle that results in pregnancy, sperm delivered through sexual intercourse will enter the uterus at its lower end, known as the cervix. Sperm are usually short-lived in the female reproductive tract, although five days is possible. Fertilization by a sperm usually occurs in the fallopian tube, and the egg then continues downward into the uterus, dividing multiple times en route so that it is no longer a single cell, and arriving as much as several days later. This tiny embryo implants into the uterine wall, known as the endometrium, in an event known as implantation. Where the fetus connects to the uterus, a vessel-rich structure known as the placenta develops. The placenta is the joint creation of the implanted embryo and the endometrium. After the child is born, the placenta is also "delivered," which is why it is called the *afterbirth*. The placenta is such a central component of most mammals' reproductive anatomy that it is used to describe the largest group of mammals on Earth, the *placental* mammals. The females of cows, bats, whales, primates, rodents, and most other familiar mammals have one, although the pouch-using marsupials (e.g., kangaroos) and egg-laying monotremes (e.g., platypuses) do not. We consider fertilization and implantation in more detail in Section 10.4.

9.5 ANOMALOUS SEXUAL PHENOTYPES IN HUMANS

Most of our examination of the biology of sex to this point has been binary in nature—male/female, sperm/egg, testis/ovary, X/Y. Yet, we know that male and female genomes are highly similar, that the gonads are initially unisexual, and that the common notion of masculine and feminine are stereotypes from which many individuals diverge. And of course, there are people who experience gender dysphoria, which is where their gender identity does not match their biological sex.

In this section, we consider the relationship between biology and sexual phenotypes that do not fit into the standard notions of male and female. In the past, the term *hermaphrodite* was widely applied in such cases, but humans are not hermaphroditic. In a truly hermaphroditic species, individuals have functional sets of male and female organs. Controversially, the medical community sometimes uses the term disorders of sexual development (DSDs, formerly known as *intersex syndromes*), but the medicalization of such conditions, especially using the word "disorder," and the associated contemplation of intervention or treatment, has made it a subject of debate.

Here, we use the terms *intersex* and *intersexual* rather than the outdated and misleading term hermaphrodite. Note that "intersexual" has two very different meanings. In this chapter, we use it for individuals who exhibit physical traits that are partway between the typical male and female patterns or that include some male and some female features. In contrast, in the context of sexual selection in Chapter 5 we used a hyphenated spelling—inter-sexual—to describe relations *between* the two sexes, as opposed to relations *within* a sex, for which we used the opposite term, *intra-sexual.*

9.5.1 Consequences of anomalies in the sex chromosomes

Sometimes, a problem in meiosis results in the failure to distribute one chromosome of each pair to each sperm or egg, a failure known as non-disjunction. This error can occur in autosomes (see Section 10.1.2) or the sex chromosomes, and it can occur in a mother's ovaries or a father's testes. It almost always has unfavorable consequences. In normal meiosis, we say that disjunction of chromosomes is the separation of homologous pairs into separate daughter cells, but remember by daughter cells we mean resulting cells, not cells that are female or are destined to be girls. When there is a non-disjunction for a pair of chromosomes, they do not separate, meaning that one daughter cell has two copies, while another daughter cell has none.

When meiotic non-disjunction occurs in the sex chromosomes, intersex individuals can be the result. Consider an egg containing no X-chromosome. When it is fertilized by an X-carrying sperm, the zygote will be XO, resulting in a zygote with 22 pairs of autosomes and one X-chromosome. This individual will have Turner syndrome. If the egg is instead fertilized by a Y-carrying sperm, the zygote will be YO; this zygote will fail to develop. A zygote without an X-chromosome simply cannot survive because the X-chromosome carries so many necessary genes.

Now contrast this with the egg containing two X-chromosomes. When it is fertilized by an X-carrying sperm, the zygote will be XXX. This individual will have triple-X syndrome. Alternatively, when fertilized by a Y-carrying sperm, the zygote will be XXY, which results in Klinefelter syndrome. Like the XO genotype, triple-X and Klinefelter genotypes also result in viable fetuses.

What are the phenotypic attributes of individuals with XXX, XXY, and XO? Individuals with triple-X syndrome occur in about 1 in 1,000 female births. We noted in Section 8.2.3 that in typical XX females, one X-chromosome is mostly inactive in any cell. It turns out that in cases of females with triple-X syndrome, two X-chromosomes are mostly inactive, and so there are not usually any noticeable phenotypic differences from XX females, although there are increased chances of some developmental issues. Most, however, have typical sexual development and are fertile.

Klinefelter syndrome (XXY) occurs at the surprisingly common rate of 1 in 600 male births. Here again, although one of the X-chromosomes

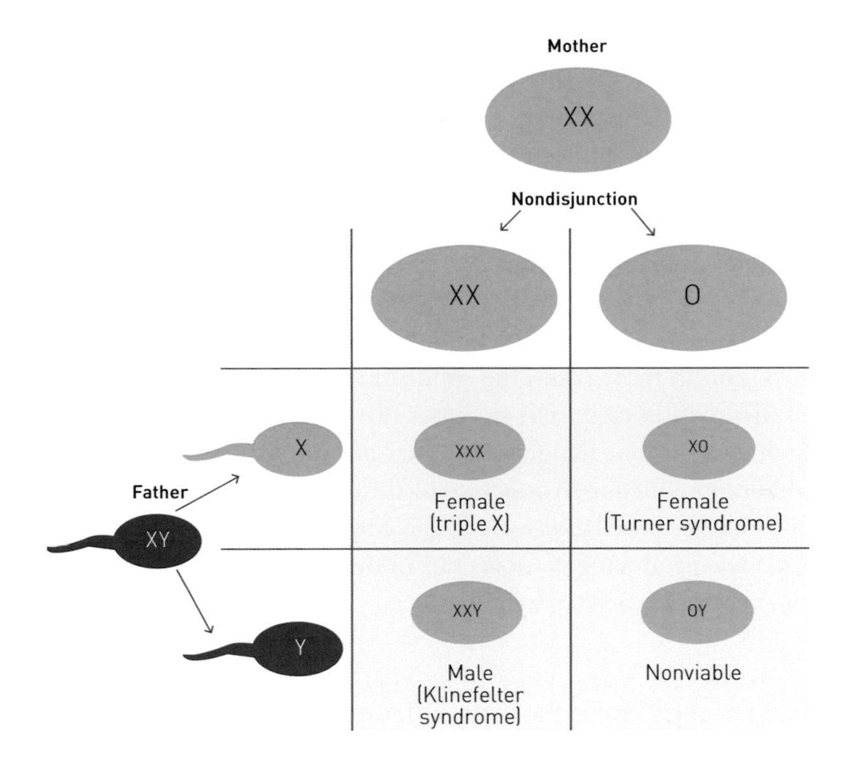

FIGURE 9.8 **NON-DISJUNCTION OF SEX CHROMOSOMES LEADS TO ANOMALIES**
This Punnett square shows the consequences of non-disjunction of the sex chromosomes during meiosis in the human ovary. The results vary depending upon whether the fertilizing sperm carries an X-chromosome or a Y-chromosome. The YO will not come to term because that individual is missing hundreds of necessary genes whose loci are on the X-chromosome. Non-disjunction in sperm production (not shown here) is less common.

is mostly inactive in each cell, the extra chromosome still has an impact on sexual development. Upon reaching maturity, males with Klinefelter syndrome exhibit *hypogonadism* (where *hypo* refers to "less than"), meaning they have testicles that are smaller than normal, as well as other symptoms of testosterone deficiency, such as breast development and very low or zero *sperm counts*.

Turner syndrome is less common than either triple-X or Klinefelter syndromes. Like males who have Klinefelter syndrome, females with Turner syndrome exhibit hypogonadism—in this case with respect to the ovaries. This results in sterility, as well as other issues, some of which can be life-shortening. This may seem surprising, given that in typical XX women, one X-chromosome is mostly inactive; clearly the presence of the second X-chromosome is important for sexual development and regulation.

Although Figure 9.8 shows these syndromes originating with non-disjunction in the egg-producing mother, it can certainly occur in the sperm-producing father. Non-disjunction in the creation of sperm can result in males with an XYY genotype. These individuals are typically taller than average but are otherwise similar to typical XY males, and in most cases, the condition remains undetected. Meiotic non-disjunction can occur in either the first division of meiosis, in which case all four daughter cells are anomalous, or in the second division, in which case only two of the four are. It also appears that some chromosomal aberrations can occur from a non-disjunction not in meiosis-producing gametes but in mitosis early in the development of an otherwise normal zygote.

9.5.2 Intersex conditions that result from dysfunctional alleles

As we've seen, sexual development usually follows a male or female "cascade" that is dependent upon the expression of a diversity of genes. Sometimes, if there is no functional allele for one of these key genes, the individual can follow a pattern of sexual development that does not match the typical pattern flowing from being either XX or XY. We'll consider two, each with an acronym or two.

The most common departure from the typical pattern of sexual development is caused by *congenital adrenal hyperplasia* (CAH; *congenital* meaning present from birth, *adrenal* referring to the adrenal gland, and *hyperplasia* meaning excessive growth of the steroid-producing cells). CAH results from a problem with one or more enzymes involved in the production of steroids in the adrenal gland. A variety of possible CAH patterns are possible, depending upon the enzyme affected, but most commonly there is a mutation in a gene on chromosome 6 that results in an excess of male hormones. In one common manifestation, XX individuals develop masculinized genitals; in extreme cases, there is no vagina and there is a functional penis that is capable of intercourse, but the gonads are female.

The second example results in feminization of genitals in XY individuals. It involves DHT, which we identified in Section 9.2.1 as being crucial for male development. Although testosterone is necessary for male development, DHT is another androgen necessary for virilization in the male genitalia. It is even more potent than testosterone in its masculinizing effects. Cells make DHT from testosterone by using an enzyme called *5-alpha-reductase* (5AR), and the gene for 5AR is on chromosome 2. Although a rare circumstance, if both alleles for 5AR carried by an XY individual have mutations that prevent the production of 5AR, the individual cannot convert testosterone into DHT, and genital development then follows the female cascade. It is not uncommon for such individuals to experience gender dysphoria, which in this case is the sense that they are male, notwithstanding that they were assigned to the female sex at birth. These XY individuals have non-descended testes, not ovaries.

9.5.3 When girls become men at puberty

In some cases, the 5AR deficiency is sufficient to produce female genitalia during fetal development, with the baby presenting as female at birth, but not so sufficient to circumvent male sexual development at puberty. These children thus present as female before puberty and male after puberty (Figure 9.9). Although rare, the mutant alleles for 5AR that produce this surprising pattern are sometimes relatively common within specific geographic areas, such as in the community of Salinas, Dominican Republic. The condition there even has its own name, *guevedoce*, meaning "penis at twelve."

How does this happen? Although there is a deficiency that fails to produce dihydrotestosterone during fetal development, resulting in apparently female genitalia, the surge in testosterone at puberty is enough to trigger the male development pattern. The "clitoris" develops into a "penis," the testes—which have been there all along—now descend into a developing scrotum, the voice deepens, and male muscle development occurs. Commonly, the gender identity of these adolescents re-aligns, as it is not uncommon for these "girls" to have identified with the male sex before puberty, notwithstanding having been

FIGURE 9.9 CHILDREN WHO APPEAR TO BE GIRLS OCCASIONALLY DEVELOP INTO MEN AT PUBERTY

These three young men are two brothers and a cousin from Dominican Republic. One was born with male genitalia and developed in the usual course into a man at puberty. The other two (cousins) are both *guevedoces*; until puberty they had apparently female genitalia and were raised as girls, but with the testosterone surge at puberty they developed male genitalia and male secondary sexual characteristics.

From Peterson, Imperato-McGinley, Gautier, & Sturla (1977). © 1977 Elsevier Inc. With permission from Elsevier

assigned as females at birth. In the Dominican Republic, at least, these individuals and their natural sex change are celebrated, although in other places their treatment has been less than benign.

9.6 SEX LINKAGE: WHY SOME GENETIC DISORDERS OCCUR MOSTLY IN MALES

Both men and women have X-chromosomes, so they both reap the benefits of the coded information from the genes of the X-chromosome. Yet women benefit by having two copies of each X-chromosome gene (although for some genes it's likely two copies of the same allele) and men have only one copy of each gene. What happens when one of the alleles on an X-chromosome is dysfunctional? That is, for a particular gene at an X-chromosome locus, what if there is a mutation in the sequence information that means the information cannot be used to make the protein that that gene codes for? If you begin your answer by saying "it depends if it is in a male or a female," you would be correct.

There are several disabilities that we say are sex-linked. To explain sex linkage, we will add a new technical term. Recall the meanings of homozygous and heterozygous (Section 3.4.1). Homozygosity means two identical alleles for a particular gene, and heterozygosity means two different alleles for a particular gene. Both terms refer to pairs of alleles. But genes on the X-chromosome exhibit a major difference when in male bodies. Because there is very little genetic information on the Y-chromosome but much information on the X-chromosome, with few exceptions the allele on the X-chromosome has no counterpart allele on the very small Y-chromosome. So, for these X-chromosome genes in males, we use the term hemizygous, where *hemi* means *half*.

Make sure you understand this sentence: *Sex-linked phenotypic traits are caused by male hemizygosity for X-chromosome genes.* If the one allele that males have for any of the many X-chromosome genes is dysfunctional, males do not have a second X-chromosome to resort to for a non-dysfunctional allele, unlike females. This means that this dysfunction is expressed in the phenotype, with notable examples being hemophilia, muscular dystrophy, and red-green color blindness. A famous example of sex linkage is the transmission of hemophilia in European royalty, where it particularly affected the offspring of Queen Victoria of Great Britain (Box 9.2).

9.7 CANCERS OF SEXUALLY DIFFERENTIATED ORGANS AND TISSUES

Cancers are a group of diseases characterized by abnormal cell proliferation, often with fatal consequences. Normally, cells behave predictably, but sometimes gene regulation fails and cells either grow and divide unpredictably or fail to die as they ought to. When this occurs in an organ, a tumor can result. In addition, non-tumorous cancers can arise in the blood system. Tumors that don't spread are referred to as benign, although they can have serious health effects regardless. Malignant tumors, on the other hand, can spread elsewhere in the body, especially via the circulatory system.

Although cancers have this capacity to spread, they are usually named after the part of the body in which they arose. Several prominent cancer types are associated with sexually differentiated tissues and organs. Among these are prostate and testicular cancers in men and breast, ovarian, and uterine cancers in women (Figure 9.11). Men can also get breast cancer, but it occurs much less frequently in men: no more than one per cent of the frequency in women.

BOX 9.2 QUEEN VICTORIA'S X-CHROMOSOME INFLUENCED EUROPEAN HISTORY

There are several dysfunctions that can lead to *hemophilia*, the disease where any injury that causes bleeding is very slow to heal due to an inability to clot, sometimes leading to death. One type is caused by a faulty clotting factor known as factor 9. Clotting factors are proteins, and so clotting factors are coded for by genes. The factor 9 gene locus is on the X-chromosome. Consequently hemophilia is a sex-linked condition, being rare in females, because females have a "backup" chromosome that likely has a functional allele.

Queen Victoria sat on the throne of Great Britain from 1837 to 1901. To understand the relationships among the X-chromosome, male hemizygosity, and hemophilia, consider Queen Victoria's pedigree in Figure 9.10. A pedigree is a genealogical chart showing multiple generations of a family, and it is customarily presented to show the pattern of inheritance of a particular trait—in this case, hemophilia.

In a pedigree, females are shown as circles and males as squares. White shapes are individuals without the trait in question (i.e., who don't have hemophilia in this case) and purple shapes are individuals who do have the trait (i.e., who have hemophilia). Sometimes, if the information is known, a shape is shown as half white and half purple to indicate a person who is a carrier of the trait (which means a person who has only one affected allele) but who does not exhibit the trait phenotypically (i.e., doesn't have hemophilia). Victoria, it turns out, was such a carrier.

In pedigrees, time travels downward, so that ancestors are to the top and descendants are to the bottom. First-born children are to the left and last-born are to the right. Lines connect parents and children. Shapes that appear in the pedigree without ancestors are people who have married into the family.

Reproducing Queen Victoria's entire pedigree would be too complicated because she lived about seven generations ago and because she had nine children, so we focus only on some of her descendants. We suggest referring to Figure 9.10 while reading the following two paragraphs. Victoria's second child and first son was Edward, who later became Britain's king. He did not have hemophilia, and because sons receive their X-chromosome from their mother, he evidently got his mother's "good" allele in the 50:50 meiotic lottery that assorted her two alleles for clotting factor 9.

Alice was the third child. She did not have hemophilia, but when she had children of her own, it became clear that she had received her mother's "bad" allele because one of her two sons, Frederick, had hemophilia. Her other son Ernest did not. Again, in the lottery of meiosis, Ernest was lucky and Frederick was unlucky. (Frederick died of hemophilia as a child following an injury.) Further proof that Alice was a carrier of the allele showed up in the Russian royal family. Alice's daughter Alexandra had married the Tsar of Russia. Their only son, Alexei, who was to inherit the Russian throne, also had hemophilia. His chronic illness resulting from hemophilia played a significant role in the political affairs of Russia as it collapsed during the Bolshevik revolution of 1917.

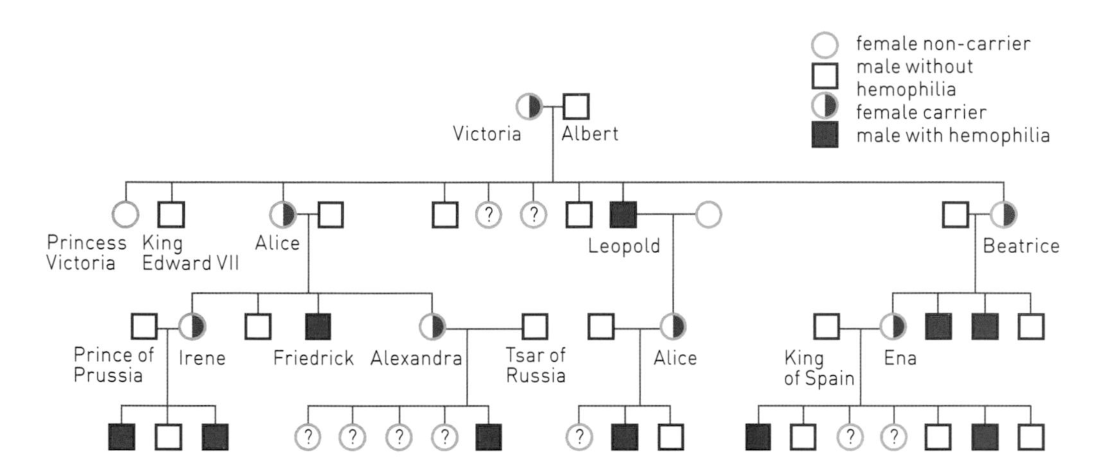

FIGURE 9.10 **QUEEN VICTORIA HAD A HEMOPHILIA ALLELE ON ONE OF HER X-CHROMOSOMES**

This pedigree begins with Queen Victoria (reign 1837–1901) and her husband Prince Albert, and it tracks the inheritance of hemophilia, a serious disease distinguished by an inability of blood to clot. Victoria's mutant allele, which was carried on one of her two X-chromosomes, continued through multiple ensuing generations, being carried by some female descendants and appearing as hemophilia in some male descendants. The "lottery of meiosis" determined whether the eggs of her and her affected female descendants carried the "good" allele or the "bad." When the mutant allele matched up with a Y-carrying sperm, the boy would be born with the disease. Her grandson Frederick died from hemophilia complications as a young child, as did her own son Leopold at the age of 30 from a slip-and-fall hemorrhage. None of the female carriers suffered hemophilia symptoms.

One can see that through yet another daughter of Victoria, Beatrice, the hemophilia-causing allele entered the Spanish royal family. Note also that no female in this pedigree had hemophilia and that approximately half the sons of carriers did.

FIGURE 9.11 CANCERS OF SEXUALLY DIFFERENTIATED ORGANS VARY IN INCIDENCE AND MORTALITY

This graph shows age-adjusted annual incidence rates (darker green bar) and mortality rates (lighter green bar) per 100,000 people for cancers of sexually differentiated organs, United Kingdom, 2008–10.

Courtesy of the Office for National Statistics, UK. © Crown copyright 2012

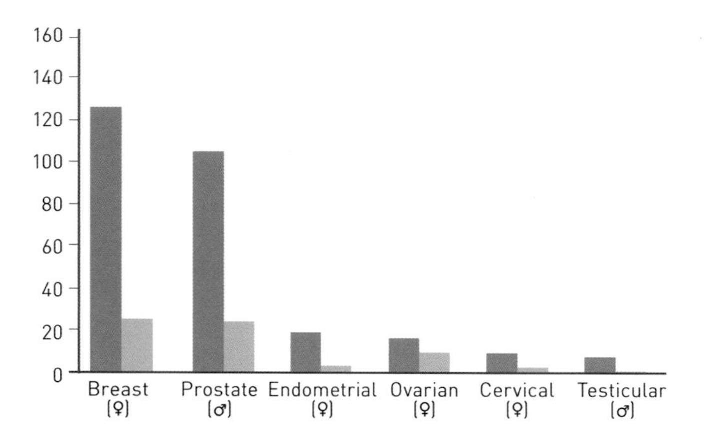

9.7.1 Cancers associated with male organs

Except for non-melanoma skin cancer, statistics show that prostate cancer is the most common cancer type experienced by men in the United States. The fluid the prostate contributes to semen counters vaginal acidity and prolongs sperm life, and small muscles associated with the prostate assist with ejaculation. The prostate approximately doubles in size at puberty, and then, beginning in the twenties, the prostate begins a process of slow growth, which is benign (non-cancerous) but can result in sizes that interfere with male urination, particularly after the age of 50. Muscle relaxants to ease urination and medications that reduce the amount of DHT are two therapies for an enlarged prostate.

Prostate cancer mostly occurs in men over the age of 65, a demographic in which the cancer grows slowly and does not usually result in death. In fact, it is very common to undertake no therapies for prostate patients who are relatively old. Eighty-four per cent of English and Welsh men who develop prostate cancer survive for at least ten years following diagnosis. In other cases, surgery, radiation, and chemotherapy, including androgen suppression drug therapies, can be prescribed. American statistics indicate that both the incidence of prostate cancer and its associated mortality rate are about twice as high in African-Americans as in European and Hispanic Americans. Asian and Indigenous Americans experience lower incidence of prostate cancer.

Testicular cancer is not particularly common, but it strikes young men disproportionately; in the United States it is the most common cancer in men in the 15-to-34 age bracket. As we've seen, sperm are made in the testicles, and almost all testicular cancers are tumors that develop in the cells that make sperm. Most testicular cancers are curable by employing surgery or radiation. Chemotherapy can be used when the testicular cancer has spread beyond the testicle or to prevent the cancer from recurring. Testicular cancer itself, as well as its treatment, can affect hormone concentrations and fertility. Usually, the cancer is in only one testicle, and the remaining testicle can produce enough testosterone and sperm for normal functioning. If therapy is expected to lower fertility, some patients will store sperm in advance of the therapy.

9.7.2 Cancers associated with female organs

Except for non-melanoma skin cancer, breast cancer is the most common cancer type experienced by women in the United States. Its detection and treatment efficacy have improved in recent decades, such that there is a high likelihood of survival. Seventy-eight per cent of English and Welsh women who develop breast cancer survive for at least ten years after diagnosis. Yet globally, it remains the main cause of death from cancer among women. Motherhood decreases the incidence of breast cancer. Indeed, early pregnancies, multiple pregnancies, and longer periods of breast-feeding are all associated with a lower incidence. As with prostate cancer among men, breast cancer incidence is disproportionately distributed among racial groups; in the United States, African-American women suffer the most and Hispanic, Indigenous, and Asian women suffer the least.

What is determined appropriate in screening for breast cancer has been contentious. Self-exams remain encouraged by some groups, although there is no demonstrated nexus between self-exams and not dying from breast cancer. The most prominent screening methodology is the mammogram, which is an X-ray of the breast. There is no consensus among medical professionals at what age mammograms should start and with what frequency they should occur, yet there is general agreement that mammograms are the best means of finding breast tumors. Accordingly, there is a demonstrated nexus between failing to have regular mammograms and dying from breast cancer.

Usually, breast cancer starts in the cells that line the ducts that carry milk from the lobules to the nipple (see Figure 10.12). Breast lumps are not always cancerous tumors, and as a result, any suspicious lump is subjected to a *biopsy*, in which a small sample of tissue or cells is removed from the breast and then tested in the lab. This is the only definite method for diagnosing breast cancer. If breast cancer is confirmed, there is then a set of treatment options that will vary in accordance with a variety of factors including age and health, as well as stage, extent, and location of the cancer. A key factor is also whether the cancer has spread to the lymph nodes, by which the cancer can *metastasize* elsewhere in the body. Most women with breast cancer will have surgery, often in combination with chemotherapy or radiation.

The most common cancer of the female reproductive system is endometrial cancer, which is a form of *uterine cancer*. In the mid-twentieth century, hormone replacement therapy (HRT) was introduced for menopausal women to combat some of the undesirable symptoms of menopause. The active ingredients in early formulations were estrogens alone, but eventually it was demonstrated that this increased cancer incidence in women, especially endometrial cancer. Now, HRT for women who still have a uterus (i.e., have not had it surgically removed) includes a combination of estrogens and *progestins* (synthetic progesterone), which counteract the carcinogenic capacity of estrogen. Women who have never been pregnant also have an increased likelihood of getting endometrial cancer.

Cervical cancer is another form of uterine cancer. There are several risk factors, but the majority of cases are caused by the human papillomavirus (HPV), a family of viruses that infects the human genitals and anus through sexual contact (Section 10.8.3). Some HPV strains are harmless and present no symptoms, but others are associated with cancer, especially cervical cancer. HPV vaccines, which are designed to be administered before exposure, can prevent most infections. Accordingly, they are usually prescribed for the pre-puberty years, especially for girls. Other, much less significant, risk factors for cervical cancer are long-term use of oral contraceptives and having multiple sex partners.

Although uterine cancers are very serious, ovarian cancer has a significantly higher death rate (Figure 9.11). Although its cause is not known, one risk factor is inheritance of an allele that is also connected to an increase in breast cancer. Other risk factors include estrogen-only HRT and having never having been pregnant.

CHAPTER 9 SUMMARY

- There are several key components of the endocrine system that regulate sexual development and function, including the hypothalamus, the pituitary gland, and the gonads, as well as numerous steroid hormones ("male" androgens and "female" estrogens) and peptide hormones.
- Hormone levels in the bloodstream vary with sex, with age (where key points of change are prenatal development, puberty, and menopause), and, in females, with the stages of the adult reproductive cycle.
- Hormone production is regulated by both negative feedback, which dampens high levels or increases low levels in the bloodstream, and positive feedback, which increases levels in the bloodstream.
- Sexual differentiation in the embryo begins to occur in about the seventh week of gestation as a consequence of changes in the levels of hormones that differ between males and females.
- The male reproductive anatomy is designed to deliver sperm to the female reproductive system, and it includes the penis, the testes, the prostate gland, and ducts for mixing sperm with seminal fluid, which together exit through the urethra.
- The female reproductive anatomy is designed not only to receive sperm but also to nurture the developing embryo and ultimately to deliver the baby at term, and it includes the vagina, the uterus (including the endometrium and cervix), the fallopian tubes, and the ovaries.
- Among the many groups of animals that employ internal fertilization, there is remarkable diversity in the structure of the male's intromittent organ.
- Several unusual genetic circumstances, such as a missing or extra sex chromosome or a mutation in a key gene, can result in the development of unusual sexual phenotypes, some of which are described as *intersexual*.
- For the many genes located on the human X-chromosome, females have two copies but males have only one; in sex-linked inheritance about 50 per cent of males manifest a dysfunctional trait inherited

from their mother if the mother has one functional allele and one dysfunctional one.

- There are numerous cancers of the sexually differentiated anatomy, including prostate cancer in males and breast cancer in females, each of which has a high survival rate.

FURTHER READING

Bendahan, S., Zehnder, C., Pralong, F.P., & Antonakis, J. (2015). Leader corruption depends on power and testosterone. *The Leadership Quarterly, 26*(2), 101–122. https://doi.org/10.1016/j.leaqua.2014.07.010

Colapinto, J. (2000). *As nature made him: The boy who was raised as a girl.* Toronto: Harper Collins Publishers.

Davis, G. (2011). "DSD is a perfectly fine term": Reasserting medical authority through a shift in intersex terminology. *Sociology of Diagnosis, 12*, 155–182. https://doi.org/10.1108/S1057-6290(2011)0000012012

Peterson, R.E., Imperato-McGinley, J., Gautier, T., & Sturla, E. (1977). Male pseudohermaphroditism due to steroid 5α-reductase deficiency. *American Journal of Medicine, 62*(2), 170–191.

Reis, E. (2007). Divergence or disorder? The politics of naming intersex. *Perspectives in Biology and Medicine, 50*(4), 535–543. https://doi.org/10.1353/pbm.2007.0054

Rogaev, E.I., Grigorenko, A.P., Faskhutdinova, G., Kittler, E.L.W., & Moliaka, Y.K. (2009). Genotype analysis identifies the cause of the "Royal Disease". *Science, 326*(5954), 817. https://doi.org/10.1126/science.1180660

Towers, B.A. (1980). Health education policy 1916–1926: Venereal disease and the prophylaxis dilemma. *Medical History, 24*(1), 70–87. https://doi.org/10.1017/S002572730003979X

Viña, J., Borrás, C., Gambini, J., Sastre, J., & Pallardó, F.V. (2005). Why females live longer than males: Control of longevity by sex hormones. *Science of Aging Knowledge Environment*, 2005(23): pe17. https://doi.org/10.1126/sageke.2005.23.pe17

Welling, L.L.M. (2013). Psychobehavioral effects of hormonal contraceptive use. *Evolutionary Psychology, 11*(3), 718–742. https://doi.org/10.1177/147470491301100315

Zietsch, B.P. & Santtila, P. (2011). Genetic analysis of orgasmic function in twins and siblings does not support the by-product theory of female orgasm. *Animal Behavior, 82*(5), 1097–1101. https://doi.org/10.1016/j.anbehav.2011.08.002

10 Human Fertility and Birth

KEY THEMES

Making sperm in humans occurs from puberty to old age, whereas making eggs begins before birth, resumes at puberty, and terminates with menopause. The female reproductive cycle begins with menstruation and features ovulation at about day 14. Infertility is reasonably common in human couples but can be countered with various technologies. People engaged in heterosexual intercourse can use contraceptives if they want to avoid pregnancy. Following fertilization, the egg implants in the uterine wall and the placenta begins developing. Sexually transmitted diseases are caused by a variety of organisms and constitute a serious public health issue.

It's not surprising that we use the same word, *fertility*, to describe both productive agricultural land and human conception and birth. Humans have always been interested and keenly attuned to the capacity of the land and animals they depend upon to be productive, as well as the capacity of their own bodies to produce offspring. The absence of a scientific approach to the subject did not stop pre-scientific cultures from formulating ideas about fertility, many of which were in the realm of the supernatural, including the belief in deities who were influential in such matters and around whom rituals were established. In most cases, worship of such deities included fertility rites, sometimes involving sexual intercourse or symbolic sexual activities. Crops, livestock, and the production of children were all of central importance to these cultures, making fertility a consuming interest. Indeed, the oldest known sculpture of a human being, the Venus of Hohle Fels excavated from a German cave in 2008, is clearly representative of female fertility, with prominent breasts and intricately carved genitalia (Figure 10.1).

Scientific advances over the past century have provided us with a high level of sophistication as well as remarkable tools for understanding and manipulating fertility, again both in ourselves and in the organisms and landscapes we depend upon. Earlier chapters in this book, particularly those dealing with genetics and the development and regulation of sexually complementary male and female reproductive systems, have already set the stage for understanding fertility and what flows from it. In this chapter, we primarily consider the production of new humans, from the creation of gametes in both sexes, through their union inside the woman's reproductive tract, and then the stages that ensue as the woman's body becomes a chamber for producing a new human being. We conclude with a survey of sexually transmitted diseases.

10.1 KEY DIFFERENCES BETWEEN MAKING SPERM AND EGGS

The fundamental goal of meiosis is the same in men and women—to produce haploid gametes that can unite through fertilization following sexual intercourse. But there are detailed differences between males and females that fall into three main themes—the pace of meiosis, the

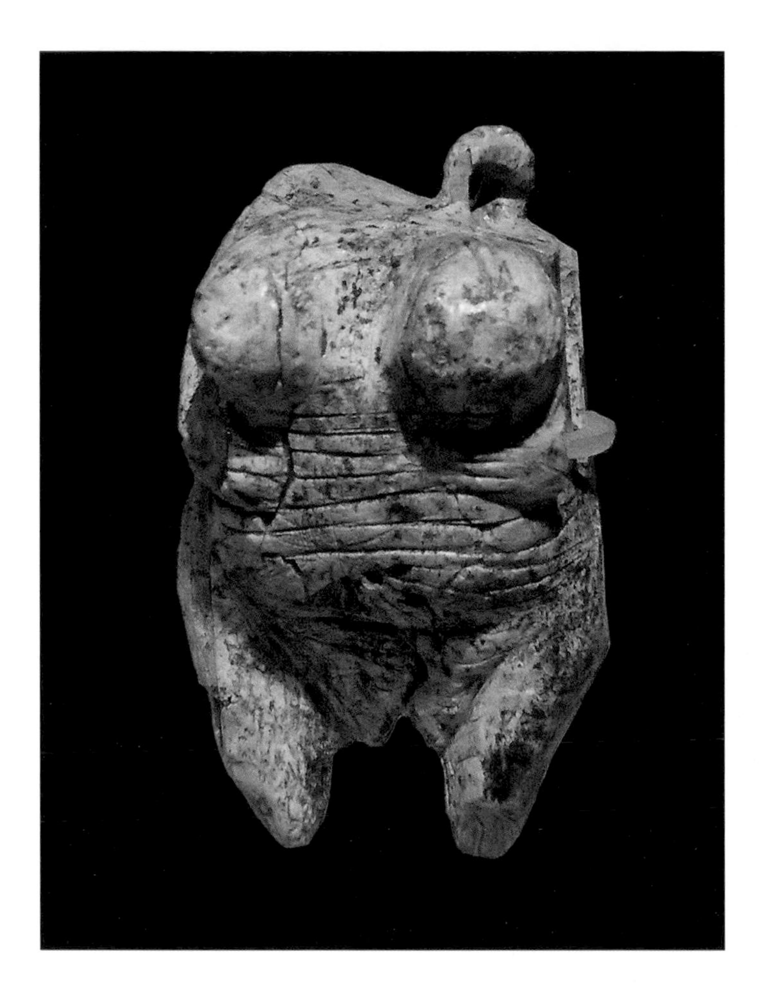

FIGURE 10.1 **FERTILITY HAS BEEN A PERPETUAL THEME IN ART SINCE PREHISTORIC TIMES**

The Venus of Hohle Fels, carved from mammoth (*Mammuthus primigenius*) tusk ivory 35,000 to 40,000 years ago, is the oldest known sculpture of a human form. Its features reveal a prehistoric interest in female fertility that has continued through history.

period of life in which it occurs, and the numbers of gametes produced (Figure 10.2). Applied to both sexes, the creation of gametes is called *gametogenesis*. We'll begin with spermatogenesis (the creation of sperm), because it matches most closely what we've reviewed so far; for oogenesis (the creation of eggs), there are several interesting twists.

10.1.1 Spermatogenesis occurs from puberty to old age

When most babies are born, they already have the reproductive equipment that will come into play later in life. Male babies have testes that contain diploid cells known as *spermatogonia* (singular,

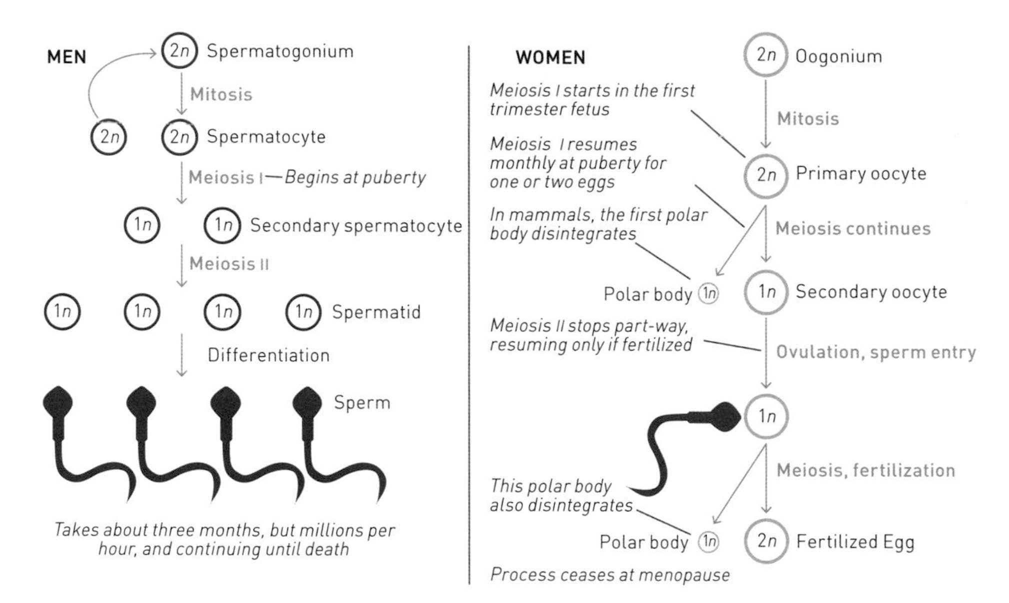

FIGURE 10.2 **MALE AND FEMALE MEIOSIS DIFFER IN SEVERAL SIGNIFICANT WAYS**
Although the result is haploid gametes in both cases, meiosis in males (spermatogenesis) and in females (oogenesis) differ dramatically. Cells that are labelled 2*n* are diploid, and those that are labelled 1*n* are haploid. Daughter cells that do not develop in female meiosis are called *polar bodies*.

spermatogonium), but they remain inactive as far as meiosis is concerned until puberty. It is only when boys have reached this developmental stage that meiosis begins. At puberty, the spermatogonia start dividing mitotically, to give rise to primary spermatocytes (the suffix *cyte* denotes "cell"). Primary spermatocytes are the cells that undergo meiosis, and from that time, meiosis is a non-stop undertaking in the male testis. When the primary spermatocyte completes the first division of meiosis, the daughter cells are called *secondary spermatocytes*, and when these divide in the second division of meiosis, the cells are called *spermatids*.

The rate of production of sperm in males is impressive. It takes about three months from the time meiosis begins to the time when the spermatids have been converted to mature sperm cells with tails, which are then referred to as *spermatozoa*. Although the process requires a significant amount of time, men have many spermatogonia. As a result, they can produce as many as 130 million sperm or more per day, which is a rate of about 1,500 per second!

10.1.2 Oogenesis occurs from the fetal stage to menopause

We contrast oogenesis and spermatogenesis in Figure 10.2. In some ways, we can simply describe female meiosis by using terms introduced for males where we substitute an *oo* (for "egg") in place of the *spermato*. But there are striking meiotic differences between males and females. One is timing. Whereas males do not start meiosis until puberty, it surprises many that females start the process long before they are born. As the ovaries develop, so too does the female version of the spermatogonium, known as the *oogonium*. Unlike the case with boys, whose spermatogonia become active at puberty, the oogonium in girls begins the meiotic process in the first trimester of development! So, by three months after pregnancy begins, the ovaries of the female fetus already contain about two million primary oocytes that have started meiosis.

Yet these primary oocytes stop before completing the first meiotic division, and they remain in that state until girls reach puberty, at which point meiosis resumes to produce secondary oocytes—but only for perhaps a dozen primary oocytes each menstrual cycle. And, even then, meiosis in the secondary oocytes stops again, this time partway through the second division. It is only if and when the secondary oocyte is fertilized by a sperm that female meiosis completes to the end. So while men produce as many as 1,500 gametes per second from puberty until death, women produce one or a few secondary oocytes per cycle, and even then none become true haploid gametes (ova) until fertilized by a sperm. Furthermore, unlike men, who produce sperm into old age, women cease to produce gametes when they reach menopause at around age 50. Menopause (Section 10.7) is the cessation of monthly ovulation, because eggs are now very few in number. This also means the end of monthly menstruation; without medical intervention, pregnancy is not possible after menopause.

Any meiotic misadventure that produces an egg with two copies of one of the chromosomes (or zero copies) is known as meiotic non-disjunction (Section 9.5.1). As women age, the likelihood of errors occurring during meiosis increases, resulting in chromosomal defects in their children, especially due to non-disjunction. This pattern is thought to be due to the age of the primary oocytes that are by that point 40 or more years old, meaning mistakes are more likely to occur. For example, the risk of having a child with Down syndrome (*trisomy 21*) increases with the mother's age, especially after 40. Trisomy 21 usually results when a sperm with one copy of each autosome fertilizes an egg that, having undergone most of meiosis, has mistakenly invested the egg with two copies of the 21st chromosome, rather than one copy. With the single copy from the sperm, the fertilized zygote now has three copies, and the child develops with all the indicators of the syndrome because all his or her cells have three copies, rather than two, of chromosome 21. This is not to say sperm does not also carry mutations. In fact, although meiotic non-disjunction is more likely in egg production, sperm are more likely than eggs to carry mutant alleles (Section 4.3.3).

10.2 THE MENSTRUAL CYCLE

The female reproductive cycle is typically 28 days, although it varies from woman to woman. It is a series of events involving the hypothalamus-pituitary-gonadal axis and the uterus. In short, each month between puberty and menopause, the body prepares for the possibility of pregnancy. Two things must occur in the woman's system for pregnancy to be possible. First, an egg must be released from an ovary so that it may be fertilized by a sperm. Second, the endometrial layer of the uterus must be prepared for the possibility of implantation of the embryo, should fertilization have occurred. The events in the ovary are strongly regulated by the gonadotropic hormones secreted from the anterior pituitary, and the events in the uterus are strongly regulated by steroid hormones secreted by the ovary (Figure 10.3).

The cycle can be divided into three sections, two of which overlap. The first five days is the *menstrual period* (also *menses*)—the bloody discharge through the vagina of the shed endometrial lining that had been built up since the woman's previous period. The menstrual period overlaps with the first 5 days of the 14-day follicular phase. Under the influence of FSH, a dozen or so ovarian follicles are activated in the ovary that will ovulate. A follicle is a spherical collection of cells within the ovary, each of which has the capacity to release one oocyte. Other

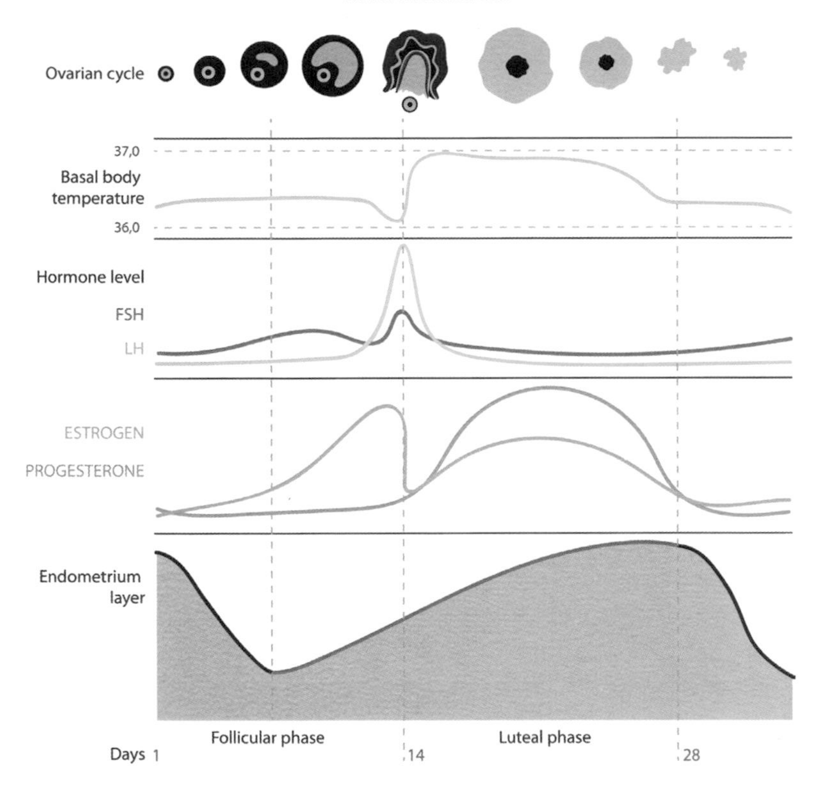

FIGURE 10.3 A NUMBER OF FEATURES VARY WITH THE MENSTRUAL CYCLE
Approximately every 28 days there is a predictable series of changes in a reproductive woman's body temperature, levels of key regulatory hormones, and endometrial lining the uterus.
Ptaha I / Shutterstock.com

cells in the follicle include those that have the capacity to produce steroid hormones. Although numerous follicles begin developing, only one continues, and it secretes increasing amounts of estradiol. This represents the positive feedback illustrated in Figure 9.3, with the result that the end of the follicular phase constitutes the peak of estradiol, LH, and FSH levels. The sudden surge of LH late in the follicular phase triggers ovulation, which is the rupturing of the follicle and the release of the secondary oocyte at the upper end of the fallopian tube. Each month ovulation occurs from either the woman's right or left ovary, and analysis of multiple cycles indicates the choice of ovary is random.

Meanwhile, under the influence of growing estradiol levels, the cells of the endometrium have been proliferating during the latter part of the follicular stage. This endometrial growth continues for most of the 14-day stage that follows ovulation, known as the luteal phase. This phase gets this label because, under the influence of LH, the ruptured follicle becomes a *corpus luteum* ("yellowish body"). This structure is a temporary gland, secreting progesterone and some estradiol. Progesterone produces a dramatic buildup of the endometrial lining. If the secondary oocyte has failed to be fertilized by a sperm, the progesterone also serves as a negative feedback system that reduces LH and results in the degeneration of the corpus luteum (Figure 9.3). Hormone levels drop, and menstruation follows.

10.2.1 Most mammals have an estrous cycle instead

A reproductive cycle is universal in female mammals, but most don't have a menstrual cycle. Many primates and bats have menstrual cycles, but in most other mammals it is an estrous cycle regulated by similar hormones. It differs in a couple of key ways. Commonly, it is characterized by a usually brief period of *heat* or estrus at or just before ovulation so that the likelihood of pregnancy is maximized (Figure 10.4). At other times, the female shows no interest in mating. Second, although there can be some discharge on a cyclical basis, the endometrial lining is almost entirely resorbed internally by most mammal species, so that menstruation does not occur. Note the differences in spelling and meaning: *estrous* is an adjective describing the cycle, and *estrus* is a noun, describing the condition or sexual readiness during a particular period in that cycle.

In cattle, for instance, females have a 21-day estrous cycle that involves the same hormones as in humans, but cows are only receptive to bulls during a period of heat—starting about a half-day before ovulation—that doesn't quite last a day. The cow in heat advertises her interest by various restless behaviors including mounting other cows, genital sniffing, and bellowing. During the other 20 days of the cycle, cows are not receptive to the attentions of bulls.

FIGURE 10.4 **MANY FEMALE MAMMALS ARE ONLY BRIEFLY RECEPTIVE ONCE PER CYCLE**

A male Thomson's gazelle (*Eudorcas thomsonii*) smells a potential female partner to assess whether she is in heat, before mating with her. Note her cooperative stance, consistent with sexual readiness.

10.4: Male gazelle checking female's receptivity to mating. Taken on safari in Tanzania. John Storr, Public domain.

10.2.2 Is women's fertile period concealed?

In humans, most pregnancies occur following sexual intercourse that occurs in the six-day period ending about a day after ovulation—a fertile period during which estradiol levels are relatively high and progesterone is low. Research analyzing sexual intercourse rates initiated by women not under the influence of hormonal contraceptives (i.e., not on the pill) indicates there is an increase in desire during this period, but female human sexual activity is obviously not limited to a period of heat as in most animals. That weekends are a strong predictor of sexual intercourse attests to this, since a woman's hormone levels can be at any stage of the cycle on weekends.

Unlike most mammals, then, humans have what is termed *concealed ovulation* (see Box 7.1). By counting days since one's period, by monitoring body temperatures (since temperatures rise abruptly at ovulation; Figure 10.3), by employing technological sampling devices that reveal when ovulation is occurring, and even perhaps by attending to levels of desire, women can increase, or decrease, their likelihood of becoming pregnant. Nonetheless, most women do not know when they are ovulating, and it is even less obvious to their male partners.

However, there has been a recent shift from the earlier dogma that human ovulation is completely concealed. The cues are subtle, but they include behavioral and hormonal changes in both the ovulating woman and potential mates with whom she interacts. In various studies, the natural scents, faces, and voices of females during their fertile period were found to be more attractive to males. In addition, females are reported to enhance their appearance more during their fertile phase, and to be more flirtatious and open to advances from potential male partners. However, these on their own do not indicate that women are more likely to unfaithfully stray from their partner. In addition, one theory about these cues is that human ovulation is essentially concealed for the purpose of promoting monogamous commitment. However, that the concealment is "leaky," meaning some subtle cues persist.

One well-publicized study examined lap dancers employed in a nightclub, correlating the tips each dancer earned with the stage each was at in her menstrual cycle. The results were rather dramatic. During the fertile period of the cycle (approximately days 10–15), dancers earned 30 per cent more tips from their male patrons than during the non-fertile luteal phase (days 18–28) and about 80 per cent more tips than during the menstrual phase (days 1–5). Such an effect could be caused either by better performances during the fertile phase or by a capacity of the male patrons to detect and prefer the fertile phase. The researchers favored the latter explanation, suggesting ovulation is not truly concealed, although it may not be consciously recognized. Interestingly, dancers on hormonal contraceptives that prevent ovulation and a fertile period showed no such peak in tips earned.

To the extent it is concealed, what is its function in humans? Perhaps it is easier to answer the converse first: why do most mammals have an estrous cycle with a period of heat that is obvious for both the female to experience and the male to notice? In most mammals (more than 90 per cent of species) males do not provide parental care. In these species, where sex appears to be exclusively for achieving pregnancy and not to form pair-bonds, an obvious period of heat is logical, because there the relationship does not continue after mating, and without heat, the opportunity for fertilization could very well be missed.

For most of our history, it has been virtually impossible for a woman to raise a child without benefit of others, particularly a committed partner. Most theories about concealed ovulation relate to this fact. The pleasure rewards associated with continuous sexual receptivity tend to keep male partners around, including during the extended period of nursing in which females have reduced fertility. Also, in "single-parent" mammal species, males mate during female heat and then move on to find another partner, but with concealed ovulation these one-night-stands would have poor prospects for success. Instead, by remaining with one partner and mating with her throughout the cycle, the male and his genes have much higher pregnancy rewards. In this sense, sexual activity among heterosexual humans is as much, or more, a feature of pair-bonding as of pregnancy.

10.3 SEXUAL AROUSAL AND RESPONSE IN MEN AND WOMEN

So far in this chapter, we have reviewed the production of sperm and eggs, as well as features of the menstrual cycle, which include a fertile period of about six days. Except where modern in vitro techniques allow for fertilization of the egg by a sperm to occur in a clinical setting, fertilization cannot occur without sexual intercourse, and sexual intercourse cannot occur without at least the sexual arousal of the male partner, leading to copulation.

Sexual arousal and response involve the whole body, ultimately focusing on the genitals. Based on the pioneering work of William Masters and Virginia Johnson, an American team who published their research in 1966 and 1970, there are four stages of sexual response in humans: excitement, plateau, orgasm, and resolution. Excitement begins when the brain receives sexually significant sensory information from vision, hearing, smelling, or touching, although auto-arousal through imagination and fantasy can also be significant.

Arousal shifts to the plateau stage with the physical stimulation of sexual activity. In both sexes, increases in heart rate, blood pressure, and blood flow to the genitals that began during excitement continue or accelerate. In some body parts this increased blood flow results in vasocongestion, a condition where blood cannot easily flow through the blood vessels, particularly the veins that drain the tissue. This engorges such tissues with blood, which warms, reddens, swells, and stiffens them. In both sexes, blood can cause skin flushing, especially in the chest and neck.

In men, the testicles elevate, bringing them closer to the body, but the most dramatic male response is vasocongestion occurring in the penis. An erection begins in the brain, when sexual stimulation—physical, imaginary, visual, tactile, etc.—prompts it to send messages to the penis via the nervous system. The receipt of these messages has the effect of relaxing the arteries that deliver blood to the penis, increasing blood flow. This increased flow compresses the veins that normally drain the penis, resulting in the trapping of blood. When not filled with blood, the two cylindrical structures on each side of the penis known as corpus cavernosa are relatively small, being made of blood vessels, tissue, and open pockets. However, the hydraulic pressure from the increased supply and decreased return of blood results in dramatic elongation and thickening of the penis—as much as a sevenfold increase in total volume. A sheath surrounding the corpus cavernosa made largely of collagen fibres prevents the shaft from bending during sexual activity. The blood engorgement does not restrict the urethra, which delivers semen. It remains uncompressed because it is wrapped by a different structure called the *corpus spongiosum.* Following sexual resolution, artery pressure declines, veins open and drain, and the penis returns to a flaccid state.

In women, vasocongestion occurs most notably in the nipples, the labia, and the clitoris. Not surprisingly, because it is the female homologue of the male penis, erection of the clitoris is achieved in a very similar manner to that of the penis: arterial blood flow is increased, venous draining is restricted, corpus cavernosa become engorged, and a collagen sheath contributes to stiffness. Other changes in women include increasing lubrication of the vaginal wall and the swelling and tightening of the outer vagina. The inner vagina expands and lengthens. The inner labia typically become a rich reddish color, the hue of which varies among women. Internally, the uterus elevates. In some women the breasts increase in size during the plateau phase, and blood also swells the *areola* (plural, *areolae*), which is the area of pigmented skin surrounding the nipple.

Not all sexual activity leads to orgasm, which in both sexes is an intensely pleasurable release of sexual tension, but it is a common goal. Accordingly, sustained sexual excitement, especially when it includes physical stimulation to the clitoris and penis, usually culminates in orgasm. Close to orgasm, the clitoris pulls back against the pubic bone and recedes, and in men a few drops of clear lubricant are released. During orgasm, coincident with the intense pleasure is a series of muscle contractions in the genital area, focused around the vagina in women and near the base of the penis in men. Subjective descriptions of orgasm by men and by women suggest it is a similar experience for both sexes, and this is true for women whether the stimulation was clitoral or vaginal.

In men, orgasm is accompanied by ejaculation, which is achieved through two stages. During *emission*, sperm from the ejaculatory ducts enter the urethra where semen is assembled. Once there, this is experienced by men as a point of no return, and *expulsion* follows. During expulsion, contractions that occur about every 0.8 seconds cause semen to be released from the tip of the penis in spurts, the first of which carries the most sperm. From a reproductive point of view, it is necessary for men to achieve orgasm because it entails ejaculation, but orgasm is not necessary for women, since pregnancy can occur without it. Some believe female orgasm promotes pregnancy, however, since the uterus experiences muscle contractions that may help move seminal fluid toward the egg, but there is no strong evidence for this. In most women, orgasm is most readily achieved through direct stimulation of the clitoris rather than by vaginal stimulation during intercourse.

Resolution follows orgasm. This is where the body returns to a more normal state, with a reduction of vasocongestion and a drop in heart rate. Resolution can occur without orgasm upon the cessation of sexual activity, although this occurs more slowly. Men require a *refractory*

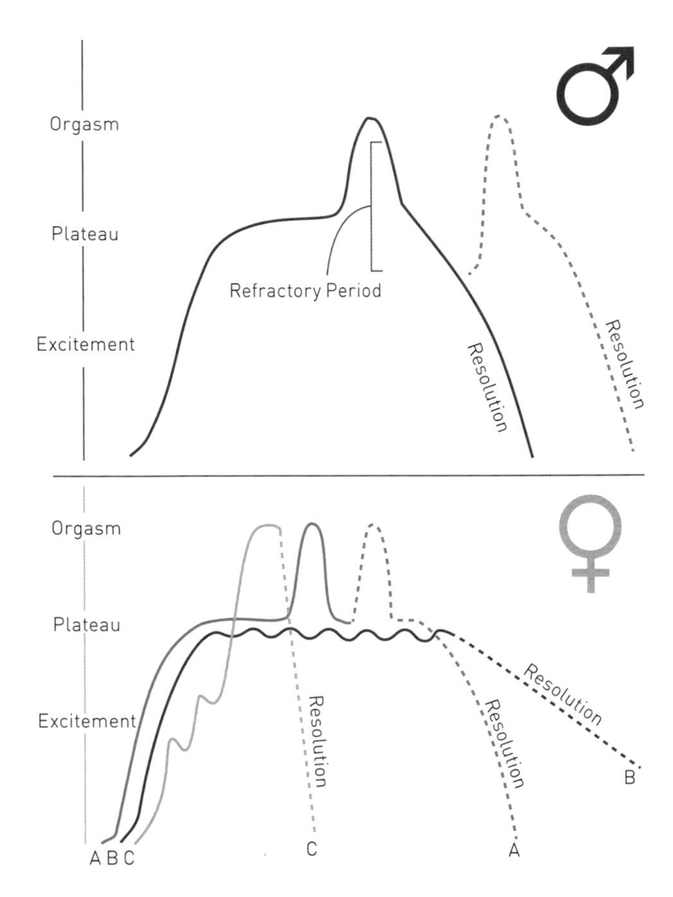

FIGURE 10.5 **THE SEXUAL RESPONSE CYCLE IN HUMANS HAS FOUR STAGES**
There are contrasting differences between men (above) and women (below). In the case of women, the pattern of the response is more variable, as indicated by alternatives A, B, and C.

period following ejaculation during which they cannot reach orgasm again, although in some cases this can be measured in mere minutes. Such a recovery period is not true of women, who sometimes are able to reach orgasm again quickly, without intervening resolution, if sufficiently aroused and stimulated (Figure 10.5).

10.4 FERTILIZATION AND THE MAKING OF A ZYGOTE

A human male's ejaculate during intercourse will, on average, contain close to 300 million sperm, and it can even be double that. This would make the odds of pregnancy seem likely, assuming it occurred during his partner's fertile period just before or at ovulation. Yet there are numerous factors that reduce the number of sperm that reach the vicinity of the egg. For one thing, only a fraction of the sperm deposited in the vagina make it to the uterus, and even that is a long way from the site of the egg. Fertilization typically occurs in the upper parts of a fallopian tube, and only a fraction of the sperm that make it to the uterus are also able to make it to the fallopian tubes. Half the sperm presumably choose the wrong side, since a woman ovulates from only the right or the left side each month. Compounding these factors is the widespread observation that large numbers of sperm appear to be abnormal, even in a healthy man's semen. Abnormalities can exist in either the head or the tail, and some can be dead or immotile, meaning they cannot move. Those few healthy sperm that do make it to the fallopian tubes, however, usually get there in less than an hour after intercourse and can live for as long as five days, although three days is probably more typical.

Technically, the egg at this point is still a secondary oocyte (Section 10.1.2). Under normal circumstances only one sperm enters the oocyte, known as *monospermy.* In humans, there are so few sperm that manage to reach the location of the oocyte in the fallopian tubes that *polyspermy* is unlikely, although there are cellular mechanisms that prevent penetration by a second sperm. In species where many sperm reach the egg at more or less the same time, a key feature that prevents polyspermy is an almost immediate change to the electric charge of the oocyte membrane, meaning that additional sperm cannot fuse with it.

Although sperm are months in the making, sperm continue to mature following ejaculation. Capacitation is the second last step in the maturation of mammalian sperm. It is a biochemical process that occurs after ejaculation, triggered by the mixing of sperm and seminal fluid that occurs during that event, and it therefore occurs in the female reproductive tract. Capacitation is primarily dependent upon a substance produced by the prostate gland, but it is also aided by secretions in the uterus. Capacitation serves to destabilize the membrane of the cap-like head of the sperm, known as the *acrosome,* making it more permeable, which prepares it for penetrating the outer layer of the oocyte. Capacitation also has the effect of increasing swimming speed.

The final step in sperm maturation, which usually occurs in the fallopian tube, is the *acrosome reaction.* Here, the sperm fuses with the outermost membrane of the oocyte. This releases enzymes needed for breaking through additional layers of the oocyte so that the DNA in the sperm may enter the cell.

As noted, we commonly refer to the sperm as fertilizing the "egg," but it is really still a secondary oocyte at that point that has not completed meiosis. So, just as sperm continue to mature upon and following ejaculation, so too does the oocyte have further maturation to do. As we discussed in Section 10.1.2, meiosis in females is a protracted event, starting in the female fetus before birth, but then delayed through pre-pubescent childhood. Upon menarche, meiosis resumes for several eggs each monthly cycle. Even then, however, the second division of meiosis does not occur unless and until a sperm fertilizes it. At that point, the second division of meiosis finishes, and the oocyte becomes a mature ovum.

10.4.1 Infertility

It is estimated that 10 to 15 per cent of American couples are infertile, where the clinical definition of infertility is that a couple cannot achieve pregnancy despite trying for at least a year. Over the longer term, the rate is lower, but is still approximately five per cent of couples. Analysis of the cause of infertility shows that where the problem is attributable to one partner, it is equally likely to be the man or the woman. This accounts for about two-thirds of cases. The remaining one-third of infertility cases are due to problems with both individuals or to unknown circumstances.

In men, causes of infertility are usually related to sperm production, including abnormal sperm, sperm with poor motility, and low numbers of sperm, or to problems related to delivery of sperm. Sperm viability problems are sometimes caused by environmental factors such as pesticide poisoning, radiation, and some medications. In women, the most common cause of infertility is an ovulation disorder, associated with an irregular menstrual cycle. *Endometriosis*, which is characterized by the growth of endometrial tissue outside the endometrium, usually in the fallopian tubes or around the ovaries, can also be a cause. So too can uterine abnormalities or fallopian tube damage.

In addition to these particular issues, risk factors for low fertility or infertility include tobacco, alcohol, or marijuana use, as well as obesity or being significantly underweight. In addition, fertility drops with age,

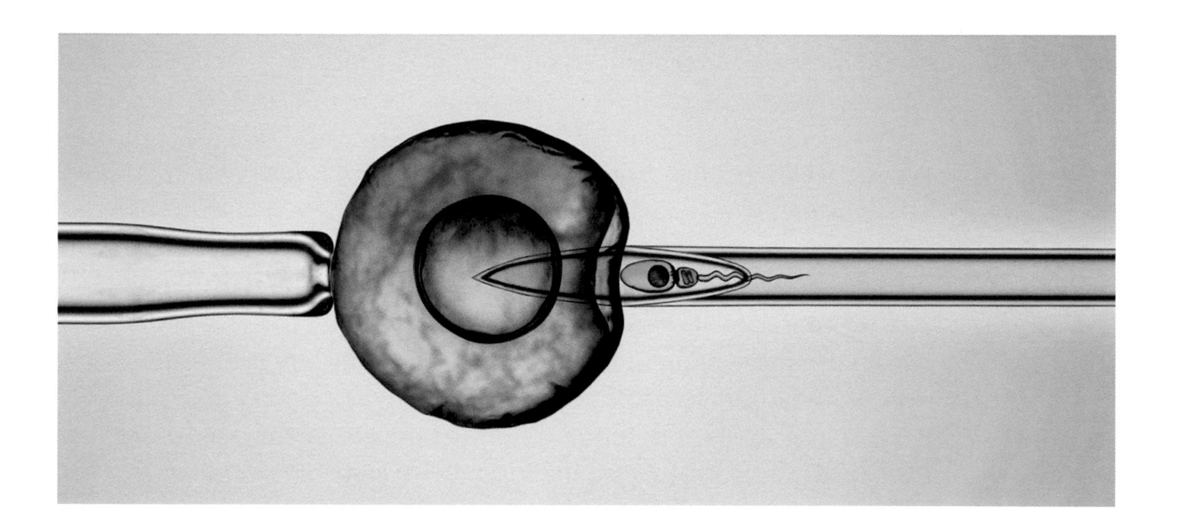

FIGURE 10.6 **SOME INFERTILE COUPLES BENEFIT FROM IN VITRO FERTILIZATION (IVF)**
In this image, a common IVF technique known as *intracytoplasmic sperm injection* (ICSI) is portrayed. A single sperm is directly injected into the oocyte, which will be transplanted into the mother after several days.
Nixx Photography / Shutterstock.com

particularly after the age of 35 in women and 45 in men.

For couples seeking help for fertility issues, the first approach is to maximize the likelihood of sperm meeting egg during the peak of the woman's fertile period. Low-tech approaches include counting days since the start of the previous menstrual period in order to identify the fertile period (about 10 to 15 days after the menstrual period) and checking for slight changes in body temperature associated with ovulation. A more sophisticated solution is by using a luteinizing hormone (LH) kit available in pharmacies that measures the level of this hormone in the woman's urine; there is a surge in LH just before ovulation (Figure 10.3), which is the most likely time for fertilization to succeed.

Other interventions can be pursued if infertility persists. These are known collectively as assistive reproductive technologies (ART). There are few options available for men if the problem relates to the quality of sperm production. For women, there are a variety of medications that either include hormones or influence hormone levels in order to stimulate ovulation, with the hopes that fertilization will then be more likely to occur following intercourse.

If it appears intercourse will not result in pregnancy, a couple's sperm and oocytes can be collected for union in a clinical setting, known as *in vitro fertilization* (IVF; literally, "in glass"; Figure 10.6). The zygote remains in the clinic under controlled conditions and undergoes cell division for several days, after which the embryo is then transferred to the woman's uterus. In cases where it appears the woman will not be able to support a pregnancy, the embryo can be transplanted into the uterus of another woman, referred to as a *surrogate pregnancy*.

In vitro fertilization was first conducted in the late 1970s, but its capacities have improved considerably during the intervening period. For instance, modern techniques can succeed even using sperm that cannot swim. In other cases, IVF has been used to correct genetic abnormalities in the mother (Box 10.1)

10.4.2 Contraception

We are almost certainly the only species that understands the connection between copulation and baby making. Armed with that knowledge, humans have a long history of employing ways to avoid fertilization when the wish has been to avoid pregnancy, a subject known as *contraception*, meaning "against conception." Conception is sometimes used synonymously with "fertilization," but it also includes implantation of the developing embryo into the endometrium. The most obvious system is abstinence, where sexual activity, if any, is restricted to methods of experiencing sexual pleasure other than through intercourse. A version of this is to only avoid sexual intercourse during the female's fertile period (natural family planning) by counting days since her period, by monitoring changes in body temperature (see Figure 10.3), or by testing methods available from a pharmacy that are designed to detect the surge in LH that occurs in the day or two before ovulation; this is the rhythm method. The gamblers' system is *coitus interruptus*, where the penis is withdrawn before ejaculation; not only can this be difficult to time, but some sperm leave the penis before ejaculation, reducing the effectiveness of the method.

The most invasive or dramatic method of birth control is abortion, where a pregnancy has occurred and the embryo or fetus is removed from the uterus by either hormonal means—using hormone-like

BOX 10.1 TWO MOMS AND A DAD

Imagine that you are a woman with mitochondrial disease caused by a deleterious mutation in the DNA of your mitochondrial chromosome. Although this is not a fatal condition, your symptoms may be many and severe, since cellular mitochondria are crucial for all sorts of metabolic functions. Imagine also that you wish to have children. However, your children will certainly get the disease, since they will only get mitochondria from you, and not from your male partner, who presumably has functioning mitochondria.

For about 20 years now, we have had the capacity for correcting this using IVF, The technique is sophisticated but the idea is rather simple (Figure 10.7), and the popular press has reported on it with variations on the phrase "two moms and a dad." The nucleus is removed and discarded from the egg of a donor who has functioning mitochondria. This donor cell consequently has no genes except the 37 genes found in the mitochondria. Similarly, the nucleus is removed from the mother's egg, but this nucleus is not discarded. Instead, it is inserted into the donor's nucleus-free egg. This egg is then fertilized with the father's sperm and inserted into the mother. The baby that develops then has three "parents"—20,000 or so nuclear genes (in duplicate) from the mother and father and 37 genes from the donor mother.

Proponents of the technique say this amount of genetic engineering in humans is defensible, since it is only to fix the energy supply system via the mitochondria in the donor's cytoplasm, and the offspring is almost entirely genetically that of the "parents." In fact, several individuals born before about 2003 in the United States had already benefited from this method before its prohibition as being unethical. Its prohibition is being reconsidered in the United States, and it was authorized in the UK in 2015.

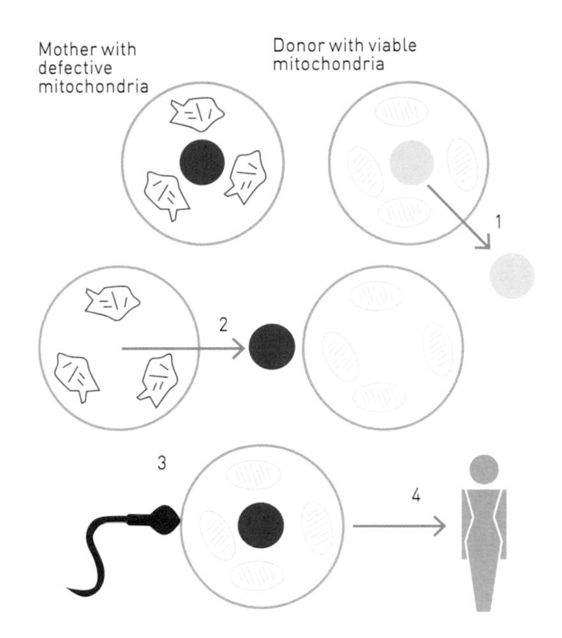

FIGURE 10.7 A BABY CAN HAVE TWO MOTHERS AND ONE FATHER

The in vitro fertilization technique shown here is one where two women contribute genetically to the egg that ends up being fertilized. The egg cell is from the donor; it contains healthy mitochondria, including the 37 mitochondrial genes, but has had the nucleus with all the nuclear genes removed (1). The nucleus is from the mother (2) and is inserted into the nucleus-free donor egg. The sperm (3) is from the father. The fertilized egg is then replaced in the mother (4), and nine months later a baby is born that is almost entirely genetically the mother and father but with a small but key genetic contribution from the donor.

prostaglandins that induce uterine contractions—or physical means. Abortion is not a contraceptive, since contraception means preventing conception, but it is a last-resort method of birth control.

Some people opt for surgical sterilization techniques (Figure 10.8), especially favored by those who have one or more children and consider their reproductive lives over. These techniques include tubal ligation in women, which closes off the fallopian tubes so that no eggs may be fertilized, and vasectomy in men, which severs the *vas deferens*, preventing the transport of sperm from the testes into the seminal stream.

All other methods of birth control allow for sexual intercourse to happen by employing either physical means to prevent sperm from reaching eggs or hormonal means to prevent pregnancy. A device whose origins go back centuries is the condom, a sheath (now of latex) unrolled over the erect penis that captures the ejaculated semen and thereby prevents sperm from entering the female reproductive tract. A secondary benefit is that condoms dramatically reduce the spread of sexually transmitted infections (Section 10.8). A less common device is the female condom, inserted into the vagina before sex.

There are other contraceptive devices designed for vaginal insertion: spermicide sponges or jellies and diaphragms designed to cover the cervical opening, or intrauterine devices (IUDs). IUDs are long-lasting methods that are inserted by a health-care provider, and they work by either a slow release of copper that kills sperm or of a progesterone-like hormone that makes the body "think" it is already pregnant.

In the 1960s, the introduction of the hormonal birth control pill—used so pervasively that it's known simply as "the pill"—proved to be a key factor in the social revolution that included increased sexual freedom. Like hormone replacement therapies, the usual formulation of the birth control pill is estrogen and a progestin. Collectively, elevated levels of these steroids inhibit follicular development, meaning that an egg is not released and so ovulation does not occur. It also alters the condition of the reproductive tract, making it less hospitable to sperm and to sperm movement. Some women report decreased sex drive or *libido* while on the pill, although this is far from universal. This observation is consistent with other research, however, that shows that women tend to experience the strongest sex drive at or about the

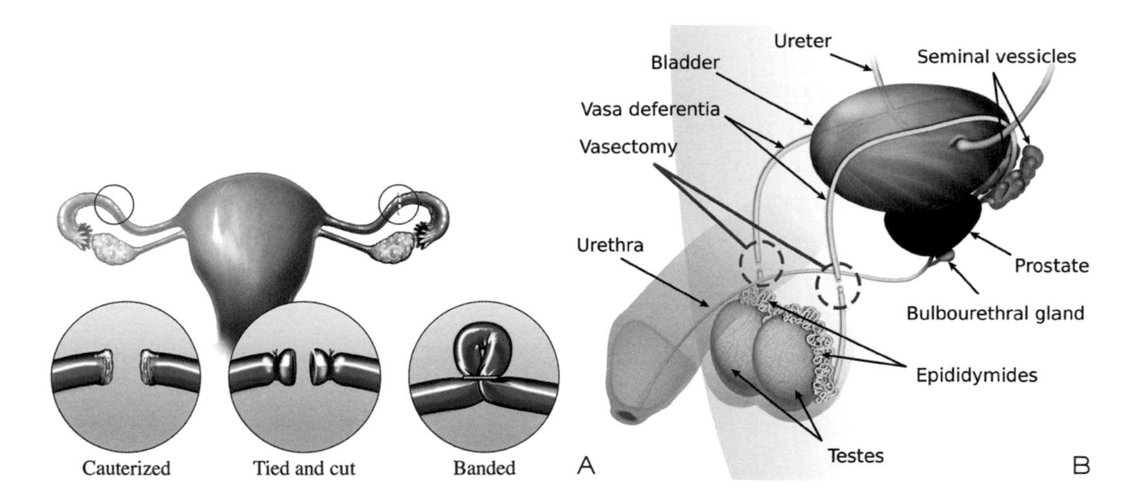

FIGURE 10.8 **SURGERY IS USED TO STERILIZE ADULTS WHO OPT FOR PERMANENT CONTRACEPTION**

Sterilization can be accomplished in the female by severing or otherwise blocking the fallopian tubes and in the male by severing and sealing the vas deferens.

[10.8a] Nucleus Medical Media Inc. / Alamy Stock Photo
[10.8b] K.D. Schroeder / CC-BY-SA 3.0

time of ovulation, and the pill eliminates ovulation. Yet the pill continues to be a very popular contraceptive method. Despite the fact that it alters a woman's hormonal cycle, making some women reluctant to take it, it continues to be used by millions.

A more recently developed oral contraceptive is the morning-after pill, designed to be taken shortly after sexual intercourse where contraception was either not used or failed, as in a torn condom for instance. Such pills primarily operate by delaying ovulation, which then makes fertilization unlikely.

It is not possible to overestimate the impact that contraception has had on the changing nature of families, as well as on population trends and patterns of immigration. It has been demonstrated repeatedly that as a society's standard of living increases, the average number of children per woman drops through various means of birth control, although there is commonly a delay of one or two generations.

10.5 PREGNANCY

If the oocyte is fertilized by a sperm that it has encountered in the fallopian tube, the resulting zygote begins to rapidly divide. After about one week, having moved into the uterus, the developing cell mass becomes a hollow ball of cells called a *blastocyst*. During *implantation*, the blastocyst lodges in the endometrial lining through a series of steps. Early in the process of implantation, the blastocyst adheres to the uterine lining, known as *adhesion*. As the process proceeds, however, the blastocyst digests some of the lining so that it embeds more deeply, a process known as *invasion*. The tiny embryo at first absorbs nutrients from the endometrium and dumps wastes into the uterus, but this changes rapidly.

10.5.1 The placenta

The placenta is the joint creation of the embryo and the mother (Figure 10.9). The first feature to develop is a circulatory system. This allows the embryo to "communicate" with the maternal blood system in the endometrium. The vessels that emanate from the embryo are in projections called *chorionic villi*, and they come so close to the mother's blood vessels that an exchange can take place: oxygen and nutrients move from the mother to the embryo, while carbon dioxide and other wastes move from the embryo to the mother. It is not until about the fifth week that the umbilical cord develops and connects with the placenta, at which point the umbilical vein receives oxygen and nutrients from the mother. The placenta also becomes the organ through which the developing baby is able to exploit the mother's immune system to fight infections. At one time, it was thought that the placenta protected the embryo against anything toxic that might be in the mother's circulatory system, but that is not true (Box 10.2).

As the mother–fetus interface, the placenta also is very much a component of the endocrine system, playing a role in hormonal feedback that maintains pregnancy-friendly levels of hormones. Recall that in the absence of pregnancy, the corpus luteum ceases activity at the end of the menstrual cycle, and this causes a drop in progesterone and estrogen levels. In turn, this causes the endometrium lining to be shed. In the early stages of a pregnancy, if this were allowed to occur, there would be a spontaneous abortion of the embryo. To prevent this, the first hormone produced by the human placenta is *human chorionic gonadotropin*. It targets the corpus luteum and plays the critical func-

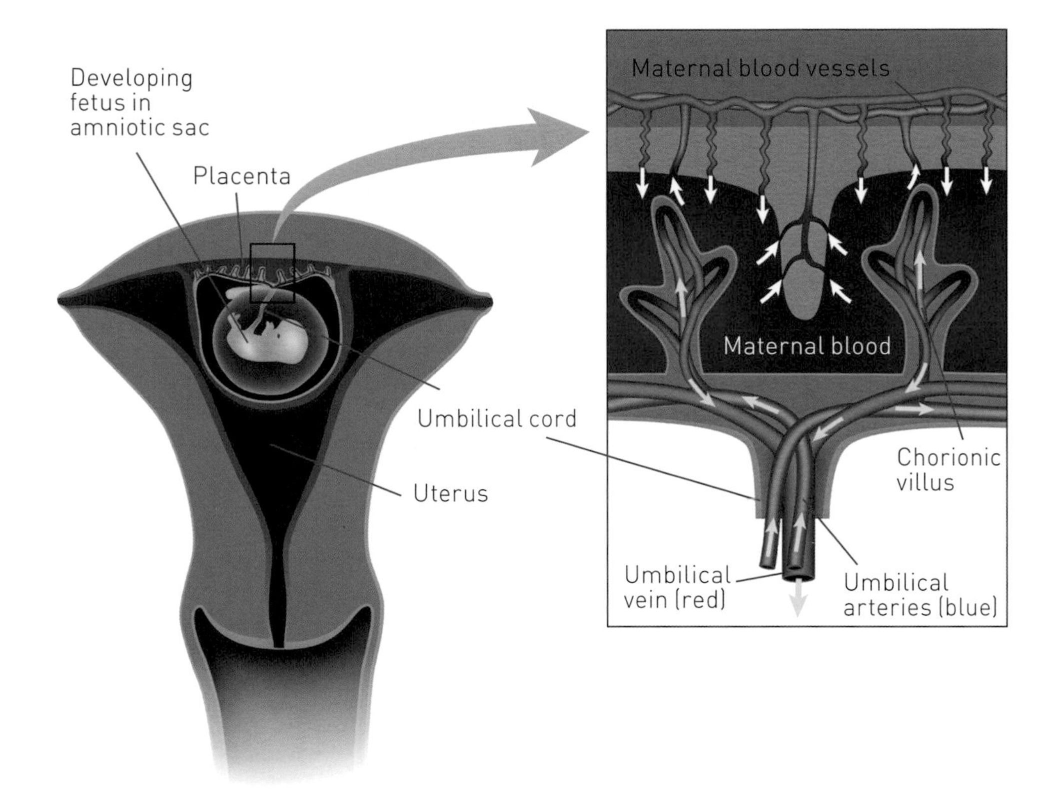

FIGURE 10.9 **THE PLACENTA IS UNIQUE TO PLACENTAL MAMMALS**
In humans, as in all placental mammals, the placenta and umbilical cord link mother and fetus. Only marsupial mammals, like the kangaroos, and monotremes, like the duck-billed platypus (*Ornithorhynchus anatinus*), do not have a placenta.
Alila Medical Media / Shutterstock.com

tion of preventing the corpus luteum from atrophying. Accordingly, the production of estrogen and progesterone remains high, and these steroids continue to support the endometrium.

10.5.2 Labor and delivery

Uterine contractions occur frequently during pregnancy. However, at or about nine months a distinct pattern of contractions begins where approximately every 15-20 minutes there is a contraction lasting about one minute. This is an indication of labor beginning. Most deliveries start slow by having a long period of early labor, lasting even more than a day, and this stage is characterized by relatively mild contractions as well as some irregularity in their spacing. The actual birth is

BOX 10.2 TERATOGENS AND THE PLACENTAL BARRIER

A teratogen is any agent that can negatively affect the development of the embryo or fetus. At one time, it was thought that drugs and similar substances couldn't cross the placenta from the mother to the developing baby, but this is now known to be incorrect. For example, alcohol drunk by the mother enters the fetal circulatory system and, depending upon amounts consumed, frequency of consumption, and stage of pregnancy, alcohol can have very serious physical and behavioral consequences to the fetus. Collectively, these disabilities are known as *fetal alcohol syndrome.*

Sometimes drugs designed to help women during pregnancy have turned out to be teratogens. Thalidomide was developed in West Germany in the 1950s to combat morning sickness in first-trimester mothers, a condition caused by the exceptional hormone levels experienced during pregnancy. Following Britain's lead, Canada licensed the drug in 1961. During that year, however, it became apparent that there were serious birth defects associated with its use, including *phocomelia*, in which the hands and feet are attached to the body without limbs or with very short limbs (Figure 10.10), and so it was discontinued by early 1962. Ironically, the Canadian who was employed by the US Food and Drug Administration in their review of the drug, Dr. Frances Kelsey, was not satisfied with the safety evidence provided by the drug company, and she never authorized its use in the United States, saving thousands from the drug's ravages.

FIGURE 10.10 **THALIDOMIDE WAS A MORNING-SICKNESS DRUG THAT CAUSED DEVELOPMENTAL ABNORMALITIES** This young girl manipulating a toy with her feet was born with phocomelia, a condition that was caused by her mother being prescribed the morning-sickness drug thalidomide. This is a case where what was circulating in the mother clearly reached the developing fetus through the placenta. Thalidomide was used only briefly, and never in the United States.
PA Images / Alamy Stock Photo

Yet another drug prescribed to pregnant mothers in the mid-twentieth century also caused widespread heartbreak. Sometimes doctors prescribed diethylstilbestrol (DES) to mothers to reduce the risk of miscarriage, which is the natural termination of a pregnancy before the fetus can survive. Many children from these pregnancies, especially girls who became known as *DES daughters*, suffered a high rate of reproductive tract cancers as young adults.

likely to still be many hours away, sometimes more than a day. With time, these contractions become more frequent and more regular. Labor is dependent upon these contractions because they ultimately push the baby downwards toward the vagina. Two molecules are key for these events. One is the hormone *oxytocin* from the posterior pituitary gland. The other is the group known as prostaglandins, which also exist in semen. Prostaglandins are hormone-like in behavior but are produced locally (i.e. in the uterus) in response to oxytocin. Both oxytocin and prostaglandins are employed by doctors when there is a desire to induce labor.

A baby is delivered from the uterus through the birth canal, which consists of the cervix and the vagina. As labor proceeds, the baby's head is pushed against the cervix, causing it to dilate. This is usually coincident with the rupturing of the uterine membranes that surround the baby, known as *water breaking*. Progressively, the cervix dilates more and contractions increase in intensity and frequency. By the time the cervix is completely dilated, known as *effacement*, contractions are much more frequent, and women report a great desire to push. Pain increases. during the final part of labor, known as *transition* and usually lasting less than an hour, contractions last longer and are even more frequent and intense. Transition tends to last longer in first-time mothers. Once the baby's head is in the vagina, the desire to push becomes even greater, and the baby is then delivered quite quickly. The placenta is usually delivered about 15 minutes later. Delivery of the baby and the placenta is called *parturition*, and the period following childbirth is the *postpartum* period.

The usual and preferable position for a baby at the time of delivery is head-down, known as *cephalic presentation* (Figure 10.11). This way the baby is delivered headfirst. Two other presentations sometimes occur. A breech presentation is head-up, meaning that the baby will be delivered head last. This occurs less than five per cent of the time, with the frequency varying with a variety of factors, two of which are twin pregnancies and fetal alcohol syndrome. The developing baby is commonly in a head-up position during the second trimester. However,

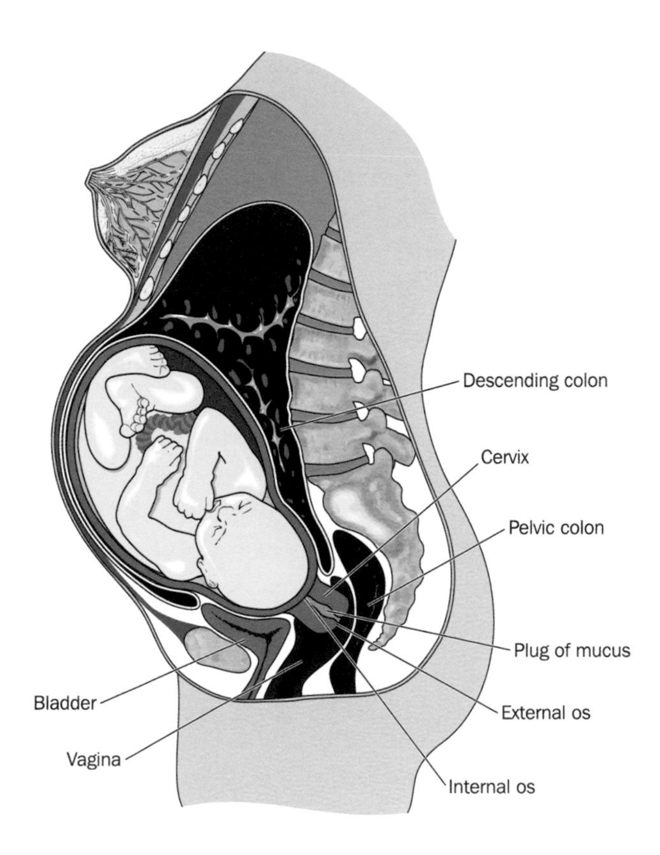

FIGURE 10.11 **CEPHALIC PRESENTATION IS THE DESIRABLE ORIENTATION OF THE BABY AT TERM**

At full term, a human baby weighs on average 3.5 kg (7.7 lbs.) and, in the usual case, is oriented head-down for delivery.
Blamb / Shutterstock.com

it moves over the course of the pregnancy, and so fetal positions up to about the 25th week cannot predict the likelihood of a breech birth. After that point, however, a breech position increases the likelihood of a breech position at the time of birth. A considerably rarer presentation at the end of term is a transverse lie, in which case the head is neither up nor down. Instead, the baby is cross-wise in presentation.

Transverse presentations delivered without intervention typically result in the death of the baby. Sometimes, manipulation of the baby can move it into a vertical orientation. Breech deliveries can work successfully, but there are more likely to be complications, so in developed nations a surgical intervention known as a Caesarean section ("C-section") is usually performed. For transverse presentations, a Caesarean section is required. Deliveries by C-section have become very

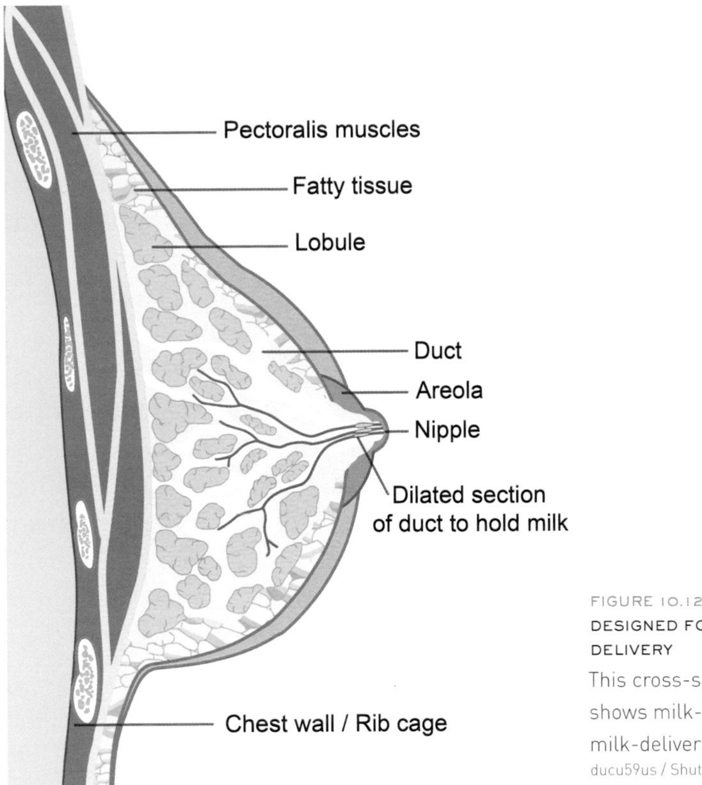

FIGURE 10.12 **THE HUMAN BREAST IS DESIGNED FOR MILK PRODUCTION AND DELIVERY**

This cross-section of the human breast shows milk-producing lobules and milk-delivering ducts.

ducu59us / Shutterstock.com

common in the developed world, likely considerably higher than is necessary but erring on the side of caution. In addition to being employed in cases of breech and transverse presentation, other considerations leading to a Caesarean section include maternal health, whether the mother has previously given birth by C-section, twin pregnancy, and umbilical cord complications.

10.6 BREAST-FEEDING

During pregnancy, the breasts increase in size in preparation for *lactation,* which is the secretion of milk by mammary glands. Breasts each contain about 20 milk-producing *lobules,* made up of many blind sacs called *alveoli* that drain into a duct. Each duct is connected to the nipple, where a suckling baby can therefore access milk (Figure 10.12).

There are two hormones directly involved in milk production: prolactin and oxytocin. Both are released by the pituitary gland in response to stimulation by the hypothalamus. Prolactin is produced in increasing levels during pregnancy, and it reaches a high concentration at the time of delivery. These levels do not result in milk production, however, as there are also high levels of estrogen and progesterone, which have the effect of inhibiting the production of milk. These steroids drop quickly following the delivery of the placenta, however, and milk production then begins, although it will be several days before it will be true milk. In the meantime, the breasts produce *colostrum,* a thin, yellowish fluid rich in proteins, including antibodies that bolster the baby's immune system.

We've already seen that oxytocin plays a role in muscle contractions in the uterus. Similarly, it induces contractions in the breast, ensuring delivery of the milk produced by prolactin. Suckling by the infant results in the delivery of some milk by suction, but more importantly, it serves a stimulus for the milk delivery machinery of the mother, since the breast areola is invested with nerves that communicate with the hypothalamus. Oxytocin induces contractions of the muscle linings surrounding the milk-producing lobules. As milk moves from the lobules to the nipples, this is experienced by nursing mothers as *letdown.*

Although standing levels of prolactin in the mother's system begin to drop shortly after birth, suckling by the infant triggers both prolactin and oxytocin production, resulting in milk production and delivery on demand. The end of breast-feeding is known as *weaning,* and this varies from mother to mother and child to child, and it is strongly influenced by culture as well. Some women nurse up until their child reaches age three, but in most developed nations children are weaned by age one.

10.7 MENOPAUSE

Recent research surprised the scientific community by revealing that the human ovary does in fact have stem cells, but it has not been shown that ovarian stem cells make new oocytes. For now, the dogma remains that reproductive aging in a woman results from the depletion of the follicle pool she produces during her own prenatal development (Section 10.1.2). This pool is known as the ovarian reserve, and when it runs out, monthly ovulation ceases and the woman enters menopause. It is a gradual process, and is usually agreed to have happened once a woman has ceased menstruating for a year, which occurs on average at the age of 51 in North America. As the hormonal system adjusts to a new norm, most women experience symptoms of discomfort such as headaches, mood swings, and hot flashes. Depression is also sometimes a consequence.

A prejudice held by young people is that sex is solely for the young. It is true that our standards of sexual beauty are best exemplified in young adults, and the intensity of sexual feelings is doubtless greatest among young adults as well. Yet many adult men and women are sexually active throughout their lives. Most men report only a gradual reduction of libido and of performance, although erectile dysfunction is widely reported among men over the age of 60.

In contrast, women experience dramatically different hormonal profiles following menopause, with attendant changes both physically and emotionally. Among frequently reported post-menopausal attributes are vaginal thinning and dryness (with consequent discomfort during intercourse), reduced orgasmic intensity and frequency, mood changes that can include clinical depression, and reduced libido. Post-menopausal women also suffer bone density losses.

Since the mid-twentieth century, hormone replacement therapy (HRT) has been available to address these symptoms. Unless a woman has had a uterine hysterectomy, which is the removal of uterus, estrogen and progestin are both typically prescribed. The estrogen is the primary therapy. However, estrogen stimulates lining of the uterus, which increases the risk of uterine cancer. Progestin, which mimics progesterone, balances this effect. For women who have had a uterine hysterectomy, the therapy can be estrogen only. Sophisticated marketing, including images of continuing youth and beauty, led to widespread use of HRT until 2002, when results of a large study were released showing significant health risks, especially in the estrogen-progestin option. These risks include an increased prevalence of breast cancer, blood clots, stroke, and even a trend to increased dementia risk in older women—costs that most agree outweigh the benefits.

BOX 10.3 GONORRHEA EVOLUTION: GENOMIC RESPONSES TO ANTIBIOTIC USE

Gonorrhea (*Neisseria gonorrhoeae*) is a dangerous but common sexually transmitted infection. In addition to often causing uncomfortable and unsightly swelling, burning, and itching in the genital areas, gonorrhea increases the risk of infertility in both sexes. In women, it also increases the risk of an *ectopic pregnancy*, which is where the fertilized egg implants in the fallopian tube instead of the endometrium, which usually means the fetus will not survive.

When *penicillin* was introduced in the 1940s to combat bacterial infections, including gonorrhea, the results were remarkably effective. But as with other bacterial infections, the efficacy of these treatments waned with time, a phenomenon known as antibiotic resistance. What happens is this: the antibiotics kill most of the bacteria in an infected host, but a few individual bacteria either (a) have some pre-existing resistance to the antibiotic or, because bacterial populations are immense, (b) randomly develop mutations that prove beneficial in resisting the antibiotic effect. This resistance can be achieved in several ways, such as being able to recognize and pump the antibiotic back out of the bacterial cell, or by changes to the bacterial cell surface that prevents the antibiotic from interfacing with it.

In any event, these survivors (as few as one individual) then populate subsequent generations, and these strains become the new normal. Nor is *Neisseria gonorrhoeae* the only kind of bacteria to develop antibiotic resistance. Other sexually transmitted bacterial infections like syphilis and chlamydia exhibit evolving resistances, as do the bacteria responsible for tuberculosis, malaria, and influenza (flu), to name just a few. This makes it difficult for public health agencies, drug manufac-turers, and medical professionals to keep up.

Neisseria gonorrhoeae gene pools have proved particularly successful in achieving resistance

10.8 SEXUALLY TRANSMITTED INFECTIONS

In 1495, an epidemic of a new and devastating disease broke out in Naples, Italy, just at the time that the French army was invading. It was quickly apparent that it was sexually transmitted, and that people of both nationalities suffered. The French called it the "Neapolitan disease" (disease of Naples) and the Italians called it the "French disease," and these weren't the only nationalities to blame the scourge on someone else. What soon came to be called *syphilis* spread quickly through Europe and beyond. It remains a serious sexually transmitted infection (STI) today, but it is treatable with antibiotics if caught in the early stages. Whether the disease arrived in Europe with Columbus and his crew upon their return from the Americas or whether it originated

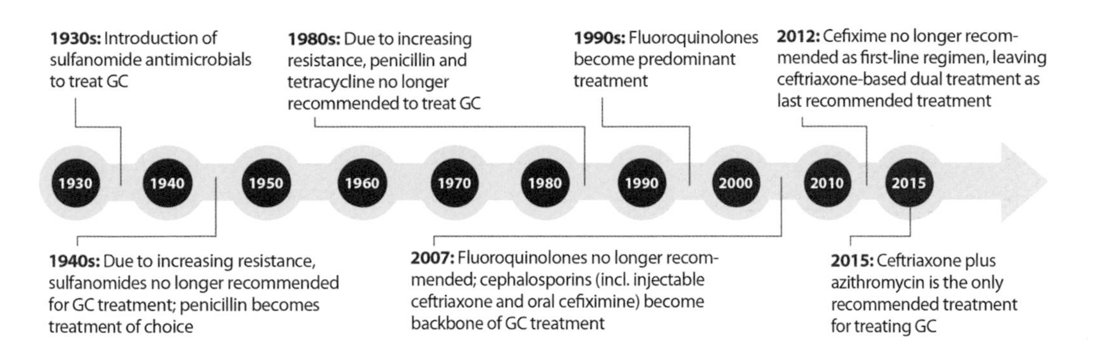

FIGURE 10.13 ANTIBIOTIC RESISTANCE IS A SERIOUS PUBLIC HEALTH PROBLEM

This is a timeline of treatment regimes and developing antibiotic resistance in the United States for gonorrhea infections. GC is a short form term for gonorrhea, based on an old name for the bacterium that causes it.

Courtesy of the United States Centers for Disease Control and Prevention

(Figure 10.13). Penicillin and tetracycline, another popular antibiotic, ceased to be prescribed for gonorrhea infections in the 1980s due to widespread resistance. During the 1990s, fluoroquinolone antibiotics were the drugs of choice, but their effectiveness quickly waned, and their recommended use ended in Asia around the year 2000 and in North America by 2007. Cephalosporin antibiotics have been the recommended medication since then, but their effectiveness too, has waned, even as treatment regimens have been developed that involve additional drugs. A crisis appears to be looming for all such drug-resistant bacterial infections, and the United States Centers for Disease Control has warned that these emerging resistant strains are likely to inevitably neutralize this last remaining first-line treatment option.

elsewhere remains a contentious debate.

Whereas pregnancy and childbirth can be a desirable consequence of sexual intercourse, STIs are certainly the ugly, undesirable face of human sexual relations. Simply reading the symptoms people suffer can generate strong discomfort. One of the early drivers for providing sex education in Western nations was to prevent the contracting and spread of STIs, perhaps as great a motivator for justifying sex education as the prevention of extramarital pregnancy. Fortunately, modern medicine has effectively tamed the impacts of many STIs, but a suite of problems associated with a variety of STIs remains an acute global health crisis, and many die annually. And, as is shown in Box 10.3, it is a battle that will likely be with us forever.

10.8.1 A diversity of organisms cause STIs

One way to think about STIs is that they are the manifestation of the opportunity that human bodies and sexual transactions present for many organisms. Whether it is nourishment, moisture, warmth, cell machinery, or genetic machinery, human body parts used in sexual transactions (genitals, mouth, anus) have much to offer parasitic or pathogenic organisms. In addition, people's tendency to engage with others sexually constitutes a highly successful conduit for getting from one host to another. That these are successful strategies for these organisms also attests to a long history among humans of sexual activity involving multiple partners. Sexually transmitted infections are caused mainly by bacteria or viruses, although one significant STI is caused by a protist, and pubic lice are tiny insects that occupy the pubic region and move from person to person through sexual contact.

10.8.2 Major bacterial STIs

Recall that bacteria are small single-celled organisms whose genes are on a circular chromosome. They are asexual reproducers, using binary fission. Many bacteria are beneficial, but others are pathogens, and there are three significant STIs caused by bacterial infections.

- Syphilis is caused by *Treponema pallidum*, a mobile, corkscrew-shaped bacterium. It is diagnosed by blood tests or by examination of fluids from the "pox" lesions called *chancres*, and a treatment of antibiotics, especially of the penicillin family, is usually prescribed. It is spread through sexual contact involving the chancres, especially during vaginal, oral, or anal sex. Without intervention, the disease can seriously affect the cardiovascular system and nervous system, sometimes leading to mental impairment and instability. Congenital syphilis results from the bacteria crossing the placenta, and such pregnancies result in stillbirths or seriously and permanently malformed babies.
- *Gonorrhea* is caused by *Neisseria gonorrhoeae*, a bacterium that commonly exists in pairs of spherical cells. In most men, it is easy to diagnose, since urination becomes painful and there is a milky

discharge from the penis. But in women, the infection is hidden until *pelvic inflammatory disease* (PID) develops, which is experienced as pain in the abdomen. PID can scar the fallopian tubes and can lead to infertility in a significant number of cases, and there can be additional internal symptoms as well as severe eye infections. *Neisseria gonorrhoeae* is a classic case of rapid evolution, as its genome continues to change in response to widespread antibiotic use (Box 10.3).

- *Chlamydia* is caused by *Chlamydia trachomatis*; it is very small, and accordingly it does its unpleasant work inside human cells—including reproducing to produce multiple copies of itself. It may be the most common STI in Western nations; it commonly goes undiagnosed in men, so it appears to be more common in women. Its diagnosis can be confused with urinary infections. Regardless, untreated it can also lead to PID (see Gonorrhea, above).

10.8.3 Major STIs caused by viruses

Viruses are incapable of independently reproducing their own genetic material, so they only reproduce inside a living host cell where they can take over the host's gene-duplicating machinery. Several major STIs are caused by viruses, as listed below.

- *Herpes simplex viruses* cause *genital herpes*, an STI in which symptoms may or may not show. Blisters often appear on the genitals within a few days to a few w1eeks following exposure, and when these rupture they cause pain, as well as secondary symptoms that include vaginal discharge. At this stage, the individual is highly contagious through sexual contact. When the lesions heal, the disease has gone into a latent stage, only to recur later, especially in the context of stress. Sometimes women with herpes elect to give birth by Caesarean section to avoid transmitting the disease to the new baby.
- *Human papillomaviruses* (HPVs) can cause genital warts. Although the immediate symptoms are usually not too severe (warts and flat lesions) and indeed sometimes go undetected, some forms of HPV are associated with cervical cancer. A vaccine that was first approved in

BOX 10.4 HIV AND AIDS

Although AIDS (acquired immunodeficiency syndrome) may have existed in humans since the 1920s, it was not until the 1980s that its impact was felt on a global scale and that it was determined to be caused by a virus (HIV; human immunodeficiency virus). During that decade, it was determined that HIV infects *T cell lymphocytes*. These are white blood cells that form a major component of the mammalian immune system because they recognize foreign cells in the body and play a role in countering these foreign cells. The human immunodeficiency virus attaches to T cells, enters them, and then uses the cell's genetic machinery for making copies of itself, eventually causing a widespread collapse of the person's immune system. Most people with AIDS die from other diseases, when their immune systems become too compromised to fight off infections.

Although one can become HIV-infected by a variety of routes, it became clear during the early 1980s that this disease was associated with both needle-sharing and sexual activity, especially among gay men. This suggested that the transfer of bodily fluids through either sex or needle sharing was the chain of infection. In 1983, it was confirmed that both men and women could get AIDS by having sex with a male partner who had HIV, and shortly afterwards it also became clear that babies could contract HIV during pregnancy, childbirth, or breast-feeding. Any sexual activity that transfers blood, semen, vaginal secretions, or rectal secretions from an infected person (HIV positive) to a non-infected person (HIV negative) can transmit the disease, especially through mucous membranes, cuts, or open sores.

Like many viruses, HIV genes are coded in single-stranded RNA instead of double-stranded DNA. The same genetic techniques used to determine the genetic sequences of DNA can be applied to RNA, allowing researchers to compare HIV genes to those of other viruses. Research along these lines shows that HIV belongs to a family of immunodeficiency viruses found in a variety of primates, with the closest similarities being to such viruses found in chimpanzees. Indeed, based on this research, a plausible scenario is that AIDS originated in Cameroon in the early twentieth century, once the virus entered humans. This perhaps occurred through the eating by humans of bushmeat that included common chimpanzees (*Pan troglodytes*) and other primates carrying similar viruses. Diseases that can be transferred from non-human organisms to humans in this manner are *zoonotic diseases*.

Between the early 1980s and 2017, about 35 million people had died due to HIV infection. No cure exists, but drugs that delay the onset of AIDS and soften its impacts have given HIV-positive individuals both hope and improved quality of life—things that didn't exist in the first two decades of the disease.

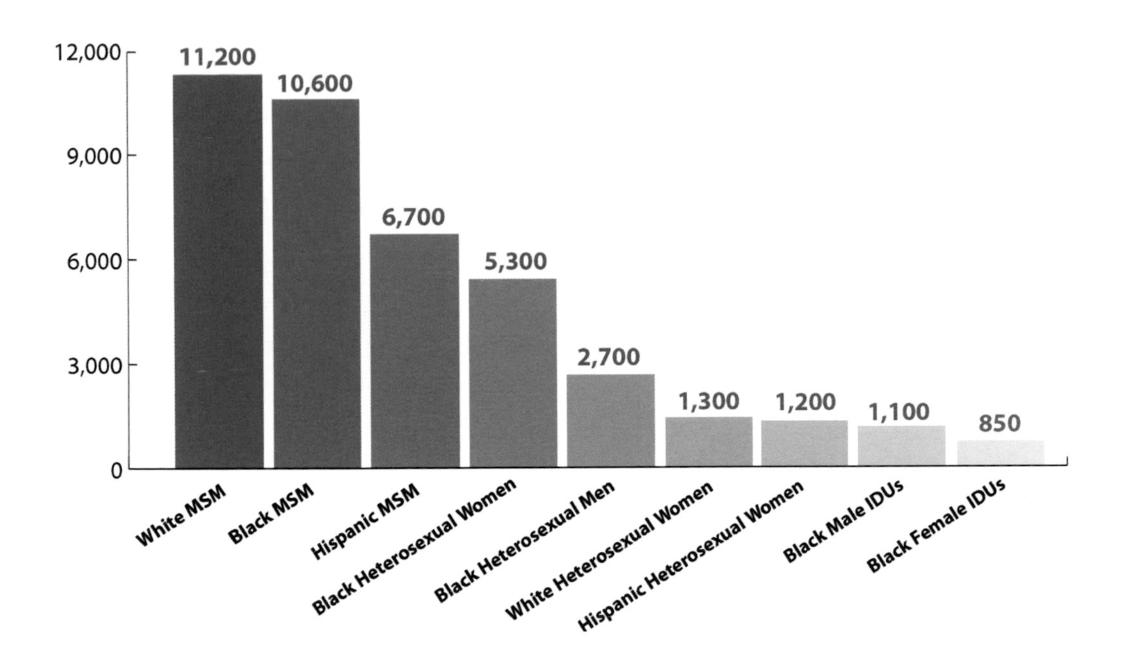

Today, men who have sex with men account for the greatest proportion of new HIV infections, and the primary reason for this is that anal intercourse is the highest-risk behavior, especially for the receiver. In the United States, new cases of AIDS are disproportionately found in non-white ethnic groups (Figure 10.14), a pattern that is attributable to a complex range of factors that include socioeconomic influences as well as different STI incidence rates within different ethnic groups.

FIGURE 10.14 **HIV INCIDENCE VARIES WITH SEX, ETHNICITY, AND RISKY BEHAVIORS**

This graph shows the estimated new HIV infections in the US in 2010, for the most affected sub-populations. African-Americans experience the most severe burden of HIV, representing 12 per cent of the population but constituting 44 per cent of new cases annually. (MSM—men who have sex with men; IDU—intravenous drug users.)

Courtesy of the United States Centers for Disease Control and Prevention

2006 and that targets girls at or before puberty has been aggressively promoted by health-care systems since then. The vaccine also has benefits that counter other types of cancer, including in males.

- *Hepatitis* comes in a variety of forms, but Hepatitis B (HBV) is most commonly spread by shared use of needles among people who inject drugs and by sexual contact between heterosexuals or gay men. It can also be transmitted across the placenta. The liver is particularly vulnerable to HBV, leading to liver disease, cirrhosis, and even liver failure that causes death. Currently, there is no cure, but there is a vaccine that is a frequent component of the vaccinations made available by most health-care systems.
- The *Zika* virus is primarily transmitted to people through the bite of an *Aedes* mosquito, but it can then be passed through sex from a person who has Zika to his or her partner. Frequent symptoms include fever, rashes, headaches, and joint and muscle pain. More seriously, Zika infection during pregnancy can cause serious birth defects.
- The *human immunodeficiency virus (HIV)* causes *acquired immunodeficiency syndrome* (AIDS). See Box 10.4.

10.8.4 Non-humans suffer sexually transmitted infections too

Australia's koala (*Phascolarctos cinereus*) is an iconic marsupial, popular for its cute features, slow-moving behavior, and apparently calm disposition (Figure 10.15). Yet numerous koala populations are in serious decline, attracting the attention of Australian wildlife managers and prompting officials to designate them *vulnerable* in some areas. A major factor driving these declines is chlamydia infections. This is a different strain from the common form of chlamydia that infects humans, but it, too, is spread sexually. In a major review of wildlife rehabilitation facility records in New South Wales, the second most common reason koalas were brought to the hospital was chlamydia; in some areas, most individuals, both male and female, suffer from the infection, making it an epidemic. Serious symptoms include blindness, urinary tract infections, and infertility.

As is the case with humans, koala chlamydia is treatable with antibiotics, but this is an expensive and labor-intensive approach to a

FIGURE 10.15 **WILDLIFE POPULATIONS ALSO SUFFER SEXUALLY TRANSMITTED INFECTIONS**
The koala (*Phascolarctos cinereus*) is an iconic Australian marsupial mammal. Some koala populations are declining due to factors that include the sexually transmitted infection chlamydia.

wildlife management problem, as it requires extended stays in captivity up to several months. It also has to be administered in the early stages of the disease to bring about a full cure. If intervention comes too late, euthanasia is called for. In other cases, the animal survives but is permanently infertile. Infertility poses a dilemma, because infertile females released into the wild may take the place of fertile ones, reducing the capacity of a population to produce young.

Queensland researchers have developed a chlamydia vaccine for koalas that requires just a single injection. Early trials on 21 free-ranging animals in a highly infected population were initially positive: vaccinated individuals that had had the disease were infection-free when they were re-tested, and those that did not have the disease did not contract it. Yet when they tested after a longer period of time, a few of the vaccinated individuals had been infected. Work to improve the vaccine continues.

Even more serious, perhaps, is a relatively new HIV-like retrovirus that has infected many koala populations, in some cases infecting all members of a population. Retrovirus infections operate very differently from conventional infections. They become inserted into the DNA of the host—in this case the koala—and then the retrovirus can travel from generation to generation through sperm and eggs, becoming, in effect, part of the genome of the host. A koala population introduced to Kangaroo Island in South Australia about 100 years ago that has remained isolated does not have the retrovirus, suggesting the retrovirus is a recent invasion of the koala genome. The impacts of the retrovirus are serious, causing cancers and immune deficiency disorders.

CHAPTER 10 SUMMARY

- Spermatogenesis occurs in human males from the age of puberty until old age, whereas oogenesis begins in human females during the fetal period, resumes during puberty, and continues until menopause.
- The female reproductive cycle is divided into two phases: the *follicular phase* (lasting approximately 14 days), the first 5 days or so of which is the menstrual period and the end of which is ovulation, and the *luteal phase* (also lasting approximately 14 days), during which the uterus is ready for egg implantation.
- Human females have a fertile period that lasts several days before and at ovulation, but it is largely hidden compared to the fertile periods of most female mammals.
- Sexual arousal and response in humans is described as moving through four stages designated *excitement, plateau, orgasm,* and *resolution*; in men, orgasm occurs with ejaculation, and in women, multiple orgasms are possible without resolution.
- Infertility related to non-viable sperm, endometriosis, or other factors is relatively common and can sometimes be resolved with assistive reproductive technologies.

- Most birth control involves contraceptive methods, including abstinence, condoms, tubal ligation, vasectomy, various intrauterine devices, and the contraceptive pill.
- The placenta is the organ connecting the developing fetus, via the umbilical cord, to the uterine wall, allowing for the exchange of nutrients, wastes, and gases with the mother.
- Labor follows a series of stages involving muscle contractions as well as the effacement and dilation of the cervix, resulting in the normal case in a head-first delivery of the baby, followed by delivery of the placenta.
- Breast-feeding is regulated by hormones and by stimulation from the suckling infant.
- Menopause represents the cessation of reproductive cycling (including menstruation and ovulation), and it results in significant changes to hormone levels.
- Sexually transmitted infections continue to be a significant public health challenge, some of which cause serious pathologies, including death.

FURTHER READING

Grive, K.J., & Freiman, R.N. (2015). The developmental origins of the mammalian ovarian reserve. *Development, 142*(15), 2554–2563. https://doi.org/10.1242/dev.125211

Haselton, M.G., & Gildersleeve, K. (2016). Human ovulation cues. *Current Opinion in Psychology, 7,* 120–125. https://doi.org/10.1016/j.copsyc.2015.08.020

Jones, M.E., Schoemaker, M.J., Wright, L., McFadden, E., Griffin, J., Thomas, D., ... , & Swerdlow, A.J. (2016). Menopausal hormone therapy and breast cancer: What is the true size of the increased risk? *British Journal of Cancer, 115*(5), 607–615. https://doi.org/10.1038/bjc.2016.231

Roney, J.R., & Simmons, Z.L. (2013). Hormonal predictors of sexual motivation in natural menstrual cycles. *Hormones and Behavior, 63*(4), 636–645. https://doi.org/10.1016/j.yhbeh.2013.02.013

Saleh, A.M., Dudenhausen, J.W., & Ahmed, B. (2017). Increased rates of cesarean sections and large families: A potentially dangerous combination. *Journal of Perinatal Medicine, 45*(5), 517–521. https://doi.org/10.1515/jpm-2016-0242

Towers, B.A. (1980). Health education policy 1916–1926: Venereal disease and the prophylaxis dilemma. *Medical History, 24*(1), 70–87. https://doi.org/10.1017/S002572730003979X

Wallen, K., Myers, P.Z., & Lloyd, E.A. (2012). Zietsch & Santtila's study is not evidence against the by-product theory of female orgasm. *Animal Behaviour, 84*(5), e1–e4. https://doi.org/10.1016/j.anbehav.2012.05.023

Welling, L.L.M. (2013). Psychobehavioral effects of hormonal contraceptive use. *Evolutionary Psychology, 11*(3), 718–742. https://doi.org/10.1177/147470491301100315
Zietsch, B.P., & Santtila, P. (2011). Genetic analysis of orgasmic function in twins and siblings does not support the by-product theory of female orgasm. *Animal Behaviour, 82*(5), 1097–1101. https://doi.org/10.1016/j.anbehav.2011.08.002

Glossary

abortion – the premature termination of a pregnancy [1.1]

abstinence – a method of birth control achieved through not having sexual intercourse [10.4.2]

adaptation – as an evolutionary process, the ongoing refinement of a trait through natural selection; as an evolutionary result, a trait with a current utility resulting in improved survival or reproductive success [1.2]

AIDS – acronym for *acquired immunodeficiency syndrome*, a sexually transmitted disease caused by the human immunodeficiency virus [10.8.3]

allele – a version of a gene, differing from other alleles of the same gene by virtue of one or more differences in DNA sequence [3.4.1]

alloparenting – a social system where parental care is provided by individuals other than the biological parents [6.6.2]

allopatric speciation – the evolutionary creation of new species where an original gene pool is divided into two physically separated gene pools that follow their own evolutionary trajectories [4.7.2]

alpha male – the male with the highest status in mating systems where there is an asymmetry among males for access to females [6.6.1]

alternative mating strategy – a male or female strategy for achieving reproductive success that differs from the prevalent strategy employed by that sex [7.3.1] anagenesis – a pattern by which an ancestral species evolves without evolutionary branching into a single new species, in contrast to cladogenesis [4.7.1]

androgen – a steroid sex hormone associated with but not limited to males [4.5.3]

anti-Müllerian hormone – a protein-type hormone associated with the suppression of female reproductive development in the male embryo and with the regulation of egg release in the reproductive female adult [9.2.1]

armament – a sexually selected character, more often expressed in males, that is used in direct intra-sexual competition [5.1.3]

artificial selection – a breeding program conducted by humans that relies on choosing breeding stock expressing desired traits, with the goal of producing, over generations, an altered phenotype that strongly expresses the desired traits [Box 4.1]

asexual reproduction – reproduction that results from means other than the fusing of egg and sperm, resulting in offspring that genetically match the parent [2.2.1]

assistive reproductive technologies – interventions used to counter infertility problems, with the goal of achieving pregnancy [10.4.1]

autosome – a chromosome that is not a sex chromosome [3.6.2]

baculum – a bone inside the penis of many mammals that assists in achieving rigidity [Box 9.1]

Bateman gradient – the relationship between an individual's degree of reproductive success and the number of mates that individual has [6.5.1] biandry – a special case of polyandry, where there is one female and two males in a reproductive trio [6.4.3]

bilateral insemination – sexual reproduction between hermaphrodites where each individual inseminates the other [Box 7.2]

binary fission – the standard means of reproduction in bacteria, whereby the DNA is first replicated and the cell then divides in two, with each daughter cell having a complete, identical copy of the DNA [2.2.1]

binomial nomenclature – the system for naming species that relies on a unique pairing of a genus name with specific epithet [Box 1.1]

biological determinism – the assertion that biology, especially at the genetic level, determines the behavior of organisms, including humans [1.2]

biological species concept – a means of defining sexually reproducing species that concludes two individuals are members of the same species if they can successfully mate together in nature [Box 4.2]

biology – the scientific study of life [1]

biparental care – care of immature individuals provided by both the mother and the father [6.2.1]

bisexuality – sexual orientation where the individual is sexually attracted to both males and females [1.1]

blending inheritance – the discredited idea that offspring must exhibit physical traits intermediate between those of each parent [3]

breed – a subset of a plant or animal species having distinctive traits and having been deliberately created by humans by the careful selection of breeding individuals [3] brood parasitism – a method of reproduction that relies on other individuals, often of other species, to provide parental care for the developing offspring [Box 7.3]

Caesarean section – the use of surgery to deliver a baby, in contrast to vaginal birth [10.5.2]

capacitation – a post-ejaculation modification of the sperm cell that prepares the sperm to fuse with the egg [10.4]

carrier – in inheritance, an individual who has a recessive allele for a particular phenotypic condition but who does not herself or himself manifest the condition [Box 9.2]

celebrity couple hypothesis – a hypothesis to explain monogamy that relies on the pairing of two superior individuals who would not increase their reproductive success by mating with any other individuals [6.2.2]

central dogma – the flow of genetic information from DNA code into messenger RNA (transcription) that is transported outside the nucleus where it is used to assemble proteins (translation) [3.6.3]

cervix – narrow outer end of the uterus [9.4.3]

character – a heritable attribute of an organism that is expressed in the phenotype, such as pea flower color [3.1.1]

chromatid – in cell division, each of the attached, duplicate strands of chromosomal DNA [2.2.2]

chromatin – the complex of chromosomes in the cell nucleus, including proteins associated with packaging [3.6.1]

chromosomal mutation – a major mutation that represents a duplication, a deletion, an inversion, or a translocation of a section of a chromosome [4.3.2] chromosome – a thread-like structure of DNA and proteins found in most cells that carries coded genetic information [2.2.1]

circumcision – the surgical removal of some of the foreskin of the penis, usually in infants; sometimes applied to the practice of female genital mutilation [9.3.3]

cisgender – a person whose gender identity matches his or her biological sex, in contrast to transgender [1.1.1]

cladogenesis – a pattern by which an ancestral species evolves into more than one new species through evolutionary branching, in contrast to anagenesis [4.7.1]

classification – a system of categorizing and naming organisms that reflects degrees of evolutionary relationships [1.3]

clitoris – a sensitive, erectile structure at the anterior end of the vulva of female mammals [2.3.1]

cloacal protuberance – a swelling of the cloaca of many male birds and reptiles that serves as an intromittent organ during copulation [Box 9.1]

coercion – the use of threats, harassment, or physical violence to force copulation [1.2, 5.5.1]

common ancestor – a species that is the shared ancestor of two or more recent species [4.7.1]

comparative approach – a method of analysis that enriches understanding by studying multiple groups to assess both shared and unique traits [1.3]

concealed ovulation – the lack of perceptible change in female behavior or appearance during ovulation, also known as hidden estrus [Box 7.1]

conception – fertilization of an egg by a sperm, followed by the successful implantation of the developing embryo into the uterine lining [10.4.2]

condition dependence – the reliance of a phenotypic expression, especially of a sexual trait, on the individual's physical state [7.3.1]

conjugation – the exchange of genetic material between two single-celled organisms [Box 2.1]

consortship – male–female bonds with recurring sexual and social interaction that does not amount to an exclusive monogamous union [5.5.2]

contraception – methods that prevent pregnancy without relying on sexual abstinence [1.1]

controlled experiment – a test of a hypothesis that draws conclusions based on comparing two groups that have been manipulated to ensure there is only one difference between them [1.5]

convenience polyandry – polyandry that results when a female yields to coercive male sexual behavior [6.4.3]

cooperative breeding – a reproductive system where parenting is shared by a group that includes parents and non-parents [6.6.2]

copulation – the insertion of a male's intromittent organ into the female's reproductive tract, including the delivery of sperm [2.1.2, 2.4.1]

corpus cavernosa – two cylindrical structures on each side of the penis and clitoris that can become engorged with blood, creating an erectile state. Singular: *corpus cavernosum* [10.3]

crossing – sexual reproduction involving gametes from different individuals, in contrast to selfing [Box 2.2] crossing-over – the exchange of genetic material between homologous chromosomes during meiosis [3.7]

cryptic female choice – mate choice by females that occurs internally by mechanisms that preferentially treat the sperm of one male over the sperm of others [6.4.2]

cuckoldry – mating by a female with a male partner other than her social mate [7.1.2]

daughter cell – in binary cell division, one of the two resultant cells; the term does not imply these cells are necessarily female [2.2.1]

developmental cascade – a pattern influencing development where one key event leads inevitably to other key events [8.2.4]

dihybrid cross – a mating experiment in which two varieties that differ in the expression of two phenotypic characters are bred with one another [3.3]

dihydrotestosterone (DHT) – a modified and more potent form of testosterone, of particular importance in male sexual development [9.2.1]

dimorphism – two different phenotypic forms; males and females within a species often exhibit sexual dimorphism [2.3.2]

diploid – describing cells that have chromosomes in pairs, in contrast to haploid [3.4.1]

direct benefits – advantages associated with mate choice that yield real, non-genetic gains, such as resources [5.3]

disorders of sexual development – unusual patterns of sexual development that deviate from the usual male–female binary [9.5]

DNA (deoxyribonucleic acid) – a long molecule in a double-helix form that carries genetic information that constitutes coded instructions for development and regulation [1.2]

dominant – an allele that is phenotypically expressed regardless of the partner allele [3.2.2]

drone – a male member of the social insects, applied especially to honeybees [6.2.2]

dual mating strategy – an approach to mating where an individual seeks two different partners for the two different values they provide [7.1.2]

egg – the female gamete (*ovum*), distinguished from the male gamete by being larger and immobile; also applied loosely to female gametes that have been fertilized [1]

ejaculation – the emission of semen that occurs at the point of male sexual climax [7.2.2, 9.3.2]

embryo – the unborn or unhatched developing offspring; in humans, applied only to about the first two months of pregnancy [8.2.4]

endocrine gland – a tissue whose function is to produce blood-transported hormones [9.1.1]

endocrine system – the collection of glands that produce hormones that travel through the bloodstream and regulate internal physiology, including sexual development, regulation, and function [9.1.1]

endometrium – the lining of the mammalian uterus that proliferates and then is shed once per reproductive cycle if there is no pregnancy [9.4.3]

endurance competition – a type of intra-sexual contest where success is measured by outlasting rivals [5.5.2]

Enlightenment – a philosophical movement centered in eighteenth-century Europe, whose foundational idea was that reason and logic are the primary source of legitimate knowledge [4.1.1]

environmental sex determination – a system where a non-genetic factor decides the sex of the developing offspring [8.3]

epigenetics – heritable changes in the expressions of genes that don't involve changes to DNA sequences [7.4]

erection – an intromittent organ in a stiffened state [7.2.2, Box 9.1]

estradiol – a steroid hormone important in the regulation of the female reproductive cycle [9.1.1]

estrogen – any of a group of steroid hormones associated with female characteristics, but also found in males [8.4.1]

estrous cycle – the reproductive cycle of most female mammals, associated with periods of fertility and uterine conditions [10.2.1]

estrus – in mammals with an estrous cycle, the period of "heat" in which the female is fertile and sexually receptive [5.2.2, 10.2.1]

evolution – change over successive generations in the genetic constitution and related phenotypic expression of living organisms [1]

evolutionary strategy – a set of tactics employed by organisms that maximize their reproductive success [7.1.1]

exaptation – the co-option of a biological feature for a novel function unrelated to its original value to the organism [2.1.1]

excitement – the first stage of the human sexual response [2.4.1] explosive breeder – a species with a brief and highly synchronized mating season, often with intense male–male competition [6.7.2]

external fertilization – sexual reproduction characterized by the union of sperm and eggs outside the female body [2.4.1]

extra-pair – in monogamy, mating that occurs outside the pair-bond; applied to mating, copulation, fertilization, offspring, young, and partner [6.2.1]

facultative sex – sexual reproduction in organisms capable of both sexual and asexual reproduction, in contrast to obligate sex [4.6]

fallopian tube – the tube through which the mammalian egg travels from ovary to uterus; mammalian oviduct [9.4.1]

Family – a grouping of organisms representing a rank in biological classification between Genus and Order [Box 1.1]

female – the sex that produces relatively large, immobile gametes known as eggs or ova [1]

female choice hypothesis – a common component of inter-sexual selection in which the female exercises mate choice from among male options, being the more common component of inter-sexual selection [5.2.2]

female mimicry – behavioral or physical imitation of females by males to gain advantages [7.2.3]

female-defense polygyny – a mating system where males directly defend more than one female mating partner from interaction with other males [6.3.1]

female-enforced hypothesis – an explanation for monogamy in which the female deserts the mating enterprise or harasses the male if her partner does not contribute resources or care [6.2.2] feminization – a process that molds a phenotype with female characteristics [8.4.6]

fertile period – that part of the female reproductive cycle in which fertilization of the ovulated egg is possible [10.2.2]

fertility insurance hypothesis – an explanation for non-monogamy that focuses on minimizing the risk to females of mating with males with shortages of viable sperm [6.4.1]

Fisher's principle – the reproductive effort expended by a population is equally divided between male and female offspring [7.3.3]

fitness – reproductive success measured as the contribution to the gene pool of the next generation [4.2.2]

fetus – the human offspring from about eight weeks following conception to birth [2.4.2]

follicle – in the vertebrate ovary, the fluid-filled sac that contains an immature egg [10.2]

follicular phase – the part of the estrous or menstrual cycle characterized by egg maturation and ending with ovulation [10.2]

foreskin – loose tissue at the tip of the mammalian penis covering the glans; removed when circumcision occurs [7.2.2, 9.3.3]

gamete – a mature haploid male (sperm) or female (egg) cell that is capable of uniting with the gamete of the opposite sex to form a zygote [1.1]

gender – the sexual identity experienced by a person with reference to psychological and social, as opposed to biological, characteristics [1.1]

gender dysphoria – the psychological experience of having one's gender identity not align with one's biological sex [1.1.1] gene – a unit of heredity made of DNA that represents a code for the manufacture of a complex molecule, usually a protein [1.1.2]

gene flow – the movement of alleles between populations through sexual reproduction [4.7.2]

gene pool – the total inventory of alleles in an interbreeding population; more accurately referred to as an allele pool [3.6.4]

genetic complementarity model – one of three explanations for the benefits of female non-monogamy, based on the desirability of finding a male partner with alleles that best complement her own alleles [6.4.2]

genetic diversity model – one of three explanations for the benefits of female non-monogamy, based on the desirability of producing more genetically diverse offspring [6.4.2]

genetic monogamy – a male–female relationship that is characterized by having no extra-pair partners, either covertly or overtly [6.2.1]

genetic sex determination – a system where a genetic factor, often associated with a particular chromosome, decides the sex of the developing offspring [8.3]

genitals – the externally visible sex organs; typically consisting of penis and scrotum in males and labia, clitoris, and vaginal opening in females [1.1]

genotype – the genetic makeup of an organism, applied to one character, many characters, or the whole individual [3.4.1]

gestation – the carrying of an offspring from conception to birth [5.2.1]

gonadotropin hormone – an endocrine substance from the pituitary gland that stimulates gonadal activity [9.3.2] gonadotropin-releasing hormone – an endocrine substance from the hypothalamus that induces the release of gonadotropin hormones from the pituitary gland [9.3.2]

gonads – organs responsible for generating gametes and regulating hormones; ovaries in females and testes in males [1.1]

good genes hypothesis – a set of ideas to explain sexual selection that relate to the providing of indirect (genetic) benefits associated with sexual armaments or ornaments [5.3, 5.3.3]

good genes model – one of three explanations for the benefits of female non-monogamy, based on the desirability of finding a male partner with superior alleles compared to the gene pool average [6.4.2]

good resources hypothesis – a set of ideas to explain sexual selection that relate to the providing of direct benefits (such as resources) associated with sexual armaments or ornaments [5.3]

handicap principle – in sexual selection theory, the idea that individuals that cope with costly traits reveal their genetic superiority, making them attractive partners [5.3.2]

haplodiploidy – a system of genetic sex determination where males are made of haploid cells and females are made of diploid cells [8.3.4]

haploid – describing cells that have a single set of chromosomes, in contrast to diploid [3.4.1]

Hardy–Weinberg equilibrium – the principle that in populations with random mating, the frequency of the alleles and the frequency of genotypes does not change unless there are evolutionary factors causing change [4.4]

harem – a group of reproductive females, access to whom is controlled by a powerful male [5.2.2] hemizygous – a chromosomal locus for which there is only one allele, rather than a pair, as in the X-chromosome in male placental mammals [9.6]

hermaphrodite – an individual that either sequentially or simultaneously has both male and female reproductive organs [2.2.4]

heterogametic sex – in species with genetic sex determination, the sex whose gametes determine offspring sex [8.3.1]

heterosexuality – a sexual orientation directed at members of the opposite sex [1.1]

heterosis – enhanced genetic quality produced by outbreeding, as opposed to inbreeding [6.4.2]

heterozygote – an individual that has inherited two different alleles for a particular gene: one allele from each parent [3.4.1]

homogametic sex – in species with genetic sex determination, the sex whose gametes do not determine offspring sex [8.3.1]

homosexuality – a sexual orientation directed at members of the same sex [1.1]

homozygote – an individual that has inherited two identical alleles for a particular gene, one allele from each parent [3.4.1]

honest indicator – a sexually selected character whose degree of expression truly reflects the genetic quality of the individual [5.3.2]

hormone – a molecule secreted by a gland into the bloodstream, which carries it to cells of target tissues or organs where it influences cell activity [1]

hormone replacement therapy (HRT) – the prescribing of hormones to adjust hormonal concentrations in the body, often to offset the impacts of menopause or ambiguous sex characteristics, or in sex change procedures [9.7.2]

hybridization – mating between two different types of organisms; sometimes but not always referring to two individuals from different populations or species [2.2.3]

hypothalamus – a component of the vertebrate forebrain involved in metabolic regulation, including regulating the production of endocrine hormones in the pituitary gland [9.1.1]

hypothesis – a proposed explanation articulated as the conceptual basis for further investigation or experimentation [1.5]

implantation – the event about one week after fertilization in which the developing embryo adheres to the endometrium [9.4.3]

inbreeding – sexual reproduction occurring between closely related individuals [4.6.2]

inbreeding depression – reductions in an individual's fitness caused by inbreeding [4.6.2]

inclusive fitness – indirect genetic contributions to the next generation through the direct reproductive success of relatives [6.6.2]

indirect benefits – non-resource, genetic benefits associated with mate choice [5.3]

indirect competition – competition among members of the same sex with respect to mating success other than by direct combat or aggression [5.5.1]

infertility – in the case of individuals, the inability to produce gametes; in the case of couples, the inability to conceive [10.4.1]

inter-sexual selection – the branch of sexual selection that involves mate choice, in contrast to intra-sexual competition [5.1.3]

internal fertilization – sexual reproduction characterized by the union of sperm and eggs inside the female body [2.4.1]

intersexual – an individual whose phenotype shows incomplete characteristics of both males and females, resulting in an ambiguous sex in the context of a male–female binary [5.1.3]

intra-sexual selection – the branch of sexual selection that involves competition between members of the same sex, in contrast to inter-sexual selection [5.1.3]

intromittent organ – the copulatory organ of male animals [2.4.1]

iteroparous – a pattern of offspring production where individuals have the potential to reproduce on more than one occasion, in contrast to semelparous [2.4.2]

K-selection – natural selection that favors a suite of characters resulting in fewer offspring but involving greater parental investment per offspring (in contrast to r-selection) [7.3.2]

karyotype – the number and appearance of the chromosomes in the cell nucleus [3.6.2]

Klinefelter syndrome – a disorder of sexual development in which the individual has three sex chromosomes per cell (two X-chromosomes and one Y-chromosome), resulting in underdeveloped male gonads and sterility [9.5.1]

labia – the inner (minora) and outer (majora) folds of the female vulva, on either side of the vagina [2.3.1]

labor – the process of childbirth, beginning with the shortening and opening of the cervix [10.5.2]

Law of Independent Assortment – one of two rules of Mendelian genetics that asserts regardless of which allele is included for one gene in a gamete, it has no effect on which allele is included for another gene [3.1.1]

Law of Segregation – one of two rules of Mendelian genetics that asserts that during the making of gametes, only one allele of each pair of alleles per gene is included [3.1.1]

lek polygyny – a mating system where males perform in an arena using competition or displays and females attend and make mate choices based on those performances [6.7.3]

lesbian – a female homosexual [1.1]

limiting sex – the sex, usually female, that limits the amount of mating that occurs on account of asymmetries between the sexes in reproductive investments and commitments [5.2]

locus – the location on a chromosome that constitutes a particular gene [3.6.1]

luteal phase – the part of the estrous or menstrual cycle between ovulation and menstruation, the first part of which prepares the uterus for pregnancy [10.2]

male – the sex that produces relatively small, usually mobile gametes known as sperm [1] male-biased – the relatively common occurrence where, among individuals prepared to mate, the number of males exceeds the number of females [5.2.1]

mammogram – a low-energy X-ray of the breast used to diagnose and locate tumors [9.7.2]

masturbation – self-stimulation of the genitals resulting in sexual pleasure [7.2.2]

mate assistance hypothesis – an explanation for monogamy that relies on the necessity of two parents for maximizing the successful production of offspring [6.2.2]

mate guarding hypothesis – an explanation for monogamy that results from the necessity for the male of preventing the female from mating with other males [6.2.2]

maternal care – parental care provided by the female parent [5.4.1]

mating plug – a glutinous secretion left by the male in the opening of the female reproductive tract following copulation that tends to reduce both sperm leakage and access to the female by other males [7.2.3]

mating strategy – a set of behaviors with matching physical traits that work together to achieve mating success [1.2.2, 5.2.2]

mating system – a social system characterized by male–female relationships with respect to mating and parenting [1]

meiotic cell division – a type of cell division in complex organisms that yields cells with half as many chromosomes as body cells have; occurs in gonads in vertebrate animals [2.2.1]

menarche – a female person's first menstrual period [9.4.2]

menopause – the time at which a woman's menstrual periods have ceased for a year [2]

menstruation – in the females of many primate species, that part of the monthly reproductive cycle characterized by vaginal bleeding resulting from the breakdown of the uterine lining [Box 1.2]

miscarriage – a natural but premature termination of a pregnancy that involves death of the embryo or fetus [Box 10.2]

mitochondrion – an abundant cell organelle that has its own DNA chromosome encoded to produce many of the proteins needed for its functioning [3.7.3]

mitotic cell division – a type of cell division in complex organisms that yields daughter cells with the same genetic content as the dividing parent cell [2.2.1]

monogamy – a mating system characterized by an apparently exclusive sexual relationship between one male and one female [1.3]

monohybrid cross – a mating experiment in which two different varieties that differ in the expression of one phenotypic character are bred with one another [Box 3.1]

Muller's ratchet – a process where the DNA of an asexual population accumulates mutations that will have a negative effect on the phenotype [4.6.3]

natural selection – a fundamental evolutionary process that favors or disfavors individuals in a population depending upon their capacity to survive and reproduce, leading to genetic changes in the population [4.1.4]

negative assortative mating – the pattern whereby individuals tend to mate with a partner who exhibits dissimilar traits [Box 5.2] negative hormonal feedback – a form of hormonal regulation where a system responds to the level of a hormone by making upward adjustments when levels are low and downward adjustments when levels are high [9.1.2]

nipple – the breast projection representing the termination of the mammary ducts in mammals which, in females, delivers milk to the suckling infant [9.7.2]

nitrogenous base – a nitrogen-based molecular structure, four of which form the building blocks of DNA [3.6.4]

non-disjunction – the failure of homologous chromosomes or of sister chromatids to separate during cell division [9.5.1]

non-sister chromatids – chromatids that come from different members of a pair of homologous chromosomes, meaning they are not identical [3.7.1]

nucleus – the cell organelle that contains the majority of chromosomes and where DNA transcription takes place [2.2.2]

nuptial gift – in courtship, an item of value offered by one sex (usually a male) to the other in return for mating access [6.4.3]

obligate sex – sexual reproduction in organisms that are incapable of asexual reproduction, in contrast to facultative sex [4.6]

oogenesis – female gametogenesis (gamete production), employing meiosis and producing mature ova [10.1]

operational sex ratio – the ratio of males available for mating opportunities to females available for mating opportunities [5.2.1]

optimal fitness strategy – a series of behaviors that collectively yield the most profitable reproductive success for the individual [7.2]

orgasm – the pleasurable climax of sexual excitement; in males, accompanied by ejaculation [7.2.2, 10.3]

ornament – a physical trait generated by sexual selection that has display value in mating [5.1.3]

outbreeding – mating between non-relatives, yielding benefits from genetically diverse alleles [4.6.2]

ovarian reserve – the pool of fertilizable ova in the ovary; declines with age [10.7]

ovary – the female gonad, in which oogenesis occurs [1.1]

oviparous – reproduction that produces eggs rather than live young [2.4.2]

ovotestis – an organ capable of producing both eggs and sperm, which in most species differentiates into either an ovary or a testis [1]

ovulation – the release of an egg or eggs from an ovary [Box 7.1]

pair-bond – a close relationship of mating and sexual activity between two individuals that is not merely of transient duration [2.1.2]

parasitic castrator – a parasite whose infectious strategy is to deprive the host organism of sufficient resources to reproduce [8.4.5]

parental care – any services provided by a parent to an offspring over and above the initial investment in the production of the egg or young [2.4.1]

parthenogenesis – the asexual reproduction of an ovum without fertilization [2.2.3] partible paternity – the biologically incorrect belief that a human child can have multiple fathers if its mother had sexual intercourse with more than one partner [6.4.3]

paternal care – parental care provided by the male parent [5.4.1]

paternity uncertainty – the doubt about which male fathered a particular offspring; associated with non-monogamous female behavior, especially in cases of internal fertilization [6.6.1]

pedigree – an ancestry chart of multiple generations [Box 9.2]

penis – the term for the male intromittent organ of various species, including mammals [2.3.1]

peptide hormone – a small amino-acid-based hormone that acts on the surface of target cells [9.1.1]

phenotype – the observable physical and behavioral traits of an individual dependent upon genetic and environmental influences [3.4.1]

pheromone – a chemical messenger released into the environment, commonly functioning to facilitate mating [4.5.2]

pituitary gland – an endocrine gland important for the regulation of other endocrine glands; located at the base of the brain [9.1.1]

placenta – an organ that develops in many mammals at the uterine interface between mother and developing fetus [9.4.3]

plateau – the second stage of sexual response in humans; follows arousal [10.3]

pollination – sexual reproduction in plants, effected by the transfer of sperm-carrying pollen to the flower's pistil, which contains the ovary [Box 2.2] polyandry – a mating system in which a female has multiple male partners [6.1]

polygamy – a mating system in which one sex has more than one partner of the opposite sex [1.3]

polygynandry – a mating system in which a breeding group includes at least two of each sex [6.1]

polygyny – a mating system in which a male has multiple female partners [4.3.3]

polygyny threshold – the point at which a female's expected reproductive success with an unmated male with inferior resources equals her expected reproductive success with a mated male with superior resources [6.3.1]

positive hormonal feedback – regulation whereby an increase in a hormone leads to a further increase in that hormone, in contrast to negative feedback [9.1.2]

post-zygotic barrier – a factor preventing the successful development of a zygote created by the fusion of gametes from individuals from different gene pools [4.7.3]

pre-zygotic barrier – any factor that prevents the successful fusion of gametes from individuals from different gene pools [4.7.3]

primary oocyte – the diploid cell created during oogenesis that is an immature egg, and which is the precursor to the haploid secondary oocyte [10.1.2]

primary sex characteristics – features distinctive to one sex that are directly related to sexual function in that sex [2.3.2]

primary spermatocyte – the diploid cell created during spermatogenesis that is the precursor to the haploid sperm cell [10.1.1] promiscuity – a mating system in which there are no pair-bonds and in which an individual is likely to mate with multiple individuals [5.5.2]

prospecting – a strategy that involves roaming for the purpose of locating potential mating opportunities [7.3.1]

prostaglandin – a hormone-like substance with reproductive and non-reproductive roles, associated with uterine contractions and also present in semen [9.3.3]

prostate – a non-endocrine gland at the base of the male genitalia that is involved in the production of seminal fluid and in the mechanics of ejaculation [9.3.3]

protandry – any patterns in which males are or appear first, such as in sexual development in sequential hermaphrodites [8.1]

protogyny – any pattern in which females are or appear first, such as in sexual development in sequential hermaphrodites [8.1]

puberty – the stage of development in which a boy or girl transitions to an adult capable of sexual reproduction [9.1.2]

Punnett square – a diagram used to determine the genotypic or phenotypic probabilities in offspring from a particular cross [3.2.2]

putative father – the assumed father of particular offspring, without genetic evidence, usually based on an association with the mother [Box 3.2]

r-selection – natural selection that favors a suite of characters resulting in many offspring but involving relatively less parental investment per offspring (in contrast to K-selection) [7.3.2]

recessive – an allele that is not phenotypically expressed unless it is present in two copies (i.e., without a dominant partner allele) [3.2.2]

Red Queen hypothesis – the idea that, because ecological circumstances change incessantly, sexual reproduction is advantageous because it produces diversity each generation [4.6.3]

reproduction – the creation of new individuals from pre-existing individuals [1]

reproductive isolation – circumstances that prevent individuals from two different populations from breeding together [4.7]

reproductive skew – the uneven distribution of breeding activity among members of a group [6.6.1]

residual reproductive value – the potential to produce offspring, diminishing over time [7.3.2]

resolution – the fourth stage of human sexual activity; follows orgasm [10.3]

resource-defense polygyny – a type of polygyny in which males secure multiple matings through control of key resources [6.3.1]

rhythm method – a birth control method that depends upon avoiding sexual intercourse near the time of ovulation [10.4.2]

runaway hypothesis – a hypothesis explaining sexual selection based on indirect benefits, in which a male ornament and a female preference for the ornament become more pronounced over evolutionary time [5.3]

satellite male – a subordinate male that associates with a higher-ranking male for the purpose of gaining mating opportunities [7.3.1]

science – a body of knowledge and a methodological system that relies on testable explanations of natural phenomena [1.4]

scramble competition – a type of mating competition where individuals achieve success by quickly exploiting mating opportunities without attempting to defend their mates from other competitors [5.5.1]

scrotum – the sac-like part of the external male mammalian genitalia that contains the testes [2.3.1]

secondary oocyte – the haploid cell produced in oogenesis that is released by the ovary and which is capable of being fertilized by a mature sperm [10.1.2]

secondary sex characteristics – traits unique to one sex that are not directly involved in reproduction [2.3.2]

selfing – sexual reproduction involving gametes from the same individual, in contrast to crossing [2.2.4]

semelparous – a pattern of offspring production where individuals produce offspring just once, in contrast to iteroparous [2.4.2]

semen – the fluid released during male ejaculation, containing sperm as well as a variety of different molecules [2.4.1]

sequential hermaphrodite – an individual that is functionally male and female but not at the same time [8.1]

serial monogamy – a type of monogamy characterized by having multiple pair-bonds, but only one at any given time [6.2.1]

sex – either of the two main categories, male and female, associated with reproductive function; also, a short term for sexual activity; also, a type of reproduction resulting from the fusion of egg and sperm, in contrast to asexual reproduction [1] sex allocation – a non-random pattern of investing in offspring in accordance with offspring sex [7.3.3]

sex assignment – the determination of an infant's biological sex at birth, based on the appearance of the genitals [1.1.1]

sex determination – a genetic or environmental trigger that governs whether the individual will manifest a male or a female phenotype [1]

sex differentiation – the accumulating changes during development that distinguish males and females [1]

sex linkage – an association between a particular trait and the sex that is heterogametic [9.6]

sex role reversal – a pattern exhibited in a minority of species where behaviors that are typically exhibited in one sex in the majority of species are exhibited in the opposite sex [5.4]

sexual cannibalism – reproduction that results in the death of the male, who is consumed by his partner [Box 5.1]

sexual conflict – differences in the optimal mating strategy for males and females [1]

sexual intercourse – sexual activity requiring the insertion of the male intromittent organ into the female reproductive tract [2, 2.4.1]

sexual orientation – the component of a person's sexual identity that determines the sex(es) to which they are attracted [1.1]

sexual reproduction – creation of new individuals through the merger of sperm and eggs [2.2.1] sexual selection – a subset of natural selection that influences reproductive success through intra-sexual competition and inter-sexual choice [1]

sexually transmitted infection – a bacterial, fungal, viral, or parasitic infection that is spread by sexual activity, especially sexual intercourse and anal or oral sex [4.5.3, 10.8]

simultaneous hermaphrodite – an individual that is functionally male and female at the same time [8.1]

sister chromatids – chromatids that come from the same member of a pair of homologous chromosomes, meaning they are identical [3.7.1]

sneaker – a male with a furtive reproductive strategy, often employed by female-mimics, characterized by quick fertilizations of females who associate with higher-status males [5.5.2]

social determinism – the assertion that socio-cultural factors determine the behavior of organisms, especially humans [1.2]

social monogamy – a male–female pair-bond that is not necessarily genetically monogamous [6.2.1]

sociobiology – the perspective that human social behavior is explainable by recourse to evolutionary biology [1.2]

speciation – the formation of new species through evolutionary processes [4.7]

species – a rank in biological classification, commonly defined for sexual species as the largest group from which two individuals can produce fertile offspring [1.1.2, Box 1.1]

sperm – a male gamete, commonly characterized as being small and mobile, more formally referred to as a *spermatozoon (pl. spermatozoa)* [1.1]

sperm allocation – the ability of some males to adjust the sperm content of ejaculates, often according to the risk of cuckoldry [7.2.2]

sperm competition – male–male competition for access to fertilizable eggs, conducted at the level of the female reproductive tract [5.5.3]

sperm depletion – the temporary drop in the number of sperm caused by recent, previous, ejaculations [6.4.1]

spermatogenesis – male gametogenesis (gamete production), employing meiosis and producing mature sperm [10.1]

spermatophore – in diverse animal groups, a vessel containing sperm and other substances of value to the female transferred during mating [2.4.1]

SRY gene – the gene located on the Y-chromosome of all placental and marsupial mammals that codes for a transcription factor that sets the developing embryo on a pattern of male, rather than female, sexual development [8.2.4]

sterilization – an intended or unintended event that renders an individual permanently incapable of reproduction [10.4.2]

steroid hormone – one of a class of endocrine hormones that play key roles in sexual function and that influence cells by acting directly on cell nuclei [9.1.1]

synapsis – the pairing of two homologous chromosomes during meiosis [3.7.1]

teratogen – an agent that disrupts embryonic or fetal development [Box 10.2]

testicle – a term applied to the mammalian testis [1.1]

testis – the male gonad, in which spermatogenesis occurs (pl. testes) [1.1]

testis-determining factor (TDF) – a transcription factor encoded by the *SRY* gene on the mammalian Y-chromosome that initiates the male sex-determination cascade [8.2.4]

testosterone – a steroid hormone found in significantly greater concentration in males, primarily associated with male sexual development and function [4.5.3]

theory – in science, an explanation of a set of natural phenomena that withstands repeated testing and which therefore forms a part of accepted scientific knowledge, in contrast to the everyday sense of the word which is akin to the word *hypothesis* [1.5]

transcription – the stage of protein synthesis that involves making a messenger RNA copy of the DNA gene for that protein [3.6.3]

transgender – a person whose gender identity differs from his or her biological sex, in contrast to cisgender [1.1.1]

translation – the stage of protein synthesis that involves making a protein from the code reflected in the messenger RNA [3.6.3]

traumatic insemination – a mating practice in some invertebrates where the male's intromittent organ inseminates the female by piercing her body cavity [Box 7.2]

triple-X syndrome – a condition in which an individual has three sex chromosomes per cell, all of which are X-chromosomes; tends to produce functional female phenotypes [9.5.1]

Trivers-Willard hypothesis – the proposition that a female can adjust the sex ratio of her offspring according to her own condition [7.3.3]

tubal ligation – a method of female sterilization that is effected by the severing of the fallopian tubes [10.4.2]

Turner syndrome – a disorder of sexual development caused by having a single X-chromosome rather than a pair of sex chromosomes, usually resulting in a female phenotype with underdeveloped gonads, sterility, and other issues [9.5.1]

umbilical cord – in placental mammals, the tube containing blood vessels that connects the developing embryo or fetus to the placenta during gestation [10.5.1]

uterus – the part of the mammalian female reproductive tract where the offspring develop before birth [9.4.1]

vagina – in mammals, the muscular tube leading from the female genitals to the cervix at the lower end of the uterus [2.3.1]

vasectomy – a method of male sterilization that is effected by the severing of the vas deferens, which are the ducts that convey the sperm from the testes to the urethra [10.4.2]

vasocongestion – the swelling of body tissues caused by changes in blood flow, with attendant local increases in blood pressure [10.3]

viviparous – a species that bears live young rather than eggs [2.4.2]

vulva – the female genitals, including labia majora, labia minora, clitoris, and vaginal opening [2.3.1]

X-chromosome – a large and gene-rich mammalian chromosome found in two copies in females and paired with a Y-chromosome in males [2.2.2]

X-inactivation – a pattern in the cells of female mammals where one chromosome from the X-chromosome pair is rendered inactive, apparently selected at random [8.2.3]

Y-chromosome – a small mammalian chromosome found in males, where it is paired with an X-chromosome [3.6.2]

zygote – a single cell that is an egg that has been fertilized by a sperm [2.3.1]

Index

ABO blood group, as gene pool, 122–25
ABO blood system, genetics and inheritance, 75–79
abortion, 340, 342
abstinence, 340
adaptation, in evolution, 115
adolescence, and puberty, 300–301, 306–7
adrenal glands, 293, 295
African jacana (*Actophilornis africanus*), conflict and sex role reversal, 169, 226–27
afterbirth. *See* placenta
AIDS (acquired immunodeficiency syndrome) and HIV, 356–57, 358
allele pool, 122
alleles
 description, 73–74
 in evolution, 113, 114, 117–18
 and gene pools, 121–25
 genetics disorders in men, 314–15
 and homosexuality, 249
 in humans, 75, 76–77, 86
 intersex conditions from dysfunctions, 312–13
 and multiple types of sex, 262
alloparenting, 210
allopatric speciation, 137, 138
5-alpha-reductase (5AR), 313
alternative mating strategies, 239, 241–44, 262
amines, 292
amino acids, 33
AMS gene, 262
anagenesis, 134–35
androchromes, 263
androgens, 291, 293
animals
 asexual reproduction, 39, 41
 developmental biology, 297
 secondary sex characteristics, 46–47, 50
 sexual dimorphism, 47–50
anisogamous gametes, 44
anomalous sex determination, 284–86
antibiotic resistance, 352–53
antigens, and blood, 76–79
anti-Müllerian hormone (AMH), 298
Archaeopteryx, 111
armaments in sexual selection, 151, 162
aromatase, in TSD, 279
arousal, stages and response, 333–36
artificial selection, 60, 111, 112
asexual reproduction
 in complex organisms, 38–43
 description, 36, 125
 and diversity, 130
 efficiency, 126–27
 and mitosis, 91
 mutations, 134
 and sexual reproduction, 41–43, 125, 129–30
assistive reproductive technologies (ART), 339–40
Atlantic puffins (*Fratercula arctica*), monogamy, 183
autogamy, 273
autosomes, 84, 261, 265–66
azure damselflies (*Coenagrion puella*), in scramble competition, 171

baby, presentation, 347–48
bacteria
 reproduction, 34–36, 38
 and STIs, 354–55
baculum, 303, 304, 305
bar graphs, 20–21
Bateman gradients, 204–5

bees. *See* western honeybee
benefits
direct benefits, 162, 201–4
indirect benefits, 162, 196–201
biandry, 203
bilateral insemination, 230
binary fission, 34, 36
binary model (male-female), 7, 43–44, 46, 309
binomial nomenclature, 16–17
biocultural and biosocial features, 30–31
biodiversity, 16
biogeography, in evolution, 111–12
biological adaptation, 115
biological classification, 13, 16–17
biological determinism, 9, 10–11
biological fitness, 115–16
biological sex, and gender identity, 7
biological species concept, 136
biology
developmental biology, 297
molecular biology, 84–85
as perspective on sex, 11
role of evolution, 102, 109
theories and hypotheses, 15, 18–19
biparental care, in monogamy, 185–86
birds
biparental care, 186
brood parasitism, 240–41
extra-pair copulation and mating, 182, 194
extra-pair paternity, 184–85
intromittent organ, 303, 304, 305
monogamy, 182–84
parenting, 168–69, 186, 250–52
reproductive organs, 234, 235
same-sex parenting, 250–52
sex determination and chromosomes, 270–72
sperm depletion, 196
birth control, 340, 342–43
birth defects, 346
birth rate, and culture, 26
Bisphenol A (BPA), 286
black-capped chickadee (*Poecile atricapillus*), extra-pair mating, 194
blastocyst, 343
blending inheritance, 60–61
blood and blood grouping (human), genetics and inheritance, 75–79
bluegill sunfish (*Lepomis macrochirus*), condition-dependent strategies, 242
bluestreak cleaner wrasse (*Labroides dimidiatus*), ESD, 282–83
blue-winged water strider (*Gerris buenoi*), coercion and harassment, 227
bonellin, 280
boys, puberty, 300–301, 313, 326
brain, and sexual selection, 174
breast, 349
breast cancer, 319–20
breast-feeding, 349–50
breech presentation and birth, 347–48, 349
breeding groups as gene pools, 121–25
broad-headed bugs (*Protenor* spp.), GSD, 270, 271
brood parasitism, 240–41
brown anoles (*Anolis sagrei*), sex allocation, 247, 248
brown garden snail (*Cantareus aspersus*), love darts, 231
brown-headed cowbird (*Molothrus ater*), brood parasitism, 240, 241
brown marsupial mouse (*Antechinus stuartii*), cost of sex and non-monogamy, 127–28, 199
burying beetles (*Nicrophorus* spp.), conflict in parental care, 238–39

Caesarean section (C-section), 348–49
cancer, in human sex organs and tissues, 315, 318–21
capacitation, 337–38
Carothers, Eleanor, 80, 81
cat (*Felis catus*), intromittent organ, 305
celebrity couple hypothesis, 188
cell nucleus, 37
cells
mitochondria, 97
reproduction, 33, 36–39
cephalic presentation, 347–48
cervical cancer, 320
cervix, in delivery, 346
Charles II of Spain, and inbreeding, 131–32
chemical warfare and sperm, 233–34
chinook salmon (*Oncorhynchus tshawytscha*), condition-dependent strategies, 242
chlamydia (*Chlamydia trachomatis*), 355, 358–60

chromatids, 37, 83, 92–93, 94
chromatin, in chromosomes, 81–82, 83
chromosomal defects, 327–28
chromosomal sex determination, 272–74
chromosomes
in birds, 270–72
defects in children, 327–28
description and discovery, 80–81
forms and division, 81–83
and gametes, 87, 90, 92–98
and genes, 37, 81
in genetics and inheritance, 80–84, 86, 265–67
as hereditary material, 80–82
in humans, 37, 83–84, 261, 265
inheritance theory, 80–81
in insects, 261, 263–64
males *vs.* females, 83, 261
and mitosis, 37–38
mutations, 118–20
and phenotype, 265–67
sex chromosomes (*See* sex chromosomes)
in sexual reproduction, 45
size and length, 265
X-chromosomes (*See* X-chromosomes)
X-like appearance and shape, 37, 83
Y-chromosomes (*See* Y-chromosomes)
circumcision, 301–2, 307
cisgender, definition, 7
cladogenesis, 134, 135, 138, 139
classification, basics of, 13, 16–17
clitoris, 304, 305, 334
cloaca in birds and reptiles, 303, 305
cloacal protuberance, 303, 304, 305
clownfish (*Amphiprion* spp.), reproduction and ESD, 2–3, 284
coercion and harassment in mating, 203–4, 219, 227–28
coitus interruptus, 340
colostrum, 350
column graphs, 20–21
common ancestor, 135
common bed bug (*Cimex lectularius*), traumatic insemination, 230
common yellowthroat (*Geothlypis trichas*), intromittent organ, 304, 305
comparative approach, 13–14
comparative genetics and morphology in evolution, 110
complementary sex determiner (CSD) gene, 275
concealed ovulation, 224, 332–33
condition-dependent strategies, 241–43
condoms, 342
conflict. *See* sexual conflict
congenital adrenal hyperplasia (CAH), 312
conjugation, in bacteria and *Paramecium*, 34–35
consortships, 172
conspecifics, 148
contraception, 340, 342–43
controlled experiment, 18–19
convenience polyandry, 203–4
cooperative breeding, 209–10
cooperative polyandry, 206–7, 209
co-option, in sexual activity, 29–32
copulation, 51, 54
copulatory plug, 236–37
corpus cavernosa, 334
correlation, in graphs, 23
creationism, 109
Crick, Frances, 84
crossing
in hermaphrodites, 272–73, 274
in Mendel's work, 65, 68–72
in plants, 52–53
crossing-over, 92–96
cruel bind, 238
cryptic female choice, 198, 201, 235
"C-section" (Caesarean section), 348–49
cuckolder phenotype, 241–42
cuckoldry, 222
cuckoo catfish (*Synodontis multipunctatus*), brood parasitism, 240
cuckoo wasps (Chrysididae), kleptoparasitism, 240–41
culture and sex, 6, 26
cyclic parthenogenesis, 42
cytokinesis, 37, 38, 95

damselflies (*Coenagrion* spp.), in scramble competition, 171
Darwin, Charles
on evolution, 102, 103, 107–9, 144
and natural selection, 113
on sexual selection, 144, 155
data, graphic presentation and sets, 19, 20–23
daughter cells, 34, 36, 38, 95

daughters, in sex allocation, 246–48
deductive reasoning, 19
delivery and labor, 345, 347–49
The Descent of Man, and Selection in Relation to Sex (Darwin), 144
desertion by one parent, 238
determination of sex. *See* sex determination
development, sexual. *See* sexual development
developmental biology, 297
developmental cascade, 269, 291, 297
Diamond, Jared, on sexual pleasure, 28
dichromatic vision, 119–20
diethylstilbestrol (DES), 346
differentiation of sex. *See* sex differentiation
dihybrid cross, 71–72
dihydrotestosterone (DHT), 298–99, 313
dimorphism, 47–50
dinosaurs and extinction, 105
diploid genes and cells, 74, 75, 274
diploid organisms, in life cycles, 90–91
direct benefits of mating, 162, 201–4
direct competition, in sexual selection, 171–72
disorders of sexual development (DSDs), 309
dispersal speciation, 137–38
divine creation, 104
DMRT1, 272
DNA (deoxyribonucleic acid)
    divisions and bases, 86
    encoding of information, 86–87
    and evolution, 116–21
    extra-pair mating, 184–85
    form, 84–85
    in genetics and inheritance, 80, 84–87
    paternity testing, 88–89
    and reproduction, 33
    transcription and translation, 84–85, 86–87, 268
DNA scissors, 184–85
dominance hierarchies, 208, 210
Dominican Republic, and *guevedoce*, 313, 314
dose-sensitive transcription factor, 272
dual mating strategy, 222, 224–25
ducks, intromittent organ, 303, 304
duplication, 118–20
eastern three-lined skink (*Bassiana duperreyi*), ESD, 281
ebony jewelwing (*Calopteryx maculata*), sexual conflict, 229, 234–35
ecotoxins, and anomalous sex determination, 286
egg dumping, 240
eggs
    cycle in women, 308
    development, 56
    and females, 44–45
    fertilization, 45, 51, 54, 308, 336–38
    and genes, 74
    and inheritance, 68–69, 74
    mutations, 120–21
    oogenesis, 325, 326, 327–28
    in sex determination, 256, 264, 265
    in sexual reproduction, 43–45
    with shell, 303
Elizabeth 1, 257–58
embryos
    in humans, 344, 346
    sex differentiation, 291, 298–99
empirical evidence and data, 18
endocrine disruptors, in environmental sex determination, 286
endocrine glands, 292
endocrine system, 289, 292–94, 295
endometrial cancer, 320
endometriosis, 338
endurance competition for mating, 172
Enlightenment, and evolution, 104–5
environmental sex determination (ESD)
    and anomalous sex determination, 284–86
    hormones in, 279, 280, 282, 284, 286
    non-social systems, 277–82
    offspring sex selection, 281–82
    overriding of GSD, 281
    overview and categories, 276–77
    social systems, 282–84
    and temperature, 277–79, 281–82
epigenetics, and homosexuality, 249–50
erection in humans, 303
ESD. *See* environmental sex determination
estradiol, 293, 294, 295, 331
estrogens, 293, 296, 351
estrogen synthase, in TSD, 279
estrous cycle, 330, 331, 333
eukaryotic cells, 38–39

evolution
alleles and genes, 113, 114, 117–18, 121
benefits of sex, 129–34
categories of evidence, 110–12
as concept, 104
costs of sex, 125–29
Darwin's theory, 109, 144
DNA and inheritance, 116–21
fitness and adaptation, 115–16
fossils and extinctions, 105, 106, 110, 111
gene pools, 121–25
historical overview, 102–9
mutations, 117–21
natural selection principles, 113–15
reproductive barriers, 139–41
role in biology and sex, 102, 109
species creation, 134–41
evolutionary psychology, 10–11
evolutionary strategy, 220
exaptation, and sexual activity, 29–30, 32
excitement, in sexual response, 333–34, 336
exocrine glands, 292
experimental evidence, 18
explosive breeders, 212
external fertilization, 51, 55, 196, 303
extra-group copulation (EGC), 193
extra-group paternity (EGP), 244
extra-pair copulation (EPC), 182, 183, 184–85, 194
extra-pair mating (EPM), 182, 184–85, 193, 194, 224, 225
extra-pair offspring, 182, 224, 225
extra-pair paternity (EPP), 182, 183–85, 224–25

factor 9 in hemophilia, 316
feathers, and exaptation, 30
female choice hypothesis, 151–52, 155–59, 163, 196
female circumcision, 307
female-defense polygyny, 189
female-enforced hypothesis, 187
female–female pairings, 250, 251
female mimicry, 236, 242–43
female reproductive tract, 51
females
biological definition, 5–6, 44–45
chromosomes, 83, 261
conflict factors, 223
costs of sex, 129
and eggs, 44–45
genitals, 44–45, 301–2, 307
as heterogametic sex, 270–72
human females (*See* women)
investment in reproduction and parenthood, 152–55, 167–70, 221
as "limiting sex" in sexual selection, 152–62
mating with multiple males, 193–204
multiple types, 262–63
non-monogamy benefits, 193–204, 221–22
in polyandry, 204
polygyny and clusters, 188–89
reproductive success, 161–62
secondary sex characteristics, 46
in sex role reversal, 167–70
ferns, reproduction, 90–91
fertility
and contraception, 340, 342–43
definition, 224
fertile period and concealed ovulation, 331–33
in history, 324, 325
infertility, 195–96, 338–40
menstrual and estrous cycles, 328–30
sexual arousal and response, 333–36
spermatogenesis and oogenesis, 326–28
fertility insurance hypothesis, 195–96
fertilization
of eggs, 45, 51, 54, 308, 336–38
external *vs.* internal, 51, 54–55, 56, 303
and sexual reproduction, 50–51, 54–55
and zygote, 51, 55, 336–40, 342–43
fetal alcohol syndrome, 346
fetus, in pregnancy, 343–44, 346
fighting, in intra-sexual competition, 170
firebugs (*Pyrrhocoris apterus*), prolonged copulation, 233
first generation (F1), inheritance, 65–67, 68–70
first sperm precedence, 231
fish
hermaphroditism, 259, 260
social systems of ESD, 282–84
Fisher, R.A., 162
Fisher's principle, 246
fitness
biological fitness, 115–16
in sexual selection, 154–55, 162–67

5-alpha-reductase (5AR), 313
flat lizards (*Platysaurus broadleyi*), she- and he-males, 243
flatworms (Order Polycladida), penis fencing, 230
flowering plants and flowers, 52–53, 61
FMF system of TSD, 278–79
FM system of TSD, 278
follicle-stimulating hormone (FSH), 301, 306
follicular phase, 328–30
fossils, and evolution, 105, 106, 110, 111
founder population, 138
frogs, external fertilization, 54, 55
fungi, reproduction, 90, 92

gametes
  in biological definition, 5–6
  and chromosomes, 87, 90, 92–98
  and inheritance, 66–67, 68
  isogamous and anisogamous gametes, 44
  production, 325, 326, 327
  in sex determination, 270–72
gametic life cycle, 90, 91
gametogenesis, 325
garden pea (*Pisum sativum*), in Mendel's work, 62–72
garter snakes (*Thamnophis* spp.), 136, 195
gender dysphoria, 7, 312
gender *vs.* sex, 5, 7–8
gene pool(s)
  breeding groups as, 121–25
  definition, 86, 121–22
  diversity, 130–31
  and species, 135, 137–39
  splitting, 137–39
generations, and inheritance, 65–67, 68–71
genes
  and chromosomes, 37, 81
  definition and concepts, 73–74, 81, 84
  and DNA, 84, 86
  and evolution, 121
  role, 37
  and sex chromosomes, 265–67
  in sex differentiation and sex determination, 257, 267, 291
  variants, 73–74
genetic code, 84–86
genetic complementarity model, 197, 200–201
genetic disorders in men, 314–15
genetic diversity model, 197, 198
genetic load, 134
genetic monogamy, 182–83
genetics
  chromosomes, 80–84, 86, 265–67
  comparative genetics in evolution, 110
  DNA, 80, 84–87
  early science of, 72–73
  and homosexuality, 31–32
  Mendellian genetics in humans, 75–79
  Mendellian genetics modernized, 72–74
  and non-monogamy benefits for females, 196–201
  paternity testing, 75–76, 78–79, 88–89
  in sex determination, 256–57
  terminology and concepts, 73–74
genetic sex determination (GSD)
  generalization, 276
  in honeybees, 274–76
  in humans, 269
  overriding of, 281
  without Y-chromosome, 269–70, 271
genital herpes, 355
genitals and genitalia
  in biological definition, 6–7, 44
  female *vs.* male, 44–45, 301–2, 307
  in sexual development, 298
genotype, 74, 75, 77, 84
genus, classification, 17
geography, and gene pool, 137–39
girls, puberty, 306–7, 313
glans, 301, 303
golden lion tamarin (*Leontopithecus rosalia*), dominance hierarchies, 208, 209
gonadotropin-releasing hormone (GnRH), 301
gonads
  in biological definition, 5–6
  hypogonadism, 311
  in sex differentiation and function, 293, 295, 297, 298
gonopodium in guppies, 303, 305
gonorrhea (*Neisseria gonorrhoeae*), 352, 354–55
good genes hypothesis/model, 162, 164–66, 197, 198–99
good resources hypothesis, 162, 166

gorilla (*Gorilla gorilla*), polygyny, 189
graphs, types and presentation of data, 19, 20–23
grasshoppers, internal fertilization, 54
gray whale (*Eschrichtius robustus*), intromittent organ, 305
green darner (*Anax junius*), contact guarding, 187
green shore crab (*Carcinus maenas*), ESD, 285–86
green spoon worm (*Bonellia viridis*), ESD, 280
GSD. *See* genetic sex determination (GSD)
Guianan cock-of-the-rock (*Rupicola rupicola*), lek polygyny, 213
guppy (*Poecilia reticulata*), sexual conflict and intromittent organ, 236, 237, 303, 305

Habsburg royal family, inbreeding, 131–32
hamlet fish (*Hypoplectrus* spp.), hermaphroditism, 259, 260
handicap principle, 165
Hanuman langurs (*Semnopithecus entellus*), sexual conflict, 223, 226
haplodiploidy, 274, 275
haploid genes and cells, 74, 274
haploid organisms, in life cycles, 90–91
harassment and coercion in mating, 203–4, 219, 227–28
Hardy, Godfrey, 121
Hardy–Weinberg Equilibrium, 121
heat (estrous cycle), 330, 331, 333
helpers, for offspring, 209–10
he-males, 243
Hemings, Eston, 88, 89
Hemings, Sally, on genetics and DNA, 88, 89
hemipenis, 303
hemizygosity, 315
hemoglobin, 117–18, 119
hemophilia, as sex linkage, 316–17
Henry VIII, male heirs, 256, 257–58
hepatitis, 358
hermaphrodites
  chromosomes and autosomes, 272–73
  definition and description, 42, 309
  humans as, 309
  in mythology, 258–59
  selfing, 273, 274
  sex determination, 257, 259–60, 272–74
  sex differentiation, 257
  sexual conflict, 229, 230–31
herpes simplex viruses, 355
heterogametic sex, 270–72
heterosis, 200
heterozygotes and heterozygosity, 74, 200–201, 314–15
HIV (human immunodeficiency virus) and AIDS, 356–57, 358
HMS Beagle, 107–8
homogametic sex, 270
homosexuality
  genetics and co-option, 31–32, 249–50
  and parenting, 250–52
  social aspects and acceptance, 5, 248
  as theoretical challenge, 249–50
homosociality, 251
homozygotes and homozygosity, 74, 132–33, 314–15
honest indicators and promises, 164, 166
honeybees (*Apis* spp.), sex determination and sex chromosomes, 274–76
  See also western honeybee
hormonal feedback, 294, 296
hormone replacement therapy (HRT), 320, 351
hormones
  and endocrine system, 292–94, 295
  in ESD, 279, 280, 282, 284, 286
  in human sex differentiation and function, 290–96, 298–99
  levels in humans, 294
  in milk production, 350
  and placenta, 344–45
  supplements, 290, 291
human chorionic gonadotropin, 343–44
human papillomavirus (HPV), 320, 355, 358
human polygyny index (HPI), 192
humans
  alleles, 75, 76–77, 86
  anomalous sexual phenotypes, 309–14
  breeding groups as gene pools, 121–22
  cancer of sex organs and tissues, 315, 318–21
  chromosomes and genes, 37, 83–84, 261, 265
  copulation, 51
  extra-pair mating, paternity and offspring, 224–25

fertilization, 55
genetics and inheritance, 75–79
genotype, 75, 77, 84
hormones in sex differentiation and function, 290–96, 298–99
investment in offspring, 57
as iteroparous breeders, 56
mating systems, 178, 181
phenotype, 75, 76–77, 84, 309–14
prenatal development, 298–300, 306
purpose of sex, 29, 30–31
rape, 227
reproductive system of men, 299–302
reproductive system of women, 299, 306–9
secondary sex characteristics, 46–47, 48
sex chromosomes, 83–84, 261, 265–67
sex differentiation in development, 297–99
sexual anatomy, 301–2, 307–9
sexual selection, 20–21, 160–61, 174
SRY gene, 267
Hume, David, *is-ought problem*, 11, 12
hybridization, 39, 41
hybrids sterility, 140, 141
hydra (*Hydra* spp.), reproduction, 39, 40
hypogonadism, 311
hypothalamus, 293–94, 295
hypothesis, definition and use, 15, 18–19

ideology
definition and use, 8–9
in sex and biology, 9–11
imperfect flower, 52
imperfect structures, in evolution, 110
implantation, 308, 343
inbreeding, 131–33
inbreeding depression, 132–33, 200
inclusive fitness, 210
independent assortment, 95, 98
law of, 64, 70–72
Indian peafowl (*Pavo cristatus*), in evolution theory, 144, 145
indirect benefits of mating, 162, 196–201
indirect competition, in sexual selection, 171–73
inductive reasoning, 19
infanticide, 223, 226
infertility, 195–96, 338–40
inheritance, sexual. *See* sexual inheritance
inheritance of acquired characteristics, 106–7
insects
in conflict, 236
kleptoparasitism, 240–41
sex chromosomes, 261, 263–64
sex determination, 274–76
interbreeding, 136
internal consistency, in evolution, 112
internal fertilization, 51, 54–55, 56, 303
intersex, 259, 309, 310–13
intersexual, definition, 309
inter-sexual, definition, 309
inter-sexual selection, 151–52, 155–59, 163, 196
intra-sexual, definition, 309
intra-sexual competition, 151, 170–73, 175
intra-sexual selection, 151, 159–62
intrauterine devices (IUDs), 342
intromittent organ
description and role, 51, 54, 303
diversity and examples, 54, 302–5
in vitro fertilization (IVF), 340, 341
isogamous gametes, 44
is-ought problem, 11, 12
iteroparous breeding, 56–57

jacanas (*Actophilornis* spp.), sex role reversal, 168–69
Jefferson, Thomas, genetics and DNA, 88–89
Johnson, Virginia, 333

karyotype, in chromosomes, 82–83
kin selection, 210
kleptogenesis, 41
kleptoparasitism, 240–41
Klinefelter syndrome (XXY), 310–11
koala (*Phascolarctos cinereus*), STIs, 358–60
K-selection traits, 245

labia majora and labia minora, 44, 307
labor and delivery, 345, 347–49
lactation, 349
Lamarck, Jean-Baptiste, 106–7
last sperm precedence, 227
law, and sex, 5
Law of Independent Assortment, 64, 70–72
Law of Segregation, 64–69, 72
Laysan albatross (*Phoebastria immutabilis*), same-sex parenting, 250–51
lek polygyny, 212–13

life cycles variants, 90–91
life history theory, 244–45
line graphs, 20
line-of-best-fit, in graphs, 22, 23
Linnaeus, Carl, 16
locus/loci for genes, 81
long-tailed widowbird (*Euplectes progne*), inter-sexual selection, 155–56, 157, 163
Lorenz, Konrad, 9
lost opportunities, 153
love darts, 231
luteal phase, 330
luteinizing hormone (LH), 301, 306, 339
Lygaeus system, 270, 271
lyonization, 266

major histocompatibility complex (MHC) genes, 160, 200–201
male children, preference for, 256, 258
male circumcision, 301–2
male-female binary model, 7, 43–44, 46, 309
males
  biological definition, 5–6, 44–45
  chromosomes, 83, 261
  conflict factors, 222–23
  costs of sex, 128–29
  genitals, 44–45, 301–2, 307
  human males (*See* men)
  in intra-sexual selection, 151
  investment in parenthood and reproduction, 152–55, 167–70
  multiple types, 262–63
  in parenting, 168–70
  in polygyny, 204
  protection of mates, 186–87
  reproductive success, 159–60
  secondary sex characteristics, 46–47
  in sex role reversal, 167–70
  and sperm, 43–45
mallards (*Anas platyrhynchos*), sexual conflict and intromittent organ, 219, 235, 304, 305
Malthus, Thomas, 113
mammals
  estrous cycle, 330, 331
  genitals, 44
  intromittent organ, 51, 54, 303, 304
  mating systems, 180, 181, 211
  parental care, 152, 211
  sex determination, 267, 268, 272
mammograms, 319
marine isopods, multiple types of sex, 262, 263
Masters, William, 333
mate assistance hypothesis, 185–86
mate guarding, in conflict, 234–36
mate guarding hypothesis, 187
mating, harassment and coercion, 203–4, 219, 227–28
mating plug, 236–37
mating strategies
  alternative mating strategies, 239, 241–44, 262
  condition-dependent strategies, 241–43
  definition and terminology, 155, 220
  dual mating strategy, 222, 224–25
  and life history theory, 244–45
  plasticity in, 239, 244–45
  same-sex parenting, 248–52
  sex allocation, 246–48
  and sexual conflict, 220–22, 239, 241–48
  for sexual success, 219–22
mating systems
  benefits, 162, 196–204
  definition and characteristics, 178
  females with multiple males, 193–204
  for humans, 178, 181
  monogamy, 180, 181–88, 193
  and parenting, 211
  polyandry, 180, 181, 204–7
  polygamy, 180, 188
  polygynandry, 180, 207–10
  polygyny (*See* polygyny)
  promiscuity, 179–80, 210–13
  and social bond, 179–80, 210–11
  types, 179–81
  in Western societies, 178
matrilineal inheritance, 97
mealworm beetle (*Tenebrio molitor*), sex chromosomes, 263–64
meerkats (*Suricata suricatta*), prospecting, 243–44
meiosis
  and crossing-over, 92, 93, 94–95
  divisions, 94, 95
  goal, 324
  life cycle variants, 90–91
  and Mendel's laws, 97–98
  men-women differences, 324–28, 338
  and mutations, 120

non-disjunction in sex chromosomes, 310–12
Meiosis I and Meiosis II, 94, 95
meiotic cell division, 36, 90, 92, 94
meiotic non-disjunction, 310–12
men
cancer of sexual organs, 315, 318–19
chromosomes, 83, 261, 265–67
erection, 303
genetic disorders, 314–15
genitals, 44, 301–2
hormonal feedback, 294, 296
infertility, 338, 339
meiosis, 324–25, 326
orgasm, 335
prenatal development, 298–300
reproductive system, 299–302
sex differentiation before birth, 299–300
sexual anatomy of adults, 301–2
sexual arousal response, 333–36
sexual development at puberty, 300–301, 313, 326
testosterone, 294, 296, 298
and women's non-monogamy, 201, 203
women turning into men, 313–14
Mendel, Gregor
background, 62
classical genetics modernized, 72–74
and human genetics, 75–79
independent assortment, 95, 98
Law of Independent Assortment, 64, 70–72
Law of Segregation, 64–69, 72, 98
and meiosis, 97–98
and rules of inheritance, 62–64, 116
menopause, 327, 350–51
menstrual cycle, 328–33
menstrual period (menses), 328
menstruation (menarche), 307
messenger RNA (mRNA), 84–85, 86–87
methodology of science, 15, 18–19, 104
MF system of TSD, 278–79
mice, sperm competition, 173, 175
milk and milk production in humans, 349–50
milkweed bugs (*Lygaeus* spp.), GSD, 270, 271
Miller, Geoffrey, 174
mitochondrial chromosome, in crossing-over, 96–97
mitochondrial disease, and pregnancy, 341
mitochondrion/mitochondria, 97
mitosis, 36–38, 90, 92
mitotic cell division, 36–38, 87, 90, 95
molecular biology, 84–85
monoecious plants, 258
monogamy
biparental care, 185–86
in birds, 182–84
definition and description, 180, 181
extra-pair mating and paternity, 182, 184–85, 193
hypotheses, 184–88
non-monogamy benefits for females, 193–204, 221–22
prevalence, 181
types, 180, 181–83
monohybrid cross, 68
monospermy, 337
morning-after pill, 343
mosquitoes, in scramble competition, 172
"mouth-brooding" cichlid fish (Cichlidae), brood parasitism, 240
mules, sterility, 140, 141
Müllerian ducts, 298, 299
Muller's ratchet, 134
mutations, 117–21, 134
mutual insemination, 230

natural selection
agent in selection, 147–48
as mechanism, 109
non-sexual matters, 146–48
observations and inferences, 113–15, 145–46, 148–49
sexual selection in, 145–49, 151–52
nature *vs.* nurture, 9
negative assortative mating, 160
negative hormonal feedback, 294, 296
neuroendocrine pathway, 294
nitrogenous bases, 86
non-disjunction, 310–12
non-monogamy in females
benefits, 195–204, 221–22
overview, 193–95
non-procreative sex, 26, 28, 30–31
non-sexual reproduction, 36, 37–38
non-sister chromatids, 93
nucleic acids, 33
nucleus, in cells, 38–39
nuptial gifts, 202–3

O blood-type, 76–79
ocellaris clownfish (*Amphiprion ocellaris*), reproduction, 2–3
off-loading parenting, 238, 240–41
older brother effect, 250
On the Origin of Species (Darwin), 102, 103, 104, 109, 144
oocytes, 327
oogenesis, 325, 326, 327–28
oogonium, 327
operational sex ratio, 153
opsin genes, and duplication, 119–20
oral contraceptives. *See* pill
orgasm, in sexual response, 333, 335, 336
ornaments in sexual selection, 152, 162–66
outbreeding, benefits, 131
ovarian cancer, 320–21
ovarian reserve, 351
ovaries, 306, 307, 308, 328–30
oviparous, 56
ovulation, 224, 308, 328, 329, 332–33
oxytocin, 346, 350

Pacific field crickets (*Teleogryllus oceanicus*), adjustment in ejaculate, 232
painted turtle (*Chrysemys picta*), nesting and ESD, 277–78
paleontology, in evolution, 110, 111
Paracerceis sculpta, multiple types of sex, 262, 263
Paramecium, reproduction, 35, 39
parasites, and anomalous sex determination, 284–85
parasitic castration, 284–85
parental generation (P generation), inheritance, 65–67, 68–70
parental investment, 152–55, 167–70
parenting
 biparental care, 185–86
 by birds, 168–69, 186, 250–52
 desertion and off-loading, 238, 240–41
 by males, 168–70
 and mating system, 211
 same-sex parenting, 248–52
 and sexual conflict, 237–39, 240–41
 and sexual reproduction, 56
parthenogenesis, 39, 41–42
partible paternity, 203
paternity testing, 75–76, 78–79, 88–89
paternity uncertainty, 221, 223
patrilineal inheritance, 96
pea, in Mendel's work, 62–72
peacock/peafowl (*Pavo cristatus*), in evolution theory, 144, 145
pedigree, description and example, 316, 317
penduline tit (*Remiz pendulinus*), desertion, 238
penis, 301, 303–4, 334
penis fencing, 230
peptide hormones, 292–93, 295
perennial monogamy, 183
perfect flower, 52
phenotype
 anomalous sexual phenotypes, 309–14
 and chromosomes, 265–67
 and conditions dependence, 241–42
 definition, 74
 in humans, 75, 76–77, 84, 309–14
 multiple types, 262
phocomelia, 346
pie charts, 21–22
"the pill"
 as contraceptive, 342–43
 and mate choice, 20–21, 160, 161
pituitary gland and system, 293, 294, 295
placenta, 308–9, 344–45, 346, 347
plants
 asexual reproduction, 39, 40
 blending inheritance, 61
 fertilization, 51, 55
 meiosis, 90–91, 92
 sex determination, 258
 sexual characteristics, 52–53
 sexual reproduction, 50, 52–53, 55
plasmids, in bacterial reproduction, 34–35
plasticity, 239, 244–45
plateau, in sexual response, 333, 334–35, 336
pleasure from sexual activity, 28
point mutations, 118, 120
pollination, 52, 63–64
polyandry, 180, 181, 204–7
polygamy, 180, 188
polygynandry
 cooperative breeding, 209–10
 definition and description, 180, 207
 reproductive skew, 207–9
polygyny
 as default system, 184–85
 definition and description, 179, 180, 181, 188, 204

human polygyny index (HPI), 192
in humans, 191–93
lek polygyny, 212
models, 188–90
polygyny threshold model, 189–90
polyspermy, 337
population
gene pool, 123–25
replacement rate, 26
positive hormonal feedback, 294, 296
post-zygotic barriers, 139, 140–41
praying mantis (*Mantis religiosa*), sexual selection, 150
pregnancy
early process, 343–44
labor and delivery, 345, 347–49
and menstrual cycle, 328
mitochondrial disease, 341
placenta, 344–45, 346
prenatal period, sex differentiation, 297, 298–300, 306
presentation of baby, 347–48
pre-zygotic barriers, 139–40
primary oocytes, 327–28
primary sex characteristics, 46, 50–51, 297
primary spermatocytes, 326
procreation, 26–27
progesterone, 330, 331
progestin, 351
prokaryotic cells, 38
prolactin, 350
prolonged copulation, 233, 236
promiscuity
definition and description, 180, 210–11
lek polygyny, 212–13
and scramble competition, 211–12
and social bonds, 179–80, 210–11
pronghorn (*Antilocapra americana*), inter-sexual selection, 158, 159
prospecting, 243–44
prostaglandins, 302, 346
prostate cancer, 318
prostate gland, 302
protandry, 259
protein
and chromosomes, 265–66
in genetics, 84, 87
and SRY gene, 268
Protenor system, 270, 271
protogynous bluehead wrasse (*Thalassoma bifasciatum*), ESD, 284
protogyny, 259
proximate answers, for sexual activity, 27–28
puberty
definition, 294
hormones, 294, 296
onset, 20, 307
sexual development of men and women, 300–301, 306–7, 313, 326, 327
Punnett, Reginald, 68
Punnett squares
for gene pools, 123–24
in inheritance, 67–69, 71, 79
meiotic non-disjunction, 311
XY system of sex determination, 264

quality–quantity dichotomy, 155

racial groups, cancer of sexual organs, 318, 319
rape, in humans, 227
reciprocal crosses, 65
recombination, 92
red deer (*Cervus elaphus*), sexual selection and parental investment, 151, 152–53, 159–62
Red Queen hypothesis, 133–34
red-sided garter snakes (*Thamnophis sirtalis*), sperm depletion, 195
regression line, in graphs, 22, 23
religion, 26–27, 104, 301, 307
reproduction
asexual (*See* asexual reproduction)
as biological concept, 33
investment in, 152–55, 167, 221
non-sexual, 36, 37–38
as reason for sex, 26–29, 30
sexual (*See* sexual reproduction)
reproductive isolation, 134, 136
reproductive skew, 207–9
reproductive success, 159–62, 218, 220
reproductive system
development, 298–99, 306
men, 299–302
women, 299, 306–9
reproductive value, 245
reptiles
intromittent organ, 303
TSD, 278–79
residual reproductive value (RRV), 245

resolution, in sexual response, 333, 335–36
resource-defense polyandry, 205–6
resource-defense polygyny, 189
retrovirus infections, 360
Rh system, 79
rhythm method, 340
ribosome, 87
royal families, inbreeding, 131–32
r-selection traits, 245
runaway hypothesis, 162–64

Sacculina barnacle, ESD, 285–86
same-sex parenting, 248–52
same-sex sexuality. *See* homosexuality
satellite males, 241–42, 243
satellite phenotype, 241–42
scatterplots, 22–23
scientific data, graphic presentation and
 sets, 19, 20–23
scientific method, 15, 18–19, 104
scorpion fly (*Panorpa cognata*), nuptial gift,
 202
scramble competition, for mating, 171–72,
 211–12
scrotum, development, 300
sea horse (*Hippocampus* spp.), sex role
 reversal, 169–70
search costs in sexual reproduction, 127
secondary oocytes, 327, 337
secondary sex characteristics, 46–51, 297
secondary spermatocytes, 326
second generation (F2), inheritance, 65–67,
 68–71
segregation, law of, 64–69, 72, 98
selection, sexual. *See* sexual selection
selection of offspring sex, 281–82
selective breeding, 60, 111, 112
selfing, 42, 52, 273, 274
semelparous breeding, 56–57
semen (seminal fluid), 51, 256, 302
 See also sperm
sequential hermaphroditism, 259–60
serial monogamy, 183
sex
 biological definition and perspective,
  5–6, 11
 comparative approach, 13–14
 definition, 5–6, 34
 and evolution, 102
 vs. gender, 5, 7–8
 ideological trends, 9–12
 interest in and knowledge of, 3–4, 13
 reasons for, 26–28, 30–31
 social aspects, 4–6, 8
sex allocation, 246–48
sex assignment, 7
sex chromosomes
 anomalies, 310–12
 discovery, 256
 genes and genetics, 265–67
 in hermaphrodites, 272–73
 in honeybees, 275
 in humans, 83–84, 261, 265–67
 in insects, 261, 263–64
 meiotic non-disjunction, 310–12
 See also X-chromosomes;
  Y-chromosomes
sex determination
 anomalous sex determination, 284–86
 chromosomal sex determination, 272–74
 definition, 255
 deviations in male-only or female-only
  sex, 258–60, 262–63
 eggs in, 256, 264, 265
 environmental (*See* environmental sex
  determination)
 gametes in, 270–72
 genes and genetics in, 256–57
 genetic sex determination, 269–70, 271,
  274–76, 281
 hermaphroditism, 257, 259–60, 272–74
 multiple types of one sex, 262–63
 social factors, 282–84
 sperm in, 256, 264, 265
 SRY gene, 267–69, 272
 systems, 269, 270
 testis-determining factor, 268–69
 without Y-chromosomes, 269–70
 with X- and Y-chromosomes, 256, 261,
  264–69
 XY system, 261, 264, 269, 270
sex differentiation
 and cancer, 315, 318–21
 definition, 255
 in embryos, 291, 298–99
 and endocrine system, 293–94, 295
 genes in, 257, 267, 291
 in human development, 297–99
 and human hormones, 290–96, 298–99
 prenatal period, 291, 297, 298–300, 306

primary and secondary sex characteristics, 297
sex education, 290
sex linkage, 314–15
sex organs and tissues
cancer in humans, 315, 318–21
men, 301–2
women, 307–9
sex role reversal, 166–70
sexual activity
and age, 351
co-option and exaptation, 29–32
and pleasure, 28
for reproduction, 26–29, 30
sexual arousal, stages and response, 333–36
sexual cannibalism, 150
sexual conflict
definition and description, 218–19
for males and females, 222–23
after mating, 234–37
before mating, 223, 226–29
during mating, 229, 232–34
in parenting, 237–39, 240–41
realms of, 222–23, 226–29, 232–39
sperm, 227, 229–34, 235, 236–37
strategies in, 220–22, 239, 241–48
sexual development
atypical patterns, 312–13
humans at puberty, 300–301, 306–7, 313, 326, 327
prenatal, 297
sexual dimorphism, 47–50
sexual function, and hormones, 290–96
sexual inheritance
blending inheritance, 60–61
and chromosomes, 80–84, 86
and DNA, 80, 84–87
dominant and recessive factors and genes, 67, 70, 76
early understanding and use, 60, 72–73, 80
and evolution, 116–21
experiments by Mendel, 62–64
and generations, 65–67, 68–71
in humans, 75–79
Law of Independent Assortment, 64, 70–72
Law of Segregation, 64–69, 72
patrilineal and matrilineal inheritance, 96
and Punnett squares, 67–69, 71, 79
rules, 62–64, 116
terminology, 73–74
sexually antagonistic coevolution, 220
sexually transmitted infections (STIs)
bacterial STIs, 354–55
causes, 354
gonorrhea and antibiotic resistance, 352–53, 354–55
in non-humans, 358–60
and sexual intercourse, 129, 353
syphilis, 352–53, 354
viral STIs, 355, 358
sexual monogamy, 182–83
See also monogamy
sexual orientation, 5
sexual reproduction
and asexual reproduction, 41–43, 125, 129–30
basics of, 43–50
benefits in evolution, 129–34
biological fitness, 115–16
costs in evolution, 125–29
cycle in women, 328–30
definition and emergence, 36, 45
and diversity, 130–31
eggs and sperm, 43–45
and fertilization, 50–51, 54–55
fundamental scaffold, 50–51
inbreeding, 131–33
meiosis, 90–91
numerical argument, 126–27
and parenting, 56
pattern variations, 56–57
primary *vs.* secondary sex characteristics, 46–51
role of sex, 102
sexual dimorphism, 47–49
speciation, 134
species creation, 134–41
sexual revolution, 4–5
sexual selection
agent in selection, 147–48
armaments in, 151, 162
Darwin's theory, 144, 155
females as "limiting sex," 152–62
fitness, 154–55, 162–67
human brain, 174
inter-sexual selection, 151–52, 155–59, 163, 196
intra-sexual competition, 151, 170–73, 175

intra-sexual selection, 151, 159–62
manifestations, 151–52
as natural selection sub-category, 145–49, 151–52
non-monogamy benefits in females, 196–201
and non-sexual natural selection, 146–48
observations and inferences, 145–46, 148–49
ornaments in, 152, 162–66
and parental investment, 152–55, 167–70
payoffs, 155–56, 158, 159–62
and reproductive success, 159–62
sex role reversal, 166–70
smell in, 20–21, 160–61
sperm competition, 173, 175, 229
and "useless" features, 174
sexual success, 219–22
"sexy sons" hypothesis, 162–64
Seymour, Jane, 257
shelled eggs, 303
she-males, 243
sickle-cell allele, 117–18
simultaneous hermaphrodites, 259, 260
single-celled organisms, reproduction, 33–35
sire-dam interaction, 200
sister chromatids, 92
size-advantage model, 259–60
smell in sexual selection, 20–21, 160–61
snails, love darts, 231
snakes, as "species," 136
social bond, and mating systems, 179–80, 210–11
social determinism, 9
social factors, in sex determination, 282–84
social insects, sex determination, 274–76
social monogamy, 182, 188, 222
sociobiology, 10–11, 12
Sociobiology: The New Synthesis (Wilson), 10, 12
sons
and ornaments in selection, 162
preference for, 256, 258
in sex allocation, 246–48
speciation, 134, 135, 137–39
species
and anomalous sex determination, 284–86
classification and diversity, 16–17
creation in evolution, 134–41
definition and concept, 136
and gene pool, 135, 137–39
patterns of evolution, 134–35, 137
reproductive barriers, 139
sperm
adjustment in ejaculate, 232–33
aggregation, 173
competition in sexual selection, 173, 175, 229
content, 256
depletion in breeding, 195–96
description, 44
ejaculation, 335, 336
fertilization of eggs, 45, 51, 54, 308, 336–38
genetics and genes, 44, 74
in human anatomy, 301, 302
and infertility, 338
and inheritance, 68–69, 74
and males, 43–45
maturation, 337–38
mutations, 120–21
and non-monogamy in females, 195–96, 197–98, 222
production, 326, 327
in sex determination, 256, 264, 265
in sexual conflict, 227, 229–34, 235, 236–37
in sexual reproduction, 43–45
sperm allocation, 232–33
spermatheca, 274
spermatocytes, 326
spermatogenesis, 325–26
spermatogonia/spermatogonium, 325–26
spermatophores, 54–55
spermatozoa, 44
sperm displacement, 229
sperm ejection, 237
spores and sporic life cycle, 90–91
spotted hyena (*Crocuta crocuta*), clitoris and copulation, 228, 304, 305
spotted sandpiper (*Actitis macularius*), resource-defense polyandry, 205–6
SRY gene, 267–69, 272
stalk-eyed fly (*Teleopsis breviscopium*), ornaments, 164–65
starlings (*Sturnus vulgaris*), biparental care, 186

"stealing a guard" strategy, 235
stereotypes (sexual), 45–46
sterilization techniques, 342, 343
steroid hormones, 291, 292–93, 295
Stevens, Nettie, 261, 263–64
strategies for mating. *See* mating strategies
subordinates, 208, 243–44
subterfuge, for mating, 172–73
success
  reproductive, 159–62, 218, 220
  sexual, 219–22
surrogate pregnancy, 340
survival, and good genes hypothesis, 164–66
survival costs in sexual reproduction, 127–29
swamping, 232
synapsis, 92, 93
syphilis (*Treponema pallidum*), 352–53, 354

taxonomy, basics of, 13, 16–17
TDF testis-determining factor, 268–69
temperature, in ESD, 277–79, 281–82
temperature-dependent sex determination (TSD), 278–79
teratogens, 346
testes, 300, 301
testicular cancer, 318–19
testosterone
  costs of, 128–29
  in sex differentiation and function, 293, 295, 298–99
  in sexual development, 294, 296, 298, 301
tetrad, 92, 93
thalidomide, 346
theory, definition and use, 15, 18–19
Thomson's gazelle (*Eudorcas thomsonii*), estrous cycle, 331
topping up system/model, 196, 197
transcription
  in genetics, 84–85, 268
  SRY and DMRT1 gene, 268–69, 272
transgender, definition, 7
translation, in genetics, 84–85
transparent roundworm (*Caenorhabditis elegans*), reproduction and sex determination, 42, 272, 273
transverse presentation and lie, 348, 349
traumatic insemination, 230
A Treatise of Human Nature (Hume), 11, 12
tree frogs, external fertilization, 54, 55
trend line, in graphs, 22, 23
trichromatic vision, 119–20
triplets, DNA, 86–87
triple-X syndrome (XXX), 310, 311
trisomy 21, 328
Trivers-Willard hypothesis, 246–47
true breeding and true-breeding varieties, 63, 65, 66–67, 69–70
tubal ligation, 342, 343
Turner syndrome, 310, 311–12
turtles, temperature and ESD, 277–78, 279
twofold cost of sex, 127
"two moms and a dad," 341

ultimate answers, for sexual activity, 27–28
umbilical cord, 343
United Kingdom, cancer of sexual organs, 318, 319
United States, 318–19, 353, 357
urogenital groove, 298, 306
uterine cancer, 320–21
uterus, 306

vagina, in sexual response, 335
vasectomy, 342, 343
vasocongestion, 334
Venus of Hohle Fels, 324, 325
vestigial structures, in evolution, 110
vicariance speciation, 137, 138
Victoria (queen), hemophilia as sex linkage, 316–17
virus, and STIs, 355, 358
vision, and genetics, 119–20
viviparous skink (*Eulamprus tympanum*), ESD, 282
viviparous, 56
vulva, 44, 307

Wallace, Alfred Russell, 107, 109, 155
water fleas (*Daphnia* spp.), reproduction, 41–42, 129–30
waterfowl, reproductive organs, 234, 235
water striders (Family Gerridae), coercion and harassment, 227–28
weaponry in sexual selection, 151, 162
Weinberg, Wilhelm, 121
western honeybee (*Apis mellifera*)
  chromosomes, 275
  mating systems, 184
  multiple types of sex, 262, 363
  non-monogamy in females, 198

- sex determination and mating, 274–75
- sperm warfare, 233–34

whiptail lizards, asexual reproduction, 40, 41
Wilson, Edmund Beecher, 269–70
Wilson, E.O., 10, 12
Wolffian ducts, 298, 299
women
- breast-feeding, 349–50
- cancer of sexual organs, 315, 319–21
- chromosomal defects in children, 327–28
- chromosomes, 83, 261, 265–67
- dual mating strategy, 222, 224–25
- estradiol, 296
- fertile period, 331–33
- fertility, 224
- gametes production, 327
- genitals, 44, 307
- hormonal feedback and hormones, 294, 296
- infertility, 338, 339
- meiosis, 324–25, 326, 327–28, 338
- menopause, 350–51
- menstrual cycle, 328–33
- non-monogamous behavior and benefits, 193, 201
- orgasm, 335
- prenatal development, 298–99, 306
- reproductive cycle, 328–30
- reproductive system, 299, 306–9
- sex differentiation before birth, 306
- sexual anatomy of adults, 307–9
- sexual arousal response, 333–36
- sexual development at puberty, 306–7, 313
- sexual selection, 20–21, 158, 160–61
- smell and mate choice, 21, 160–61
- turning into men, 313–14

X-chromosomes
- in crossing-over, 96
- in DNA, 84, 88–89
- duplication, 119–20
- genes and genetics, 265–67
- genetic disorders in men, 314–15
- in humans, 83, 261, 265
- in sex determination, 256, 261, 264–69
- size, 265
- *See also* sex chromosomes

X-inactivation, 266
XO system, 270, 271
XX pairing, 83, 261
XX/XO system, 270, 271
XX/XY system, 270, 271
XY pairing, 83, 261
XY system of sex determination, 261, 264, 270, 271

Y-chromosomes
- in crossing-over, 96
- discovery, 264
- in DNA, 84, 88–89
- genes and genetics, 265–67
- in humans, 83, 261, 265
- in sex determination, 256, 261, 264–69
- sex determination without, 269–70
- size, 265, 267
- SRY gene, 267
- *See also* sex chromosomes

Zika virus, 358
zygotes
- definition, 45
- and fertilization, 51, 55, 336–40, 342–43
- and genes, 74
- heterozygotes and heterozygosity, 74, 200–201, 314–15
- homozygotes and homozygosity, 74, 132–33, 314–15
- and inheritance, 68–69
- meiosis and mitosis, 90

zygotic barriers, 139–41
zygotic life cycle, 90, 91